Architectural Structures

An Introduction to Structural Mechanics

Henry J. Cowan

Professor of Architectural Science
University of Sydney, Australia

SECOND EDITION

ELSEVIER

New York/Oxford/Amsterdam

AMERICAN ELSEVIER PUBLISHING COMPANY, INC.
52 Vanderbilt Avenue, New York, N. Y. 10017

ELSEVIER SCIENTIFIC PUBLISHING COMPANY
335 Jan Van Galenstraat, P. O. Box 211
Amsterdam, The Netherlands

Library of Congress Cataloging in Publication Data

Cowan, Henry J
  Architectural structures.

  Includes bibliographical references.
  1. Structures, Theory of. I. Title.
TA646.C68 1976   624'.171   75-26330
ISBN 0-444-00177-8

Designed by Ellen Louise Blissman

Manufactured in the United States of America

Architectural Structures

To Peter Smith,
a friend and colleague for many years

Contents

Sections marked with an asterisk (*) may be omitted at first reading

Contents

Preface to the Second Edition

Since the publication of the first edition there has been a renewed interest in the design of tall buildings, and new and sophisticated structural systems have been devised which made revisions to Chapters 1 and 8 appropriate. The 1971 ACI Code established ultimate strength as the main basis of reinforced concrete design, and Chapter 7 has accordingly been rewritten. Some of the material has been rearranged for greater clarity. Many of the illustrations have been redrawn, and photographs have been added.

Great Britain, Australia, New Zealand and South Africa have revised their structural design codes in accordance with the SI metric system; Canada is in the process of conversion, and the American Society for Testing and Materials has drawn up a Metric Practice Guide. All the numerical examples have been worked out a second time in metric units in Appendixes G and H. Readers and instructors who wish to employ metric units are asked to turn to these appendices when they reach examples or problems in the text. For ease of identification, such examples and problems are marked with a dagger (†).

I am particularly indebted to Dr. E. L. Buckley and Mr. R. Powell of University of Texas in Arlington, and to Dr. G. S. McClure of Montana State Univeristy, for drawing my attention to a number of errors and ambiguities which have been corrected; to Dr. Valerie Havyatt for checking the new edition, and to Mrs. Rita Arthurson for typing it.

Sydney, Australia   H. J. C.
January 1975

Preface to the First Edition

The study of structures forms part of the education of every architect, and has done so for more than a century. The emphasis, however, is changing. Structures are becoming more complicated, and calculations are increasingly done by automatic computation. The architect produces the original design concept which embodies the form of the structure; but an engineer carries out the structural calculations and detailed design. What kind of structural course, then, does the architecture student require?

Few architects have a profound knowledge of mathematics and physics, at any rate by comparison with engineers. Some curricula therefore confine themselves to the most elementary structures, which even engineers solve without advanced mathematics. This does not satisfy students, who tend to use a much wider range of structures in their design. Unless the structural course can be applied to architectural design, and unless the student can apply it himself without a consulting engineer (or instructor) at his elbow, it has not succeeded.

Another approach omits mathematics altogether. The course covers descriptively the whole field of structures, including complex space frames and shells which engineering students hear of only in a graduate program; but it omits all theory and calculations. Although this is delightfully easy, it does not really help the student to understand structures. He cannot choose between alternative solutions unless he understands their mechanics, nor can he judge structural scale without some calculations. Buildings 20 ft high and roofs spanning 20 ft, however unusual their shape, present no structural problems; buildings 200 ft high and roofs spanning 200 ft need to be correctly conceived.

I believe that an understanding of structural behavior is possible only through a study of theory. To be useful to architects, the course must cover the whole range of available structures, and deal with the principal structural sizes, because big structures differ from small structures. It need not go into design details, or deal with the full calculations, which have become the responsibility of the engineer. I have attempted to achieve this objective by deleting inessentials, and by deriving theories in the simplest terms, with frequent reference to the physical behavior of structures.

Instructors who teach a junior college course may find it helpful to cover Chapters 1 to 6, omitting sections marked with an asterisk

(*), in a first course, and use the remainder of the book in a second course. When structures are first taught at senior undergraduate or at graduate level, sections marked* should be within the competence of all students; Chapter 3, which deals with statics, has probably been covered previously and could be omitted.

Integration of structural teaching with studio work and the use of physical demonstrations help students to follow the lectures and seminars. For the benefit of instructors, some suggestions are given in Appendix A. I hope that they will save time, rather than add to the length of the course.

I am greatly indebted to Dr. Carolyn Mather, and to Mr. Graham Morris, who drew the illustrations from very rough sketches; to Mrs. Rita Arthurson, who typed the manuscript; to Professors Day Ding, Peter Smith, and John Gero for permission to use problems set by them; and to my daughter Judy for helping with the proofreading. Finally, I would like to express my appreciation to the heads of architecture schools and teachers of structural subjects in more than forty universities, who gave me their time to discuss the proper way of teaching this subject, and to my own students, past and present, for their critical comments.

Sydney, Australia   H. J. C.
September 1970

Choice of Structure 1

*Strange it is to think how building do fill my mind
and put all other things out of my thoughts.*

Samuel Pepys

In this chapter we describe the principal structural types. Later some are considered in detail, and formulas are derived for determining the principal structural dimensions. *The force diagrams used in this chapter are intended only to illustrate principles; we will consider the mechanical solutions of the simpler cases in later chapters.*

1.1. The Problem

The enclosure of space for human activities is a fundamental objective of architectural design. With a few exceptions, such as open-air theaters, buildings must have roofs, and most also have floors suspended above the ground. Bridging the gap between the supporting walls or columns is the more difficult problem in structural design; except for tall buildings (see Section 1.13), the design of the columns or walls is simpler.

The length of the gap to be bridged is of decisive importance. A good craftsman lacking in theoretical knowledge can design a structure over a span of a few feet (or metres). Great skill and elaborate calculations are required to build a roof with a clear span of several hundred feet (or more than a hundred metres). In this chapter we will therefore concentrate our attention on roofs and floors, i.e., on the various ways in which a horizontal gap between supports may be bridged. Long-span bridges introduce special problems, whereas very short spans present no structural problems at all.

Structures are designed to carry loads. In the first place, they must carry their own weight, and in the design of medium- and long-span roofs the weight of the structure itself is by far the most important consideration. In the second place, they must carry the loads superimposed on the structure by the roof covering, the floor surface, the ceiling finish, the furnishings, and the people who use the building. All these loads act vertically. For the present, we will ignore horizontal loads imposed by wind on tall buildings and horizontal loads caused by earthquakes or moving machinery. These can be very important for some structures; but the design of the majority is dominated by vertical loads, due to the weight of the building and its contents.

Let us now consider a structure spanning between two supports, and carrying vertical loads (Fig. 1.1). The supports stop the structure and the loads from falling, and there is thus a reaction at each support. Let us imagine that the structure is cut along a vertical line A–A. The (shaded) part of the structure to the left of the imaginary cut (i) tends to rotate under the combined action of the left-hand reaction R_L and the loads W_1, W_2, and W_3, and (ii) tends to move up or down, depending on whether R_L is greater or smaller than the three loads W.

If the structure is strong enough and does not collapse, it resists both tendencies. At the section A–A, the loads tending to rotate the shaded portion of the structure set up a *bending moment M* (Fig. 1.2), which tends to bend the beam downward (and indeed causes a measurable deflection—Fig. 1.3). The tendency to move the shaded portion of the structure vertically relative to the unshaded portion sets up a *shear force V* (Fig. 1.4), which tends to cut or shear through the beam.

Vertical loads acting on a structure bridging a horizontal gap set up bending moments and shear forces, and their magnitude is independent of the form of the structure. However, the ability of

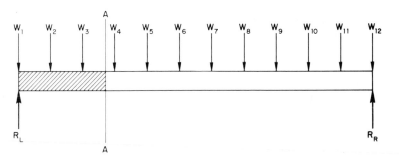

Fig. 1.1.
Bridging a horizontal gap. The problem is how to carry loads W_1, W_2, W_3, etc. between the left-hand and right-hand reactions R_L and R_R.

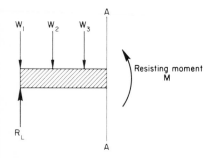

Fig. 1.2.
The loads W_1, W_2, W_3 and the left-hand reaction R_L produce a bending moment which is equal and opposite to the resisting moment M of the beam at the section A–A; this depends only on these loads and not on the type of structure.

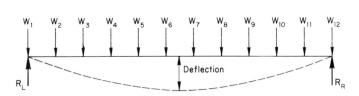

Fig. 1.3.
Bending moments must not be allowed to become so large that they cause damage to brittle finishes, such as plaster, or allow people to notice movement in the building (see Section 6.5).

Choice of Structure

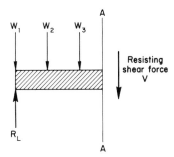

Fig. 1.4.
The load W_1, W_2, and W_3 and the left-hand reaction R_L produce a net upward force which is resisted by the shear force V of the beam at the section A–A; this depends only on these loads, and not on the type of structure.

the structure to cope with large bending moments and shear forces greatly depends on its form, and the choice of structure is therefore particularly important for large spans.

1.2.   Post and Beam (see Chapter 5)

The simplest structure is a beam supported on two posts. It is easily built either by laying a block of stone across two others or by placing a tree trunk across two vertical posts. Such structures go back to prehistoric times; although none survive in timber, there are several of large size built in stone several thousand years ago (Fig. 1.5). In historical times, the post-and-beam construction formed the

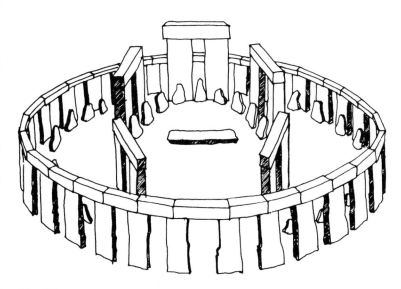

Fig. 1.5.
Stonehenge. A giant post-and-beam structure erected in Southern England between two thousand and three thousand years ago.

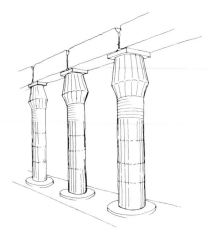

Fig. 1.6.
Egyptian temple colonnade. Post-and-beam structures were employed in Egyptian architecture for more than three thousand years.

Fig. 1.7.
Greek (Doric) temple colonnade. Greek post-and-beam construction originated about 700 B.C., and through imitations and revivals its use continued into the twentieth century.

basis of Egyptian and Greek temple architecture (Figs. 1.6 and 1.7).

The bending moment produced by the weight of the beam, the loads on the beam, and the reaction of the post, is resisted by an equal and opposite stabilizing moment within the beam. Let us further consider the beam of Fig. 1.2 at section A–A (Fig. 1.8). The bending moment M tends to bend the beam downward, and thus compresses the top and extends the bottom of the beam. The compression C near the top and the tension T near the bottom are separated by a resistance arm a (whose magnitude we shall determine in Chapter 7). These form a resistance moment which equals M and thus prevents the beam from collapsing.

Stone is strong in compression, but weak in tension. This is easily proved by pulling a piece of one of the weaker stones (such as a soft sandstone or chalk); it is easily pulled apart. However, it can be broken in compression only by putting one's full weight on it, crushing it underfoot. Stone is therefore unable to span large distances as a beam, because it cannot resist the tensile component of the resistance moment.

Timber is not as strong in compression as stone, but its tensile and compressive strength are about equal. It is therefore a good material for short beams and is still used in that way, for example, in the timber floors of small houses.

Steel, like timber, has about the same strength in tension and compression, but it is a much stronger material. When it became available in large quantities in the middle of the nineteenth century, steel beams replaced timber beams for the larger buildings.

The problem of the low tensile strength of stone has been solved by introducing steel reinforcement into artificial stone, or concrete. Reinforced concrete (see Section 7.10) is cheap, durable, and fire-resistant, and it has therefore become the predominant material for floor construction in all but the smallest buildings. The concrete resists the compression, C, and the steel reinforcement the tension, T (Fig. 1.9).

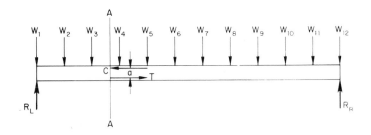

Fig. 1.8.
Resistance moment of a beam in post-and-beam construction. The loads W and the reaction R produce a bending moment M at the section A–A. This gives to rise to an internal compressive force C at the top of the beam, and an internal tensile force T at the bottom. The resistance arm a separates the two forces, so that they form a resistance moment which equals M.

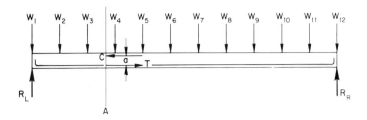

Fig. 1.9.
Reinforced concrete. Since concrete is weak in tension, its bending resistance is low, unless it is reinforced with steel. In reinforced concrete, the concrete resists the compression due to bending, and the steel resists the tension due to bending.

Further economies are achieved in modern structural design by making the beams continuous across intermediate supports. This significantly reduces the maximum bending moment (see Section 8.3), and continuous beams consequently require less material than those simply supported without any restraints at their ends, as in Figs. 1.5 to 1.9.

1.3.   Cable (see Section 9.1)

Post-and-beam construction is well-suited to small- and medium-sized spans. The straight members are easily made and assembled, and the finished structure is horizontal, which is essential for floors.

For long spans, however, other solutions must be found, and for medium spans other solutions may be more economical. From Fig. 1.8 it is evident that a beam must have depth to give a reasonable distance, or resistance arm, between the tensile component of the moment, T, and the compressive component of the moment, C. As the spans get larger, the members get bigger and heavier. But the main weight carried by long-span members is the weight of the structure itself. The stage is thus reached where a beam can no longer support its own weight, however deep it is made, because the greater the depth, the greater the weight it has to carry. In practice, the limit of economic usefulness is reached long before that.

We can produce a large resistance arm without using a thick piece of material by curving our member. Let us consider a cable hanging freely under its own weight (Fig. 1.10). The cable is held by two vertical springs, which represent the same vertical reactions also required in a beam carrying the same load, and by two horizontal springs, which are additional horizontal reactions required to support the cable because it is curved and tends to pull inward.

Let us now consider the equilibrium of the cable at mid-span, where it is horizontal by symmetry (Fig. 1.11). The cable tension T and the horizontal reaction R_H are separated by the sag of the cable s. This is much greater than the resistance arm a in a beam (see Fig.

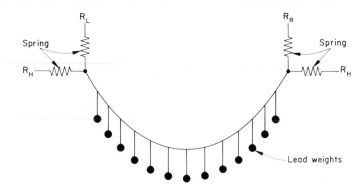

Fig. 1.10.
The cable. A cable carrying its own weight requires the same vertical reactions, R_L and R_R, as a beam over the same span. In addition two horizontal reactions, R_H, are required, because of the curved shape which pulls the cable inward. In the model the weight of the cable is increased by lead weights. The reactions are then readily illustrated if a cable is hung from two vertical and two horizontal springs.

1.8), so that the resistance moment is correspondingly increased. There are several additional reasons why cables are efficient for long spans:

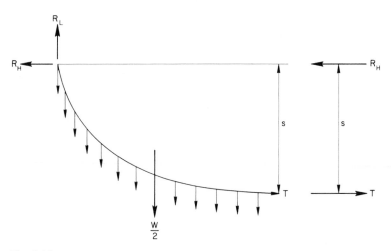

Fig. 1.11.
The cable provides a resistance moment because it sags. The resistance moment is formed by the horizontal reaction R_H, the cable tension T, and the sag s. Since the sag s is far greater than the resistance arm a of the beam in Fig. 1.8, the cable can span farther for the same tension T.

1. The resistance arm of a beam lies inside the beam, and increasing it means increasing the thickness, and therefore the weight, of the beam. The sag of a cable lies outside the cable section, and increasing the sag does not increase the thickness of the cable.

Choice of Structure

2. The steel in cables can be made much stronger than in beams. Steel used in beams may be stressed to between 20,000 and 35,000 p.s.i. (pounds per square inch) or, in metric units, between 140 and 240 MPa (megapascals). Steel which may be stressed up to 200,000 p.s.i. (1,400 MPa) can be made by pulling a bar through a wire-drawing die. The distortion of the crystals in the steel increases not merely the strength, but also the hardness, and the material is so difficult to cut that it cannot easily be used for anything other than cables, which are made by gathering the strands of wire together. Because the steel is so strong, a much higher load can be carried by the same weight of steel, and over long spans the weight of the structure is the biggest load to be carried.

3. Materials used in tension straighten out, while materials in compression buckle under load if the shape is long and slender. A sheet of paper or a thin bar can carry a good load when either is pulled, but both buckle sideways when their ends are pushed together (see Section 7.1). Similarly, a piece of rope can carry a big load in tension, but it collapses in compression (Fig. 1.12).

Cables are therefore invariably used for bridges of very long span (Fig. 1.13). The Verrazano-Narrows Bridge in New York has a span of 4,260 ft (1,300 metres), and the 12 longest bridges in the world are all suspension bridges.

Cables are not nearly so useful for long spans in buildings. The flexibility of the cables is an advantage in bridge design, because the cables adjust themselves to the loads placed upon them. In buildings, the cables are required to support a roof which has to keep the rain out, and waterproofing a flexible roof is not easy. The large area of a flexible roof may vibrate or flutter unless steps are taken to prevent it (see Section 9.1).

Moreover, spans in buildings rarely exceed a few hundred feet, while suspension bridges may span several thousand feet. The

(a) No buckling in tension

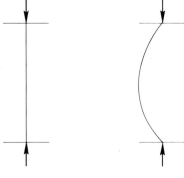

(b) Buckling in compression

Fig. 1.12.
Compression members may buckle. Tension members do not buckle.

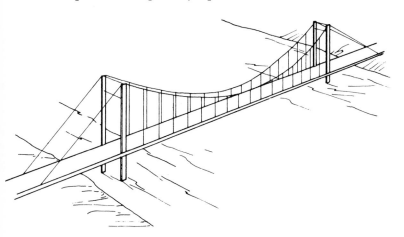

Fig. 1.13.
Long-span suspension bridge.

relatively small spans used in buildings can be roofed economically by stiffer curved structures.

Suspension roofs are therefore still comparatively rare, although with increasing knowledge of their behavior they are becoming more common. The simplest type of suspension roof is one in which the cables are parallel, as in a suspension bridge (Fig. 1.14). A large number of cables carried down to the ground for anchorage takes up valuable space, and the separate anchorages add to the cost. They can be economically replaced by buttresses (Fig. 1.15), which may serve as supports for rising banks of seats. Suspension roofs are most commonly used to cover sporting arenas, and the high seats surrounding a low arena are more functional than a flat roof or a dome because less space needs to be enclosed (Fig. 1.16).

Transferring the horizontal reactions of the suspension cables to the ground far below adds to the cost when the cables are arranged

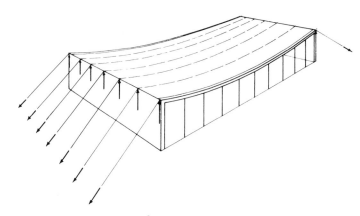

Fig. 1.14.
Suspension roof with parallel cables, anchored to the ground.

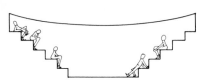

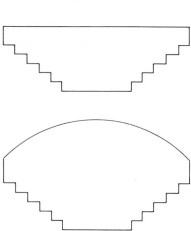

Fig. 1.16.
A suspension roof is better suited to covering the banked seats surrounding an arena than either a flat roof or a dome.

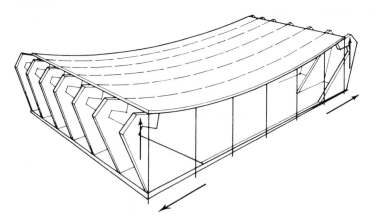

Fig. 1.15.
Suspension roof with parallel cables, anchored to banked seats.

along parallel lines, and it can be unsightly (Fig. 1.14). The anchorage problem is simpler in a circular building with tiers of seats arranged around a central arena. The cables are supported between an inner tension ring and an outer compression ring (Fig. 1.17), and the reactions are completely absorbed because each circular ring is self-balancing.

Another self-balancing support is provided by two crossed arches (Fig. 1.18). If the shape is properly designed, the reaction of the cables is absorbed by the arches in pure compression.

1.4. Arch (see Section 5.9)

If a piece of string, acting purely in tension, is made rigid (say by treating it with a lacquer, such as hairspray) and turned upside down, we obtain an arch which spans the same gap and carries the same load in pure compression. A flexible arch is evidently impracticable, since it would collapse under the slightest irregularity of loading.

The rigidity of the arch is, however, an advantage in building, where movement causes trouble with waterproofing of the roof and with brittle finishes. Shifting the emphasis from tension to compression helps materials which are strong in compression and weak in tension, such as concrete.

Before the nineteenth century, timber and masonry (including in that term concrete and brick as well as natural stones) were the only structural materials available. Timber lacked durability, because it was sooner or later destroyed by fire or by timber pests, and the oldest timber structures in existence are only a few centuries old. Masonry structures more than two thousand years old are still in use. Masonry is not used to the best advantage in post-and-beam construction (Fig. 1.7), because of its low tensile strength, and it cannot be used at all in pure tension.

From a very early age, masonry has been used in the form of arches, particularly in regions such as Mesopotamia, where natural stone and fuel for burning hard bricks were scarce, and a weak mudbrick was consequently employed. Limitation in the choice of materials enforces by necessity their most effective use.

The arch, being rigid, cannot adjust its shape to loads placed upon it, and the shape must therefore be designed for the loads which it is to carry. To be in pure compression, the arch must be the exact reverse of the shape assumed by a cable under the same loads. When carrying a single central load, a cable forms two straight lines (Fig. 1.19). Under its own weight, with that weight distributed uniformly along its length, it forms a catenary (Fig. 1.20). If it carries a load that is distributed uniformly along its span, it forms a parabola (Fig. 1.21). Although mathematically quite different, the two curves are geometrically similar (Fig. 1.22).

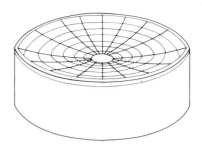

Fig. 1.17.
Suspension roof with radial cables. The cables are supported between an outer compression and an inner tension ring, and the resulting shape is a dished dome.

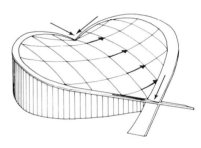

Fig. 1.18.
Suspension roof with parallel suspension cables, supported by crossed arches.

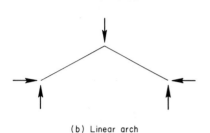

(a) Linear cable

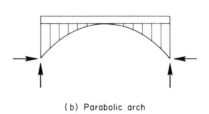

(a) Parabolic cable

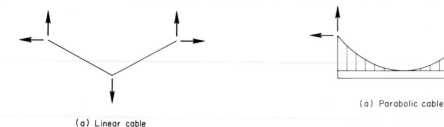

(b) Linear arch

(b) Parabolic arch

Fig. 1.19.
The linear cable and the linear arch. A flexible cable hangs in pure tension in two straight lines if loaded with a single heavy weight (Fig. 1.19a). Hence a linear arch carrying a single concentrated load is in pure compression if its own weight is, by comparison, negligible (see also Fig. 1.25).

Fig. 1.21.
The parabolic cable and the parabolic arch.
A flexible cable loaded uniformly along its span hangs in pure tension in a parabolic curve (a). Consequently a parabolic arch uniformly loaded *along its span* is in pure compression (b). A cable or arch uniformly loaded along its length carries more load near the supports and less load near mid-span than a cable or an arch loaded uniformly with respect to span.

(a) Catenary cable

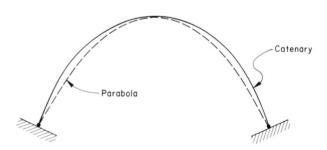

Fig. 1.22.
The catenary and the parabola, although mathematically quite different, are geometrically similar.

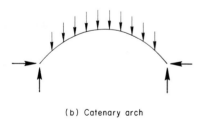

(b) Catenary arch

Fig. 1.20.
The catenary cable and the catenary arch. A flexible cable uniformly loaded along its length adopts a curve known as a catenary (Fig. 1.20a). Consequently, a catenary-shaped arch uniformly loaded *along its length* is in pure compression.

The Persians built at Ctesiphon in 550 A.D. a 112-ft (34 m) high arch in mudbrick, which is still standing in a suburb of Baghdad, and this conforms closely to the catenary shape (Fig. 1.23). The same shape is also to be found in primitive mud huts in Central Africa (Fig. 1.24). The brick cone which carries the dome of St. Paul's Cathedral in London, built in 1698, is closer to two straight lines, because in addition to the dome and its own weight, it carries the heavy lantern on top (Fig. 1.25). The precise shape is less important when steel or reinforced concrete is used, because both are able to resist a substantial amount of bending (Fig. 1.9). However, for very large spans the shape must conform to the load

10

Choice of Structure

system. The longest arches are still, at the time of writing, the Sydney Harbour Bridge and the Bayonne Bridge in New York, both completed in 1931 in steel with almost identical spans of 1,650 ft (504 m). The longest concrete arch is the Gladesville Bridge, completed in 1964 four miles upstream from the Sydney Harbour Bridge, with a clear span of 1,000 ft (305 m) (Fig. 1.26).

The traditional Roman and Romanesque semicircular arch is evidently not well-adapted to its structural purpose, and the Gothic arch, in spite of its pointed crown, is much closer to the structurally advantageous catenary (Fig. 1.27).

If the arch is in pure compression by virtue of its shape, then its mechanics is exactly the same as that of the cable, except that the force within the arch is compressive (Fig. 1.28). The arch compression C and the horizontal reaction R_H are separated by the rise of the arch r, and these provide the resistance moment. Evidently, the

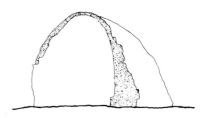

Fig. 1.23.
Ruin of the Palace of Ctesiphon built by the Persians in 550 A.D. This is a catenary arch of mudbrick, 112 ft (34 m) high.

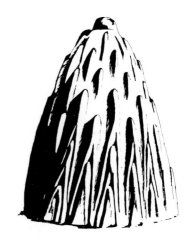

Fig. 1.24.
Primitive mud hut in Central Africa.

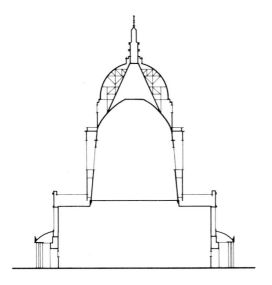

Fig. 1.25.
The outer dome of St. Paul's Cathedral in London, Christopher Wren's masterpiece, is actually a timber truss resting on a brick cone, visible neither from the outside nor the inside. The shape of the cone is determined by the heavy concentrated load imposed by the lantern.

Fig. 1.27.
The Gothic arch is closer to the structurally advantageous catenary than the Roman semicircular arch.

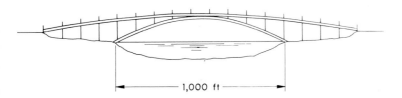

Fig. 1.26.
The Gladesville Bridge, Sydney, completed in 1964, has the world's longest concrete arch. It spans 1,000 ft (305 m).

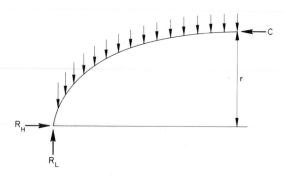

Fig. 1.28.
The arch provides a resistance moment because of its rise. The moment is formed by the horizontal reaction R_H, the compressive force at the crown of the arch C, and the rise of arch, r. It is the exact opposite of the suspension cable (see Fig. 1.11).

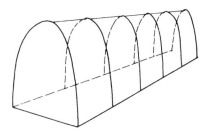

Fig. 1.29.
Parallel arches form a vaulted roof over a rectangular plan.

greater the rise, the smaller the compressive force C, and the less material is required in the arch to provide this force. On the other hand, too great a rise may provide an awkward shape for the building, and an inconveniently high space. Since arches are stiff, they need not conform to the "inverted cable" shape. The more they depart from it, however, the greater the bending in the arch (see Section 1.6).

Over rectangular or square plans it is best to place the arches parallel (Fig. 1.29). In Gothic cathedrals, arches were normally crossed (Fig. 1.30), and this is still occasionally done today. Over a circular or oval plan, arches may converge onto a common crown (Fig. 1.31). However, the interaction of crossed arches presents additional problems, which are avoided in parallel arches. Some quite complex shapes may be formed by a series of arches, as in the case of the Sydney Opera House (Fig. 1.32)

Fig. 1.30.
Crossed arches in a Gothic roof.

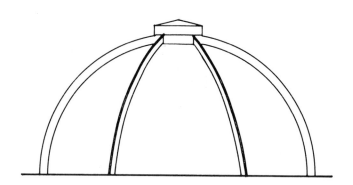

Fig. 1.31.
Arches converging radially to a common crown.

Choice of Structure

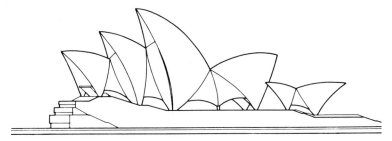

Fig. 1.32.
Complex shapes formed by arches (Sydney Opera House).

1.5. Plane Truss (see Chapter 4)

We may absorb the horizontal reactions of the linear arch (Fig. 1.19) with a tie, and we then have a simple truss (Fig. 1.33). Provided that the joints of a truss act like hinges which allow rotation, the forces in the members are either tensile or compressive. Bending within the members is avoided if the weight of the members themselves can be neglected. There is therefore no need for very thick members capable of accommodating the resistance arm of the bending moment. (However, compression members may buckle if they are made too slender: see Sections 2.6 and 7.1.)

Fig. 1.34 shows a model of a truss in which the forces can be measured with tension and compression spring balances. Ideally the members should be joined with pins as in the model. In practice the cost of machined pin joints is warranted only in special cases, such as the supports of large bridges, and any joint which is sufficiently flexible may be used in "pin-jointed" trusses (see Section 4.1). However, the joint must not be made rigid; otherwise bending is introduced into the individual members.

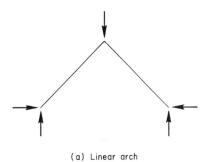

(a) Linear arch

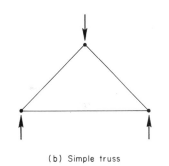

(b) Simple truss

Fig. 1.33.
A simple truss in a linear arch, in which the horizontal reactions are replaced by a tie member, and all joints are flexible.

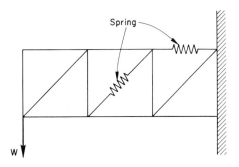

Fig. 1.34.
The members of a simple truss are either in tension or in compression. None of the members is subjected to bending or shear, although the frame as a whole resists bending and shear. This is demonstrated with a model, in which tension or compression spring balances are inserted in the individual members.

In a truss the resistance to bending (Fig. 1.8) is provided by the top chord, acting in compression, and the bottom chord, acting in tension. The depth of the truss provides the resistance arm, and since this consists mostly of empty space, the truss is much lighter than a solid beam carrying the same load over the same span (Fig. 1.35).

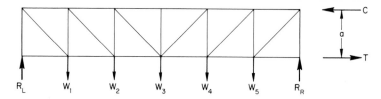

Fig. 1.35.

The resistance moment of a parallel-chord truss is provided by the compressive force in the top chord, the tensile force in the bottom chord, and the depth of the truss, which provides the resistance arm a.

The shear resistance (Fig. 1.4) is provided by the vertical and diagonal members.

Parallel-chord trusses may have only diagonal members between the chords, as in the Warren truss (Fig. 1.36), or a combination of diagonal and vertical members, as in the Pratt truss (Fig. 1.35). However, to ensure that the individual members of the truss are in tension or compression without bending, the truss must be triangulated—i.e., built up entirely from triangles (see Section 4.1). Eliminating some of the diagonals and replacing them by stiff joints induces bending in the members and thus requires heavier members (see Section 1.6).

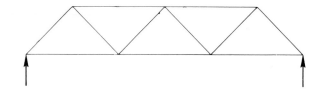

Fig. 1.36.
Warren truss.

Trusses are well-suited for sloping roofs (Fig. 1.37), and they can be adapted to indirect lighting by putting the glazing on the north side (northern hemisphere) or south side (southern hemisphere). In the temperate zone a sloping north rafter is used, but in the tropical and subtropical zone the shaded side is usually made vertical to prevent admission of undesirable heat. The same applies to monitor roofs, which admit top light from both sides. The rafters (top

Choice of Structure

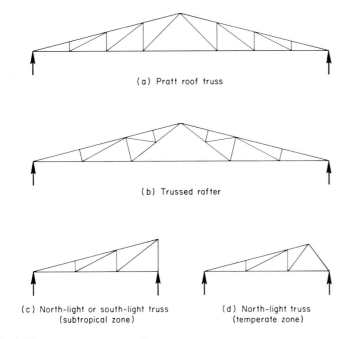

(a) Pratt roof truss

(b) Trussed rafter

(c) North-light or south-light truss
(subtropical zone)

(d) North-light truss
(temperate zone)

Fig. 1.37.
Trusses for sloping roofs.

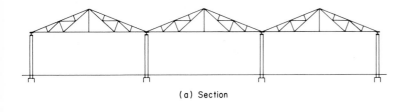

(a) Section

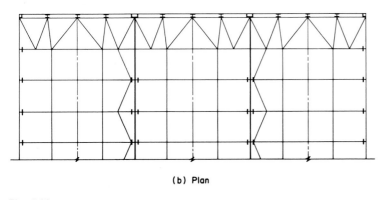

(b) Plan

Fig. 1.38.
Wind bracing for factory structure covered with "pin-jointed" roof trusses. The bracing (shown in the plan) consists of bars placed under the roof sheathing.

members) are always in compression in simply supported trusses, whatever their shape, and the bottom members are in tension, and are usually called *ties*. Approximately half the interior members are in compression, and the remainder in tension (see Chapter 4).

Trusses can be used over quite large spans; e.g., the Severin Bridge over the Rhine in Cologne has a clear span of 990 ft (300 m). In architecture, plane trusses are particularly useful for single-story utility buildings requiring column-free spaces, such as factories and gymnasia. Triangulated trusses are not sufficiently attractive for large, formal rooms, for which rigid frames, shells, or folded plates (see Sections 1.6, 1.11, and 1.12) are preferable. Plane trusses are normally arranged in parallel lines, joined by purlins, which carry the roof sheathing and additional diagonal bracing is provided at the ends to resist the wind (Fig. 1.38). However, it is difficult to fireproof triangulated trusses economically, and this limits their use to those building types which need not be fireproof.

1.6. Rigid Frame (see Section 8.6)

Rigid frames have some of the characteristics of post-and-beam construction (Fig. 1.39), of arches (Fig. 1.40), and of trusses (Fig. 1.41). By departimg from the arch shape, which produces pure compression (the parabolic arch for a uniformly distributed load), we induce bending (Fig. 1.40). By joining the columns to the beam we induce bending in the columns and horizontal reactions at the supports (Fig. 1.39). Bending is induced by the elimination of the triangulation of the truss, and the horizontal restraint provided by the tie (bottom chord) is transferred to the ground (Fig. 1.41).

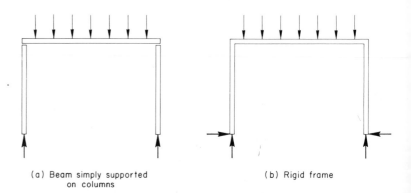

(a) Beam simply supported on columns

The beam is subjected to bending only.

The columns are subjected to compression only.

No horizontal reactions required.

(b) Rigid frame

The beam and the columns are subjected to bending *and* compression, but the maximum bending moment is less than in (a).

Horizontal reactions are required.

Fig. 1.39.
Comparison of rigid frame with post-and-beam construction.

Choice of Structure

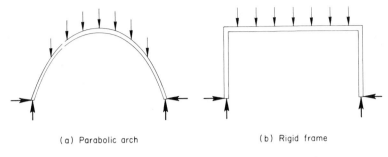

(a) Parabolic arch

A uniformly distributed load subjects the arch to compression only.

(b) Rigid frame

The beam and the columns are subjected to bending *and* compression.

Fig. 1.40.
Comparison of rigid frame with arch. The support reactions are identical for both (a) and (b).

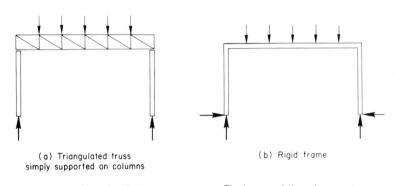

(a) Triangulated truss simply supported on columns

All members subjected *only* to either tension or compression.

No horizontal reactions required.

(b) Rigid frame

The beam and the columns are subjected to bending *and* compression

Horizontal reactions required.

Fig. 1.41.
Comparison of rigid frame with simple truss.

Rigid frame analyses involve a much more complex theory than cables, pure compression arches, or pin-jointed trusses (see Section 8.7). On the other hand, they present a better appearance than triangulated trusses, and their rigidity results either in greater economy or in increased load-bearing capacity.

Rigid frames may be used singly (Fig. 1.42), multibay (Fig. 1.43), or multistory (Fig. 1.44). North (south) lights or monitors can be inserted into rigid frames, but every sharp corner creates high local bending stresses, with appropriate increase in structural depth. If a multistory building is designed as a rigid frame, the columns may be taken to ground level (Fig. 1.44a) or cut off to give more open space at ground level (Fig. 1.44b and c). Eliminating some ground-floor columns, however, increases the size of the remaining columns.

Although the rigid frame is more economical, the beams can be

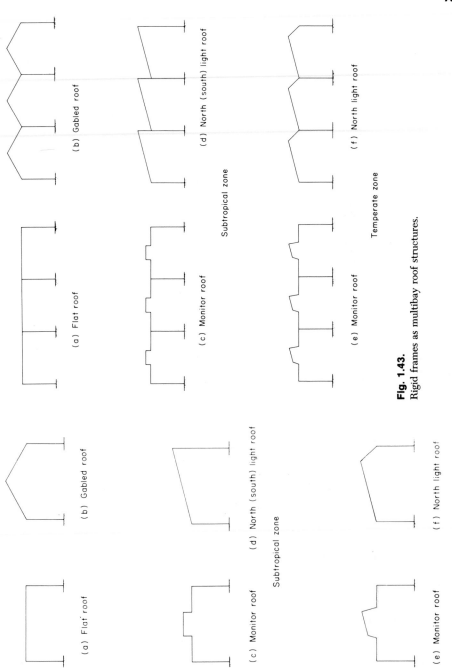

(a) Flat roof

(b) Gabled roof

(c) Monitor roof

(d) North (south) light roof

Subtropical zone

(e) Monitor roof

(f) North light roof

Temperate zone

Fig. 1.43.
Rigid frames as multibay roof structures.

(a) Flat roof

(b) Gabled roof

(c) Monitor roof

(d) North (south) light roof

Subtropical zone

(e) Monitor roof

(f) North light roof

Temperate zone

Fig. 1.42.
Rigid frames as single-bay roof structures.

Choice of Structure

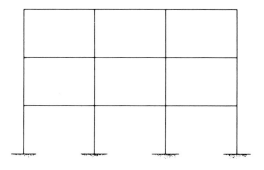

(a) Normal multistory frame

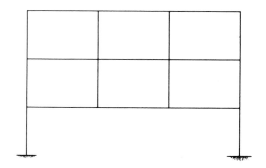

(b) Multistory rigid frame without interior ground-floor columns

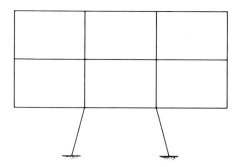

(c) Multistory rigid frame on splayed legs (pilotis)

Fig. 1.44.
Rigid frames as multistory structures.

designed for simple support on the columns, if all columns are taken to the ground (Fig. 1.45). Separate provision must, however, be made for the lateral loads (see Section 8.5).

On the other hand, a rigid frame is appropriate if some of the ground-floor columns are to be eliminated. A parallel-chord truss as in Fig. 1.35, but without diagonals, to give space for window or door openings (Fig. 1.46), is particularly useful for supporting a multistory building. The two lowest floors can be joined into a *Vierendeel truss*, a whole floor in depth, to support the building over its entire length, while still leaving space for window openings unobstructed by diagonals (Fig. 1.47). The diagonals are replaced by rigid joints at the corners of the Vierendeel truss.

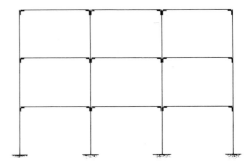

Fig. 1.45.
Nonrigid multistory frame, built with beams simply supported on beam connectors.

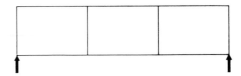

Fig. 1.46.
Vierendeel truss. The diagonals are omitted to allow space for corridors or windows, and all "pin joints" are made rigid.

1.7. Horizontal Grid (see Section 7.10)

The simplest method of framing a floor is to support it on beams which in turn rest directly on the columns (Fig. 1.48). If the span is too great, the beams, or joists, carrying the floor are supported on larger, primary beams, or girders, which are supported on the columns (Fig. 1.49). The floor may either be rigidly connected to the beams, which is normal practice in reinforced concrete cast in one piece, or each may be simply supported, which is common practice in steel construction.

Choice of Structure

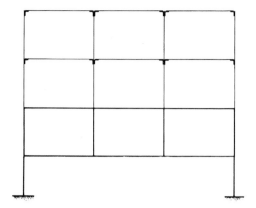

Fig. 1.47.
Nonrigid multistory frame without interior ground-floor columns. The beams on the upper floors are simply supported on column connectors. The two lowest floors are joined into a rigid Vierendeel truss, a full floor in height.

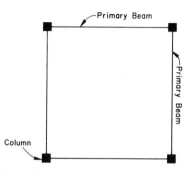

Fig. 1.48.
Floor slab supported on primary beams, which span directly between the columns.

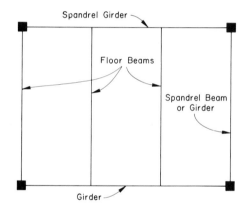

Fig. 1.49.
Floor slab supported on secondary floor beams or joists, which in turn are supported on primary floor beams or girders.

When the mesh of the grid is reduced, it is worthwhile to allow for its two-way action, whether the beams are simply supported or rigid. Consider two simply supported beams, carrying a load W over a square plan (Fig. 1.50). Since the two beams are joined together, or one rests on the other, they deflect by the same amount under the load W, and because they are identical, *each* carries half the load. The two-way system therefore saves material. The same argument can be used if the plan is rectangular, or the beams are of different size; the stiffer beam then absorbs a larger share of the load, the exact proportion depending on the relative stiffness of the two beams. Two-way grids are therefore shallower than one-way beams over the same span.

Two-way grids can be arranged parallel or diagonal to the column

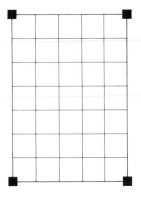

(a) Parallel-to-column lines

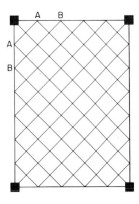

(b) Diagonal-to-column lines

Fig. 1.51.
Horizontal grid.

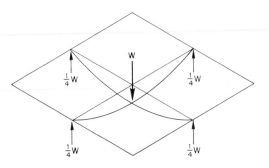

Fig. 1.50.
Distribution of load between two beams at right angles.

lines (Fig. 1.51). The diagonal system costs more to fabricate, because the members have different lengths and do not all join at right angles. On the other hand, the structure is stiffer for the same overall depth. The members crossing diagonally at the corners (A–A, B–B, etc. in Fig. 1.51b) are shorter, and therefore deflect less than the members in Fig. 1.51a. The greater stiffness of these end members increases the relative stiffness of the entire grid, so that a smaller structural depth is attainable.

The rigid interconnection of horizontal members in two directions introduces a form of deformation which is new to our vocabulary (see Section 6.7). If one member of a rigid horizontal grid is bent, those connected to it at an angle, except the one at the center, are *twisted* (Fig. 1.52). The members of a horizontal rigid grid, whether they run parallel to column lines or diagonally to them, are therefore subjected to combined bending and *torsion* (or twisting).

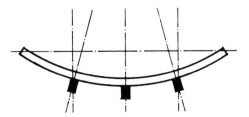

Fig. 1.52.
The rigid connections between the members of a horizontal grid introduce torsion into most members. In this figure, the central cross-member is not twisted; but the two outer members are subjected to torsion.

1.8. Space Frame (see Section 4.7)

The lamella roof is a three-dimensional variation of the diagonal grid (Fig. 1.53). It can be formed into a vault or a dome, but the connections must be rigid, and the members are therefore subject to a combination of bending and torsion (because it is a rigid diagonal grid) and compression (because each member acts like an arch—see Fig. 1.20). The members consequently need considerable depth to accommodate the bending stresses.

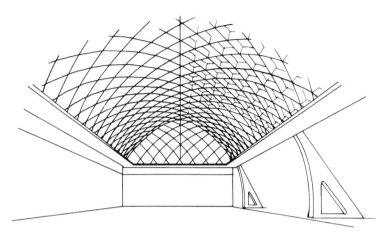

Fig. 1.53.
Lamella roof. This is the space-frame version of the diagonal grid, and it requires rigid joints for stability.

If additional members are added, so that the frame of the vault or dome becomes triangulated (Fig. 1.54), and if the joints are made flexible, the forces in the individual members of the structure are reduced to simple tension or compression. The space frame then behaves similar to the triangulated plane frame (Fig. 1.35) except that it forms a three-dimensional shape.

Because of its shape, a triangulated dome is stable with a single layer of triangles assembled with flexible ("pin") joints. A horizontal frame with flexible joints would fall down, and triangulated horizontal frames must therefore have at least two layers interconnected by triangularly arranged members (Fig. 1.55). The members of triangulated space frames are light, since they are only subjected to either tension or compression. On the other hand, the roof structure is not necessarily economical, because jointing in space is complicated, and because the frame still requires a roof covering. Flat roofs, vaults, and domes can be built by other means, which at present are generally simpler and cheaper (see Sections 1.9, 1.11, and 1.12).

Space frames have other and more important uses. Towers and tall masts are usually designed as space frames, since they cannot easily be built in any other way.

In conclusion, it is worth pointing out that the traditional domestic pitched timber roof is a space frame of some complexity. The sizes of the members are determined by empirical rules, established

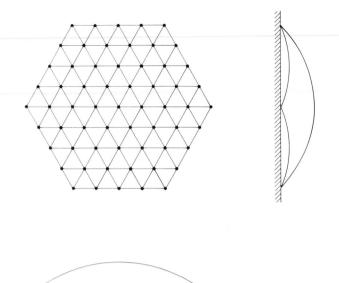

Fig. 1.54.
Geodesic dome on a hexagonal plan. This is the space-frame version of the triangulated truss, and all joints are "pin joints."

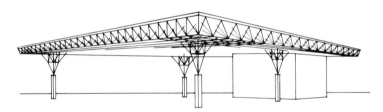

Fig. 1.55.
Space frame for flat roof. Two layers are needed for a triangulated "pin-jointed" flat roof.

over long periods by trial and error. However, if the size of the rafters is worked out as for a plane truss, it generally comes out bigger than the empirical rules (which are quite satisfactory in practice) require. Evidently the structure behaves sufficiently like a space frame to warrant the reduction in size.

1.9. Plate (see Sections 7.10 and 8.7)

We have seen how a horizontal grid carries the loads in both directions by two-way action (Fig. 1.50). If the mesh of the grid is progressively reduced, we eventually get a solid plate. Thus the plate carries the loads placed upon it by two-way action (Fig. 1.56), and it is therefore stronger than a series of separate planks.

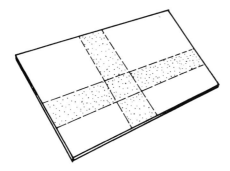

Fig. 1.56.
Two-way action in plate. The square forms part of the two shaded strips spanning at right angles. Part of the load is carried by one strip and part by the other. The proportion depends on the location of the element, and on the ratio of the long to the short span.

A plate carrying a load W by two-way action is subjected to bending (like the individual strips of the plate), to shear between the strips, and also to torsion (Fig. 1.57). The torsion causes the corners of the plate to curl up (Fig. 1.58) unless they are held down by additional reactions at the corners. Consequently, the corners of a two-way reinforced concrete plate require torsional reinforcement at the corners if cracking is to be avoided.

Plates may be supported on primary girders and secondary floor beams (Fig. 1.49). If the ratio of the long to the short span is greater than 1.5, the two-way action is negligible, and the plate may be assumed to span entirely in one direction. If the ratio of long-to-short span is between 1.5 and 1 (i.e., square or nearly so), the plate spans between the beams by two-way action, as shown in Figs. 1.56 to 1.58.

Reinforced concrete plates may also be supported directly on the columns, with or without enlarged column capitals (Fig. 1.59). To distinguish between the two, the plate supported directly on the columns is called a *flat plate*, and the plate supported on enlarged

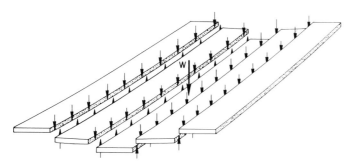

Fig. 1.57.
Two-way action in the plate, as in the horizontal grid, produces torsion. It also produces shear between adjacent strips of the plate.

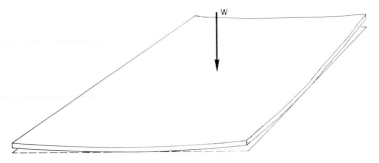

Fig. 1.58.
Because of torsion, the corners of a two-way plate curl up.

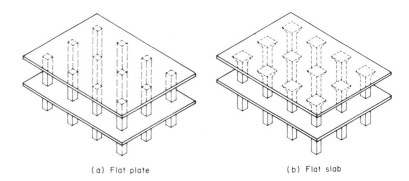

(a) Flat plate   (b) Flat slab

Fig. 1.59.
The flat plate is directly supported on the columns; the flat slab is supported on enlarged column capitals to reduce the shear around the joint between the column and the slab.

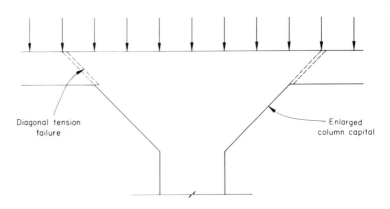

Diagonal tension failure

Enlarged column capital

Fig. 1.60.
Flat slabs (as shown, with enlarged column capitals) and flat plates (without enlarged column capitals) are liable to fail in diagonal tension due to the shear force around the periphery of the column.

column capitals is called a *flat slab*. If the plate is supported directly on the columns without beams, a shear failure around the column must be considered, the column punching through the loaded slab (Fig. 1.60). Without column capitals the surface resisting the punching shear is further reduced, and special reinforcement is sometimes needed in the flat plate around the column.

The flat plate has great economic merit for multistory buildings with moderate spans between columns. No floor structure could be simpler than a series of concrete plates supported directly on columns. For longer spans, however, the depth of the plate becomes excessive.

1.10.   Pneumatic Membrane (see Section 9.3)

To be thin and light, surface structures should be free of bending and transverse (i.e., punching) shear, like a balloon or a soap bubble. If any attempt is made to bend or punch through a soap bubble, it bursts immediately. A balloon resists longer, but only by deforming, say, under the pressure of a pencil to minimize bending and tranverse shear. It can be blown up, which proves that it can resist tension and shear within the membrane.

It is perfectly feasible to use a balloon to form a pneumatic roof over a small space (Fig. 1.61). Collapsible pneumatic structures can be formed of plastic or impregnated cloth and be held up by a very small internal pressure, which is unnoticeable to the people inside, and is maintained by an air-blower. Such buildings are useful for temporary exhibitions, and perhaps for turning an outdoor swimming pool into an indoor one in winter. In theory they could also be used as large-exhibition buildings, but the practical problems are too formidable at the present time to make this economical.

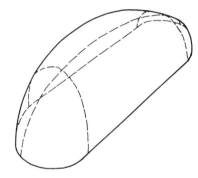

Fig. 1.61.
Roof formed by pneumatic membrane, supported by a small internal excess air pressure.

1.11.   Shell (see Sections 9.3 to 9.5)

The pneumatic structure is the *surface* corresponding to the *linear* suspension cable. Like it, it is very light, but too flexible for widespread practical use in architecture. The shell dome corresponds to the arch, and under the most favorable conditions it is entirely in compression.

Although we think of an eggshell as particularly thin and fragile, reinforced concrete shells are, in proportion to their span, even thinner. Let us compare the dome of St. Peter's Cathedral in Rome, built in 1590, which is a large Renaissance masonry dome; the first thin reinforced concrete shell, the Planetarium at Jena, East Germany, built in 1923; and a large, modern concrete shell, the CNIT Exhibition Hall, built in Paris in 1958 with a span of 720 ft (216 m) (see table on the following page):

	Ratio of span to thickness of shell
St. Peter's Cathedral (taking the total thickness of the double shell)	13
Average hen's egg	100
Jena Planetarium	420
CNIT Hall, Paris (taking the total thickness of the double shell)	1,800

The thinness of a well-designed shell saves a great deal of material and, therefore, weight. Since the structure carries mostly its own weight, the saving is progressive.

The thinness of the shell is, however, dependent on the elimination of bending and transverse shear over most of the shell, since neither can be resisted without depth. Bending, in particular, requires depth to accommodate a resistance arm between the tensile and compressive components of the moment (Fig. 1.8). Tension, compression, and shear *within* the surface of the membrane (*membrane forces*) can be resisted by a thin shell, if the designer provides correctly designed supports.

There is a great variety of practicable shell forms, and we will discuss these in more detail in Sections 9.3 to 9.5. In this introductory chapter we can only look at the principal classifications. The simplest shell form is the cylindrical vault. We can make this easily by taking a piece of paper, forming it into a half-cylinder, and tying the ends with a string to prevent them from spreading (Fig. 1.62). The paper is much stiffer now than when it was flat, which indicates

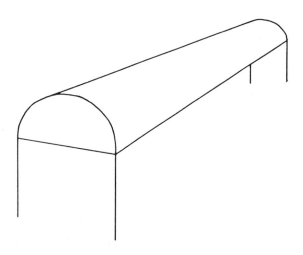

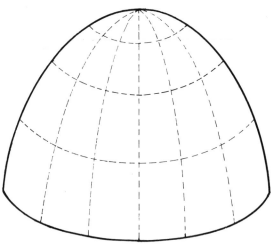

Fig. 1.62.
Cylindrical shell. The cylindrical shell may be formed from a flat sheet, but it requires ties at the ends to prevent them from spreading.

Fig. 1.63.
Dome. The dome has double curvature, and spreading is prevented by a tie which is integral with the dome. It cannot be formed from a flat sheet, or generated by straight lines. This complicates the formwork for casting a dome in concrete.

Choice of Structure

the structural superiority of the shell over the plate. The tie at the ends is essential to absorb the horizontal reaction of the curved shape (see Fig. 1.20); without it the paper would flatten out. The cylindrical vault was used in Roman and Romanesque structures and has had a modern revival in reinforced concrete.

The cylindrical shell is curved in only one direction. If curvature improves the structural performance so much, then curvature in both directions may be expected to be even more favorable. We thus obtain the dome, which has a long and honorable history in architectural design (Fig. 1.63). The dome has one important disadvantage in terms of modern technology: it cannot be formed by a series of straight lines. Practically all architectural shells are now made of concrete, and the formwork of the concrete comes in straight pieces (usually timber, sometimes steel). Since we cannot draw straight lines on a dome, however we try to arrange them, the formwork is expensive.*

If we look for shells that can be formed entirely from straight pieces of timber, there is only a small range to choose from. The most important is the *hyperbolic paraboloid* shell, or *hypar* for short (Fig. 1.64). This is a remarkably versatile form, which can be used in a variety of ways. While the dome is like the top of a hill, the hypar shell is like the saddle of a mountain pass (Fig. 1.65). Hypar shells can be used as saddles or they can be used with straight boundaries, either alone or in combination (Fig. 1.66).

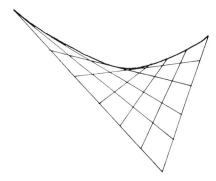

Fig. 1.64.
Hyperbolic paraboloid (*hypar*) shell. The hypar shell has double curvature, so that it cannot be formed from a flat sheet. It is, however, one of the few surfaces generated entirely by straight lines, which greatly simplifies formwork for casting the shell in concrete.

---

*In a recent innovation by Dante Bini, the reinforced concrete is cast flat on the ground on a dome-shaped plastic membrane. The dome is then raised with an air compressor before the concrete has set. The reinforcement consists of springs, which extend as required. As soon as the concrete has gained sufficient strength, the membrane is collapsed, and the necessary openings are cut into the concrete. The method has been successfully and economically used in Italy, the USA and Australia. The completed structure is a dome, but the construction is based on the pneumatic membrane (see Section 1.10).

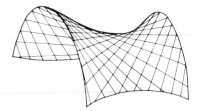

Fig. 1.65.
The hypar shell as a saddle. This shape is obtained if the shell of Fig. 1.64 is cut at 45°.

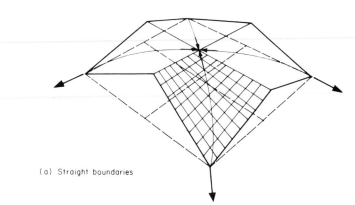

(a) Straight boundaries

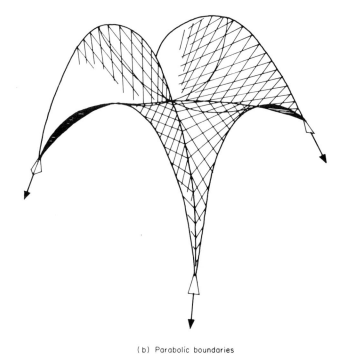

(b) Parabolic boundaries

Fig. 1.66.
Combination of hypar shells as roof structures.

1.12. Folded Plate (see Section 9.6)

Most structural forms are of great antiquity, although the modern version is usually much lighter and stronger. Beams and cables have been used since prehistoric days. Wooden trusses, stone arches, cylindrical vaults and domes were well-established in Roman days. The hypar shell is a modern invention, and so is the folded plate.

Choice of Structure

Its effectiveness has been illustrated by Curt Siegel in a simple experiment (Fig. 1.67). A single sheet of paper has insufficient strength to carry its own weight (a). If it is folded in the direction of the span in a series of parallel folds, it acquires stiffness, and will easily carry a hundred times its own weight (b). If the load is increased, the structure fails as the folds straighten out (c). This can be prevented by gluing a stiffener to the ends (d). Transverse stiffeners of this type are essential in folded plates, just as ties are essential in cylindrical shells (Fig. 1.62).

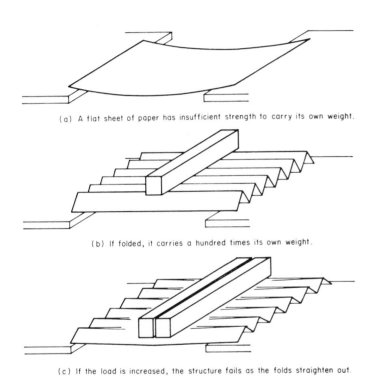

(a) A flat sheet of paper has insufficient strength to carry its own weight.

(b) If folded, it carries a hundred times its own weight.

(c) If the load is increased, the structure fails as the folds straighten out.

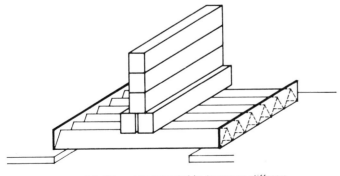

(d) This can be prevented by transverse stiffeners.

Fig. 1.67.
Folded plate.

Since the folded plate consists of straight pieces joined with sharp edges, it is subjected to bending, and it cannot be made as thin as a shell. We have remarked that the pneumatic membrane is the space version of the cable, and the dome the space version of the arch. Similarly the folded plate is the space version of the rigid frame (Fig. 1.68).

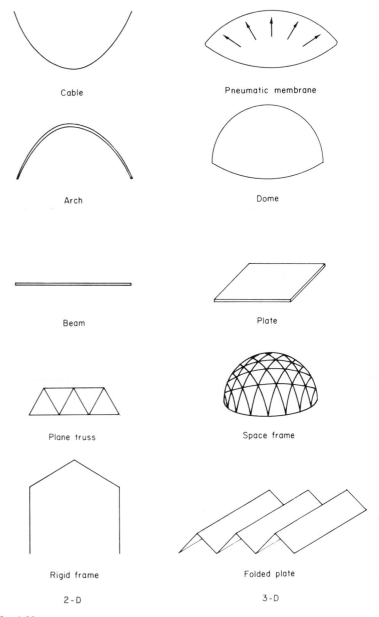

Cable	Pneumatic membrane
Arch	Dome
Beam	Plate
Plane truss	Space frame
Rigid frame	Folded plate
2 - D	3 - D

Fig. 1.68.
Comparison of structural types in two and three dimensions.

Choice of Structure

Folded plates are difficult to make in timber because of jointing problems, and metal is liable to buckle in compression. However, the flat surfaces are easily formed for casting, and most folded-plate structures are therefore made of reinforced concrete.

An attractive variation of the folded plate is obtained by triangular folding (Fig. 1.69) which, like the previous example, provides an essentially flat roof. Folded plates can also be formed as frames or domes (Figs. 1.70 and 1.71).

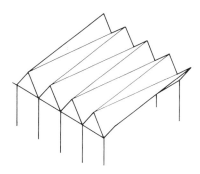

Fig. 1.69.
Flat folded-plate roof formed from triangles.

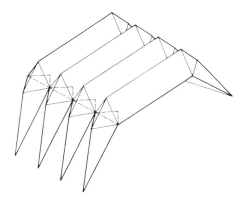

Fig. 1.70.
Rigid frame formed from folded plates.

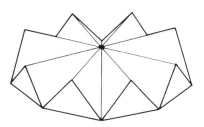

Fig. 1.71.
Dome formed from folded plates.

1.13.   Tall Buildings (see Sections 8.7 and 8.8)

We have so far confined our discussion to horizontal spans carrying vertical loads. For tall buildings, the horizontal loads due to wind and earthquakes (see Section 2.7) may be as important as or more important than vertical loads.

There are two main structural problems to be solved in the design of a tall building. First, the vertical loads, which were considered in Section 1.6, act on each floor; and the more stories there are, the greater are the loads at the base of the columns. In a two-story building, the upper columns carry the loads of only one floor, and the lower columns carry the loads of both floors. In a hundred-story building the columns of the lowest story carry the vertical loads of all hundred stories. The columns, therefore, must increase in size with the height of the building for that reason alone.

Second, the horizontal loads bend the building sideways. It behaves like a vertical cantilever (Fig. 1.72), and the overall bending moment reaches a maximum at ground-floor level (see Section 5.2). The frame-design of multistory buildings illustrated in Figs. 1.44 and 1.45 is therefore not economical for tall buildings.

A better solution is to design the frame of the tall building like a perforated tube (Fig. 1.73). The tube consists of horizontal members (the spandrel beams at each floor level) and vertical members (the

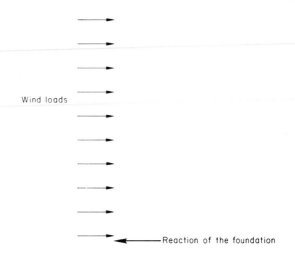

Wind loads

Reaction of the foundation

Fig. 1.72.
The wind loads acting on a tall building are resisted by the reaction of the foundation, and they bend the building as a vertical cantilever.

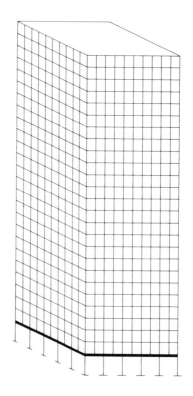

Fig. 1.73.
The tall building as a perforated tube, formed by the horizontal spandrel beams at each floor level and the vertical columns, with spaces between for window and other openings.

Top
Residence at Highland Park, Illinois, with exposed steel frame.
Photograph by courtesy of AISC.

Bottom
Federal Reserve Bank, Minneapolis, Minnesota. The building is supported over a clear span of 275 ft (82.5m) by two catenaries formed by cables and steel beams.
Photograph by courtesy of the architects, Gunnar Birkerts and Associates.

Top

The Knights of Columbus Headquarters, New Haven, Connecticut. The structure consists of reinforced concrete faced with brick, and an exposed, weather-resistant steel frame.

Photograph by Courtesy of AISC.

Bottom

Myer Music Bowl, Melbourne, Australia. An open-air concert shell formed by aluminum-faced plywood supported by steel cables.

Photograph by courtesy of Yuncken Freeman Architects Pty. Ltd.

Centrepoint Tower, under construction in Sydney, Australia. A cable-stayed steel tower 885 ft (265m) high, atop a 15-story reinforced-concrete framed building.
(*Top*) Architectural model superimposed on location photograph.
(*Bottom*) Scale model in boundary-layer wind tunnel to test dynamic behavior of structure.
Photographs by courtesy of the structural engineers, Wargon, Chapman and Associates.

Fig. 1.74.
The John Hancock Center in Chicago uses the structural windbracing as a visible feature.

columns), and there are spaces between them for window and other openings. The tube is loaded by the vertical loads, which produce compressive forces in the columns and, in addition, relatively small bending moments in all the members of the tubular frame, as was discussed in Section 1.6. The tube is bent sideways by the horizontal loads, producing much bigger bending moments and shear forces in all the members of the tubular frame.

Buildings present a special problem if they are very tall. The longest span in a building (as distinct from a bridge) is at present 720 ft (220 m). The tallest building is at present 1,450 ft (442 m) high. Since it is loaded as a cantilever, this is equivalent to a horizontal beam of double the length spanning between two end supports (see Section 5.2 or Table 6.2), i.e., a simply supported span of 2,900 ft (884 m). We are thus dealing with bending moments that are much larger than those for any horizontal span in an architectural structure.

We can join the horizontal and vertical members rigidly, as shown in Fig. 1.73, or employ the truss concept of Section 1.5, and use diagonal bracing members (Fig. 1.74). The diagonals, are, however, a dominating visual feature that is not always acceptable.

Suggestions for Further Reading

H. J. Cowan: *An Historical Outline of Architectural Science*. Elsevier, Amsterdam, 1966. Chapters 1-4, pp. 1–95.

M. Salvadori and R. Heller: *Structure in Architecture*. Prentice-Hall, New York, 1963. Chapters 6-12, pp. 97–357.

C. Siegel: *Structure and Form in Modern Architecture*. Reinhold, New York, 1962, 302 pp.

F. Wilson: *Structure—The Essence of Architecture*. Van Nostrand-Reinhold, New York, 1971. 96 pp.

Design Criteria 2

Accident: an inevitable occurrence due to the
action of immutable natural laws.

Ambrose Bierce

In this chapter we examine the criteria of structural design. What loads does a structure carry? How can it fail? Which of the various modes of failure is likely to be critical? How do we establish whether the structure is safe?

2.1. Building Regulations

Practically every structure built today must conform to the appropriate building regulations. At first sight this may seem a restriction imposed by an ill-informed local council on a highly skilled designer who is trying to create more imaginative structures. This impression is reinforced when one observes that designers working for the central government do not normally have to conform to the regulations of the local council in whose area the structure is to be erected.

In reality the safety of structures is not so much a matter of opinion as of ascertainable fact. People's opinions may differ on the question of erecting very tall buildings in the center of the city. Some feel that it is unwise to generate in already overcrowded streets the additional traffic that is certain to result from a high building accommodating many persons. Others think that the central area of a city must develop or decay, so that rigid restrictions stand in the way of progress. Neither opinion could be proved right or wrong by experiment, so that' regulations restricting the height of buildings (or lack of such regulations) will always be attacked by those who hold opposite views. Most regulations affecting structural safety, however, are based on established facts, and those that are open to questions can mostly be resolved by experimental research.

There are some exceptions to this general rule. Prior to the eighteenth century, all structures were designed by empirical rules, i.e., traditional rules established by experience, and slowly modified by experience and changing circumstances. These rules are still used for some traditional construction, particularly in brick and timber.

As we have mentioned (Section 1.8), the traditional domestic pitched timber roof is a space frame of some complexity, whose members are sized by empirical rules established over long periods by trial and error. If we design it by mechanics, resolving the space frame into a series of parallel-plane frames, we often get larger members, so that the traditional rules are retained for the sake of economy. The same applies to the small brick wall, which is a complex assembly of small blocks (see Section 2.3). The other exception is the very large and complex structure, which sometimes goes beyond the range of our knowledge of the theory of structures. Since it may not be worthwhile or feasible in the time available to conduct a research program just for this one structure, it is necessary to make simplifying assumptions. These generally err on the side of safety, but opinions may differ on just how conservative they should be, or even on the precise way in which the structure will behave under load.

The great majority of structures are designed in accordance with mechanically based rules, and the designer working for the central government does not really "get away with it." He uses a code approved by his department, which is probably based on the same model code as that adopted by the local council.

Model codes are documents prepared by professional committees. In Great Britain, Australia, and most British Commonwealth countries the committees are established by bodies such as the British Standards Institution or the Standards Association of Australia. These are institutions chartered by the government but supported mainly by voluntary contributions from industry. Their councils invite professional bodies, such as engineering and architectural institutions and industrial research or development organizations, to nominate the committee members, and the drafts are made available for public comment before the code is enacted. In the United States of America, committees for drafting model codes are commonly appointed directly by the professional institutions—e.g., the model concrete code is written by the American Concrete Institute—but the procedure is otherwise the same. Model codes are amended with changing circumstances, and rewritten from time to time. Few structural codes remain unchanged for more than 10 years.

Model codes have no legal force, but in practice, local authorities adopt them without change, or with only minor alterations. Innovations in important model codes generally receive attention throughout the world. For example, a major change in the American Concrete Code is likely to lead to similar changes in many other model codes within a few years.

Structural building regulations are thus not as arbitrary or as varied as they may seem at first sight. The wording often differs considerably, but the resulting structural sizes are much the same by any code. A designer may be restricted to erecting a building 150 ft (50 m) high in one city, and permitted to go five times as high elsewhere. However, it is most improbable that the structural sizes required for the same building would differ by as much as 50% in any two cities.

2.2. Overstressing the Structural Material

There are several ways in which a structure can fail. The entire structure may overturn. The structure may come apart at the joints and thus collapse. The entire structure or a part of it may buckle, and the building then collapses. Deformation or cracking may reach proportions that render the building useless for its assigned purpose. The principal criterion for mechanically designed structures is, however, the maximum stress.

If a material is loaded in tension, compression, bending, or torsion, or by a combined loading, it eventually fails. Failure may mean permanent deformation, as in structural steel, or fracture, as in concrete (see Section 6.1). If we could control materials perfectly at a reasonable cost, we could predict their strength precisely. In practice, we can only specify a minimum.

The designer determines the *minimum* strength of the materials on which his design is based, and the builder (sometimes under the designer's supervision) must ensure that this strength is achieved. Taking concrete as an example, we require that test samples be taken (usually 6-in. diameter cylinders, 12 in. long; 150 mm × 300 mm cylinders; 6-in. cubes; 150 mm cubes; or 200 mm cubes). These are broken, and the maximum load which they carry in compression is measured.

We cannot break every piece of concrete, otherwise there would not be any left. However, by a judicious choice of samples we can determine its minimum strength with a high probability, say 95%. The remaining 5% is then covered by a *factor of safety*.

Structural steel and reinforcing steel for concrete are made in a large factory, and the manufacturer undertakes the testing to ensure that the steel conforms to the standard specification. Steel is made in different strength grades, each having a guaranteed minimum strength.

We now design the building for these specified steel and concrete stresses, with a factor of safety. This takes account of the fact that we have tested only selected samples; it allows for uncertainties in the loading (see Section 2.7); and it provides a margin for the simplifications made in every structural theory to keep the amount of calculations commensurate to the importance of the structure.

The determination of this factor is one of the most important tasks for any committee drafting a building code. The factor has been reduced progressively with increasing knowledge of structural theory, materials technology, and improved supervision on the building site. At present it ranges from 1.4 to 2.4.

The factor can be introduced in one of two ways. We can say that the service load (normally carried by the structure), W, multiplied by a load factor must not produce stresses anywhere in the structure which exceed the ultimate strength of the material, f'. This is called *ultimate strength design*.

Alternatively, we may specify that the *actual* stresses anywhere in the structure caused by the service load W normally carried by the structure must not exceed the *maximum permissible stress*. We thus get the basic design criterion

$$f_{\text{act.}} \leqslant f_{\text{perm.}} \qquad (2.1)$$

The maximum permissible stress, in turn, is defined as

$$f_{\text{perm.}} = \frac{f'}{S_f} \qquad (2.2)$$

where f' is the ultimate strength of the material and S_f is the factor of safety. This is called *elastic design*.

If the stresses were always directly proportional to the loads, then the two methods would always give the same answer, and the load factor should equal the factor of safety; however, this is not necessarily so. Although ultimate strength design is being used to an increasing extent, we shall in this introductory text confine ourselves to elastic design, except for reinforced concrete in Sections 7.9 to 7.11.

There are elastic solutions for all structural problems which can be solved by ultimate strength. The reverse is not true at present; in particular, there are no satisfactory ultimate strength theories for shells.

2.3. Failure of Weak Joints

Although the design of modern architectural structures is generally based on the stresses in the members, this has not always been the criterion.

Traditional masonry structures almost invariably fail through opening up of the joints. Let us consider a flat arch, first in reinforced concrete (Fig. 2.1a) and then in stone blocks joined by a weak mortar (Fig. 2.1b). A reinforced concrete arch, carrying a concentrated load as shown may fail through overstressing (crushing) of the concrete on the compression (top) face, or through opening up of a crack on the tension (bottom) face, followed by overstressing (plastic yield) of the steel. In a traditional masonry arch, failure is initiated when the line of thrust caused by combined

Design Criteria

bending and compression (see Section 7.12) touches the outside (extrados) or the inside face (intrados) of the arch. A joint then opens up. Four open joints are necessary to turn the arch into a "mechanism," so that it can move freely under load (see Section 4.1). It then collapses, without overstressing the stone. The blocks of stone are not damaged before the arch fails. Evidently we have not made full use of the strength of the stone.

Modern masonry structures still fail mainly by the same mechanism. Thus, most brick structures fail through opening up of the joints between the bricks, before any of the bricks are crushed.

Similarly, the members of traditional timber structures were often far larger than necessary, and failure occurred in the joints. Traditional timber joints, which are still used in many pieces of wooden furniture, were quite complicated before metallic jointing materials, such as nails and timber connectors, became freely available, and the size of the member was partly determined by the space needed for the joint.

The joints in modern steel and concrete structures are almost invariably designed to be stronger than the members they connect, so that design calculations concentrate on sizing the members. This is easily achieved in concrete, where additional reinforcement can be inserted into the joints without being visible from the outside. In steel, welding and bolting has been perfected to the stage where it is easy to produce strong joints.

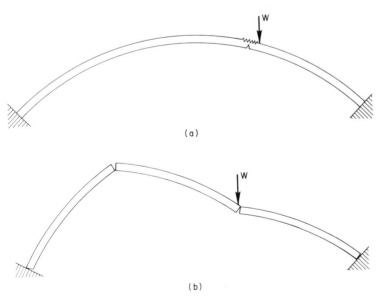

Fig. 2.1.
Collapse of a flat arch in reinforced concrete and in traditional masonary. (a) The arch fails because of crushing of the concrete on top, or because of cracking of the concrete below followed by tension failure of the steel reinforcement. (b) Collapse occurs after four mortar joints have opened up sufficiently. The stone blocks are *not* overstressed.

Making the joints stronger than the members is convenient, but not necessarily good practice. For example, if a brick wall shows signs of failure through opening up of the joints, it can be strengthened, and the damage repaired by scraping the joints and repointing them, provided the bricks have not been damaged. If traditional lime mortar is used, the bricks are rarely cracked if the joints open up. When the much stronger cement mortar is used, cracks often run through the bricks, and if the damage is serious, there is no alternative to demolishing the wall.

2.4. Excessive Deformation and Cracking

Every material under load deforms elastically, i.e., it elongates in tension or shortens in compression, and recovers its shape when the load is removed (see Section 6.2). Consequently, beams deflect simply because they are loaded. Under a uniformly distributed load, the bending moment is proportional to L^2 and the deflection to L^4 (where L is the span). Consequently if beams are designed for stress, then they deflect more as they get longer.

Elastic deflection is, in itself, harmless, and it is rarely so large as to be unsightly. If, however, a steel beam or truss carries a ceiling made of a brittle material, such as plaster, this will crack as the structure deforms elastically. Load tables for steel structures (see Section 7.5) are based on maximum stresses; but they also indicate loads that cause deflections in excess of, say, $\frac{1}{250}$ of the span, so that brittle materials should not be used in conjunction with excessively deflecting structures.

Concrete is unlikely to show excessive elastic deflections, since the sections are much stiffer. In addition to its elastic deflection, however, concrete deforms further as water is squeezed out of the fine pores of the cementing material; this is called *creep*, and it occurs slowly over a long period of time. The total creep deflection may be two or three times as much as the elastic deflection. Concrete structures with a low ratio of depth to span, such as flat plates (see Section 1.9), are therefore liable to show significant deflection some months after the building is complete, and this may damage brittle materials, such as plaster finishes and brick partitions, or cause doors and windows to jam. Creep deflection of concrete cannot be controlled by the simple device of drawing a limiting line on a load table, as for the elastic deformation of steel. Careful attention must be paid to the design details of concrete structures, which are liable to show excessive deflection (see Section 8.9).

Cracking occurs when the tensile stress in a brittle material reaches the limit of fracture, and this is liable to happen in concrete, brick, plaster, and other brittle finishes. It can be caused by loads, but also by temperature and moisture movement. Concrete expands

Design Criteria

approximately $\frac{3}{4}$ in per 100 ft for a 100°F change of temperature, or 6.3×10^{-6} per °F (3.5×10^{-6} per °C).* A change in temperature of 100°F (55.5°C) is likely to occur in every structure during its lifetime, so that temperature movement must be allowed for. Concrete also moves as a result of drying shrinkage, which occurs as the wet concrete dries out after casting. This ranges* from 3 to 10 $\times$ 10^{-4} (i.e., $\frac{3}{8}$ to $1\frac{1}{4}$ inches per 100 ft or 3 to 10 mm per 10 m).

Concrete must be designed with sufficient reinforcement to control the cracking. It must also be separated from other parts of the building, which it may damage by its own movement. Thus, reinforced concrete joined firmly to brick often causes cracks in the brickwork.

Deformation and cracking rarely cause the collapse of a building. However, they frequently cause unsightly damage, which seriously disfigures both the exterior and the interior. Cracks may also permit water to enter the building, or reach the steel reinforcement of concrete and thus cause it to rust. Both must be controlled, whether they are due to the loads or due to movement caused by temperature, moisture, or settlement of the foundations (see Sections 6.9 and 8.9).

2.5. Overturning of the Entire Structure

A building could overturn under the action of wind, if it is made too tall for its base (Fig. 2.2). The mistake is so gross, that it is most unlikely to occur. The answer lies in either spreading the basement

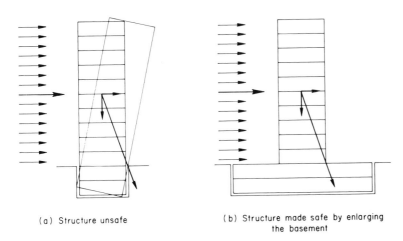

(a) Structure unsafe

(b) Structure made safe by enlarging the basement

Fig. 2.2.
Failure by overturning of the entire structure under wind pressure.

*Temperature expansion and shrinkage are ratios of length per unit length, so that they have the same numerical value in inches and in millimetres.

or taking it deeper. Evidently it is expensive to erect a very tall building on a restricted site, which does not allow room for a wide pedestal or basement, since it requires a deep excavation. Digging a deep hole is more expensive in soft soil (particularly in waterlogged clay) than in hard rock, which does not need retaining walls.

Buildings occasionally overturn partly or completely due to settlement of the foundations (see Section 8.9). The most spectacular recent failure was that of the São Luis Rei Building, an 11-story reinforced-concrete building under construction in Rio de Janeiro, in 1958. This was inadequately founded on soft clay, toppled over because of slow, uneven settlement during construction, and was a total loss.

Many old buildings had inadequate foundations, and settled unevenly. The best-known example is the Leaning Tower of Pisa, founded on volcanic ash; it, too, started to settle unevenly during construction, but has survived for more than eight centuries.

2.6. Buckling

Buckling does not necessarily damage the material. Let us, for example, take a long, thin piece of spring steel (or the long, thin rod used for pointing to objects on a projection screen) and lean on it. It will bend sideways but recover its shape when the load is removed. Although it has not been damaged, it is evidently useless as a column in the buckled condition. Buckling therefore constitutes a structural failure.

Pure buckling is independent of the strength of the material, but is proportional to the modulus of elasticity and is influenced by the geometry of the cross-section of the compression member (see Sections 7.1 and 7.3). Members under axial tension and short compression members are affected only by the strength of the material and the cross-sectional geometry, because buckling is not a problem for these members. Thus, a high-strength steel carries a much larger tension or short-compression load than ordinary structural steel (see Section 1.3), but both buckle at the same load because the modulus of elasticity is the same for all steels. On the other hand, it is possible to produce high-strength aluminum alloys which are as strong as structural steel, but the alloying does not alter the modulus of elasticity of aluminum, which is only one-third of that of steel; consequently the aluminum alloy buckles at one-third of the load.

Deformation due to elastic buckling usually produces a bending failure in practical structures, so that elastic recovery does not occur.

Buckling occurs only where there is compression (see Fig. 1.12). It is an important design criterion in aircraft frames and other very light structures. In concrete structures, buckling causes little trouble

Design Criteria

because they are massive; but it has to be considered in steel structures. It is never a major design criterion in buildings. The design of columns is modified to allow for the effect of buckling (see Section 7.3), and stiffeners are used to prevent thin structural members (such as the webs and compression flanges of steel girders, or sheet metal used in floors and roofs) from buckling (see Sections 7.5 to 7.7).

2.7   Loads

The structure first of all carries the *dead load*, which includes its own weight, the weight of any permanent nonstructural partitions, built-in cupboards, floor surfacing materials and other finishes. It can be worked out precisely from the known weights of the materials and the dimensions on the working drawings; but it can also be predicted with some accuracy before design commences from previous experience with similar buildings. Although the dead load can be accurately determined, it is wise to make a conservative estimate to allow for changes in occupancy; e.g., the next owner might wish to demolish some of the fixed partitions and erect others elsewhere. The dead load always acts vertically (Fig. 2.3).

The *live load* includes removable partitions, removable furnishings, removable machinery, materials in storage, and the weight of

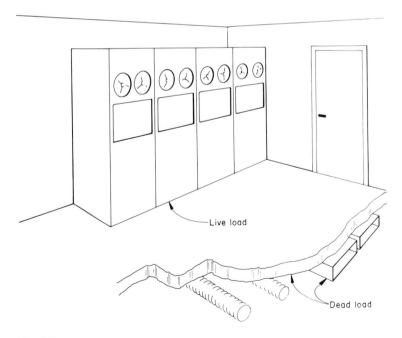

Fig. 2.3.
Dead loads and live loads in a building.

the people in the buildings. With rare exceptions, only the vertical component of the live load need be considered, although some machines produce a horizontal force.

Building codes always give rules for the dead and live load. These rules tend to be conservative, but it must be remembered that excessive live loads are particularly liable to occur when the building is already in a weakened condition, as after an outbreak of fire. The people in the building may then crowd together in some parts of it (which are largely determined by the unpredictable location of the fire) and produce live loads far in excess of normal.

The designer has little freedom of choice in the determination of the dead and live loads, since building authorities usually require strict compliance with their codes.

The effect of *wind* becomes particularly important in tall buildings. (Fig. 2.4). Wind applies a pressure to the windward side of buildings, and suction to the leeward side. It also applies suction to a flat or slightly pitched roof (Fig. 2.5). Wind forces on buildings are thus largely horizontal. The magnitude of wind forces depends on the location of the building. They are higher in the tropics than in

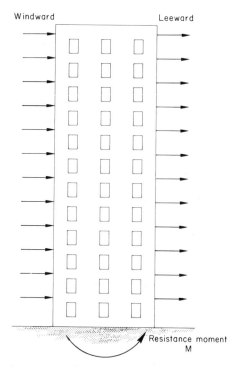

Fig. 2.4.

Wind pressure on the windward side of a tall building, supplemented by suction on the leeward side, produces a bending moment, and the building behaves like a giant cantilever (see Section 5.2.) The taller the building, the greater the resistance moment required for stability.

Design Criteria

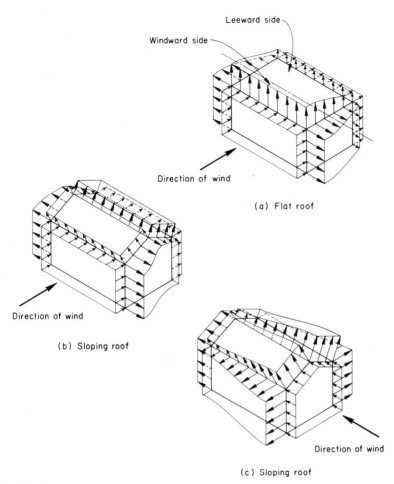

Leeward side

Windward side

Direction of wind

(a) Flat roof

Direction of wind

(b) Sloping roof

Direction of wind

(c) Sloping roof

Fig. 2.5.
Wind pressure and suction on buildings. (a) Flat roof. Pressure on windward wall; suction on other three walls and on roof. (b) Sloping roof with gable end parallel to direction of wind. Pressure on windward wall; suction on other three walls and on leeward roof surface. Windward roof surface is subject to pressure if windward slope of roof is more than 30°, and subject to suction if slope is less. (c) Sloping roof with gable end perpendicular to direction of wind. Pressure on windward wall; suction on other three walls and on both roof surfaces.

the temperate zone. They are higher near the sea than inland. They are higher on the top of a hill and lower when sheltered by surrounding buildings. In particular, they are higher above the ground, so that tall buildings have high wind loads not merely because they have a bigger surface.

Structures must be provided with adequate horizontal bracing to resist wind forces (Fig. 1.38). This is usually hidden within the cladding, although occasionally it is made a structural feature (Fig. 1.74). Rigid walls (*shear walls*) may be used as an alternative to diagonal bracing, and it is generally possible to cut small holes for windows and doors through these (see Section 8.8).

Most building codes specify the wind velocity v for a given zone. The wind velocity is converted into pressure from the equation

$$p = \tfrac{1}{2}\rho v^2 \tag{2.3}$$

where ρ is the density of the air. This conversion is normally performed with the aid of a table in the code.

The velocity v is, in some codes, taken to be at the reference elevation of 30 ft (10 m) above finished grade or the ground floor elevation. The increase with height (Fig. 2.6) is obtained from the empirical formula

$$\frac{v_1}{v_2} = \left(\frac{h_1}{h_2} \right)^k \tag{2.4}$$

where k is a constant specified for different conditions; it ranges from 0.07 to 0.2. For buildings of unusual shape or tall blocks partly sheltered by surrounding buildings, the precise distribution of wind pressure and suction is usually determined by a scale model test in a wind tunnel (Fig. 2.7).

In cold climates, snow may impose heavy loads on roof structures. No allowance is required in Sydney, Australia; 30 lb/sq ft (146 kg/m^2) is specified for London, England; 40 lb per sq ft (195 kg/m^2) for New York; and 60 lb/sq ft (292 kg/m^2) in Northern Canada. Snow loads act vertically.

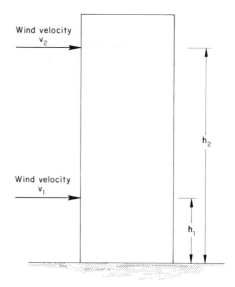

Fig. 2.6.

Experiments show that wind velocity increases with height above ground according to the formula $v_1/v_2 = (h_1/h_2)^k$.

Design Criteria

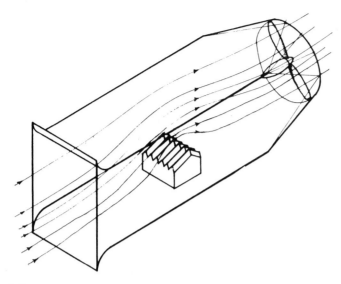

Fig. 2.7.
The distribution of wind pressure and suction on buildings of unusual shape is determined by a scale model test in a wind tunnel.

Major earthquakes occur in some parts of the earth's surface, including the north and south American west coast, New Zealand, Japan, and several Mediterranian countries. Only minor disturbances have been recorded in eastern America, the British Isles, and Australia. An earthquake is a jerky motion of the ground; it takes the foundation with it, but leaves the body of the building behind because of its high speed and the inertia of the building. It thus imposes a horizontal force on the structure. Although the ground moves very quickly, we can design small- and medium-sized buildings for earthquakes by considering their effect as an equivalent static horizontal force. Building codes in earthquake zones stipulate a minimum value for this force, which must be included in the structural calculations.

Only in rare instances is it necessary to allow for the dynamic effects of loads. Moving machinery, wind on flexible buildings, or earthquakes on tall buildings are liable to cause stress conditions which magnify those obtained by static calculations; but these are exceptional conditions in architectural structures, which do not enter into the design of simple buildings.

Suggestions for Further Reading

ARCHITECTURAL INSTITUTE OF JAPAN: *Design Essentials in Earthquake Resistant Buildings*. Elsevier, New York, 1970. 295 pp.

J. E. GORDAN: *The New Science of Strong Materials, or Why We Don't Fall Through the Floor*. Penguin Books, Baltimore, 1948. Chapters 1 and 2. pp. 15–60.

P. SACHS: *Wind Forces in Engineering*. Pergamon, Oxford, 1972. 392 pp.

The following works describe some case histories of modern failures:

T. McKAIG: *Building Failures*. McGraw-Hill, New York, 1962. 262 pp.

J. FELD: *Lessons from Failures of Concrete Structures*. American Concrete Institute, Detroit, 1965. 179 pp.

J. FELD: *Construction Failure*. Wiley, New York, 1968. 399 pp.

C. SZECHY: *Foundation Failures*. Concrete Publications, London, 1961. 141 pp.

Statics 3

A state of balance is attractive only when one is on a tightrope; seated on the ground there is nothing wonderful about it.

André Gide

In this chapter we start with the basic laws of statics. Although many readers will already be familiar with them, this brief account may be helpful for revision.

We consider the composition and resolution of concurrent and parallel forces and also the conditions for static equilibrium, which form the basis of all structural theory. We then determine the location of centers of gravity for simple and composite sections, and the reactions for beams, trusses, foundations, retaining walls, and buttresses.

3.1. The Definition of Force

In classical physics a *force* is defined as anything that changes, or tends to change, the state of rest of a body or its uniform motion in a straight line. Since buildings do not, or at least should not, move, we may confine ourselves to a definition of force as something which tends to change the state of rest of a body.

Thus, the weight of a body is a (gravitational) force which acts vertically downward (Fig. 3.1); the reaction of the structure is equal and opposite, and prevents the weight from dropping. The two forces are thus *in equilibrium*.

A force has direction as well as magnitude. Thus a force can be represented by a straight line: the length of the line denotes the magnitude of the force; and its angle of inclination, the direction of the force (Fig. 3.2).

3.2. Composition and Resolution of Forces

We can thus add and subtract forces in the same direction by simple arithmetic (Fig. 3.3). Forces not in the same direction can be added by force triangles (Fig. 3.4) or, as we shall see presently, by force polygons. This method was first used by Leonardo da Vinci in the

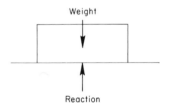

Fig. 3.1.
To every action there is an equal and opposite reaction.

(a) Vertical force of 1 lb.,
acting in the direction of gravity

(b) Vertical force of 1.5 lb.,
acting opposite to the direction of gravity

0.5 lb

(c) Horizontal force

(d) Inclined force

Fig. 3.2.
Force can be represented by the length, direction, and location of a straight line. (The arguments are equally true when metric units are substituted; that is when the forces are 1 N, 1.5 N, etc., instead of 1 lb, 1.5 lb, etc.)

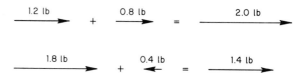

Fig. 3.3.
Addition and subtraction of forces acting along the same line. The same applies if the forces are in newtons.

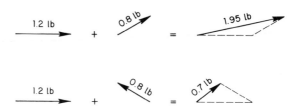

Fig. 3.4.
Addition of forces not acting along the same line. The same applies if the forces are in newtons.

fifteenth century, and formally published by Stevinus of Bruges in 1586.

It is easily proved by experiment (Fig. 3.5). Let us draw on a sheet of paper the two forces to be added and their vector sum, complete the parallelogram, and draw the diagonal. Then let us join three strings together, and arrange them so that they are in line with the directions of the three forces. Let us now place spring balances at the end of the three strings, and measure the forces. These will be proportional to the three forces represented by the strings.

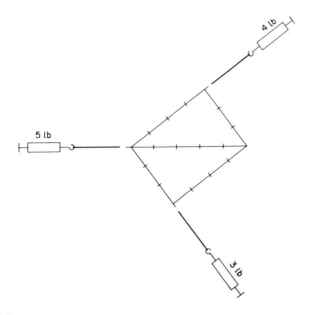

Fig. 3.5.
Experimental demonstration of the triangle of forces. The strings are arranged so that they follow the lines of the triangle of forces. The forces in the strings are measured with spring balances pinned to the baseboard. The springs balance at loads directly proportional to the corresponding lengths of the sides of the triangle of forces. (The argument is equally true if the forces are 3, 4, and 5 N respectively.)

We may therefore conclude that *if two forces, acting at a point, be represented by two sides of a triangle, then their resultant is represented by the third side of the triangle.* This theorem of the *Triangle of Forces* is the fundamental theorem of statics.

We can also resolve a single force into two components by the same means. In structural mechanics, forces are usually resolved horizontally and vertically, and we will therefore confine ourselves to this case (Fig. 3.6).

In Stevinus' time, arithmetic was made difficult by the lack of computers, slide rules, and even logarithmic tables, so that solutions were performed graphically wherever possible. Graphic statics remained popular for the solution of structural problems well into the twentieth century. Indeed, graphical solutions of statics, performed on a large drawing board with a finely sharpened pencil, are far more accurate than those carried out on a standard slide rule, although not as accurate as those performed with an electronic digital calculator.

With the growing use of computational aids the labor of arithmetic has been reduced to the point where it requires no serious consideration in structural mechanics, and solutions are now normally performed analytically by computation rather than graphics.

Let us consider the right-angled triangle shown in Fig. 3.7. By

Fig. 3.6.
Resolution of forces. The same applies if the units of the forces are newtons.

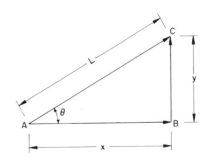

Fig. 3.7.
Definition of *sine* and *cosine*.

55

trigonometry

$$\cos \theta = \frac{AB}{AC} = \frac{x}{L} \qquad (3.1a)$$

and

$$\sin \theta = \frac{BC}{AC} = \frac{y}{L} \qquad (3.1b)$$

If AC represents a tensile force T_{AC}, both in magnitude and direction, then it can be resolved into two components: a horizontal component T_{AB}, represented in magnitude and direction by the length AB; and a vertical component T_{BC}, represented in magnitude and direction by the length BC.

But we have found that

$$AB = AC \cos \theta$$

and

$$BC = AC \sin \theta$$

It therefore follows that the horizontal component of the tensile force T_{AC} is the tensile force

$$T_{AB} = T_{AC} \cos \theta \qquad (3.2a)$$

and the vertical component of the tensile force T_{AC} is the tensile force

$$T_{BC} = T_{AC} \sin \theta \qquad (3.2b)$$

To obtain horizontal and vertical components of forces we need therefore only multiply by $\cos \theta$ and $\sin \theta$

$$(\text{Horizontal component of } T) = T \cos \theta \qquad (3.3a)$$

$$(\text{Vertical component of } T) = T \sin \theta \qquad (3.3b)$$

Finding horizontal and vertical components by sines and cosines is still the normal method of solving statical problems. Components can also be expressed in terms of Cartesian (x and y) coordinates.

From eqs. (3.1) and (3.2)

$$T_{AB} = T_{AC} \frac{x}{L} \qquad (3.4a)$$

and

$$T_{BC} = T_{AC} \frac{y}{L} \qquad (3.4b)$$

†Example 3.1. *A force of 10 lb is inclined to the horizontal at an angle of 30°. Determine the horizontal and vertical components.*

Horizontal component = 10 cos 30° = 10 × 0.866 = 8.66 lb

Vertical component = 10 sin 30° = 10 × 0.500 = 5.00 lb

The composition and resolution of *more than two* forces is accomplished by the same means. Let us consider four forces which keep a structural joint in equilibrium; these might consist of a load and three structural members converging at one joint (Fig. 3.8a). The joint is in equilibrium if the quadrilateral of forces closes (Fig. 3.8b); if the quadrilateral does not close, the joint would move in the direction of the resultant force, and the building would probably fall down. By the same reasoning, if we know four of the five forces acting on a joint, and know that the joint is in equilibrium because it does not move, we can determine the remaining force; it is the closing line of the pentagon (Fig. 3.9).

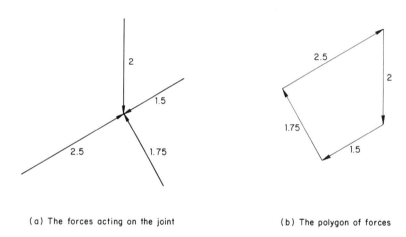

(a) The forces acting on the joint   (b) The polygon of forces

Fig. 3.8.
Quadrilateral (or four-sided polygon) of forces for a joint held in equilibrium by four forces.

We may therefore say that the forces converging on a joint in equilibrium must form a closed *polygon of forces*, and one unknown can thus be determined for each joint which is known to be in equilibrium; it is the closing line of the polygon (Fig. 3.10).

We can similarly determine the *horizontal and vertical components of several forces* converging on a joint, and determine any two unknown components if we know that the joint is in equilibrium. Considering the joint of Fig. 3.10, the horizontal and vertical components are shown in Fig. 3.11b, illustrating one of the

†This and all following examples are worked in metric units in Appendix G.

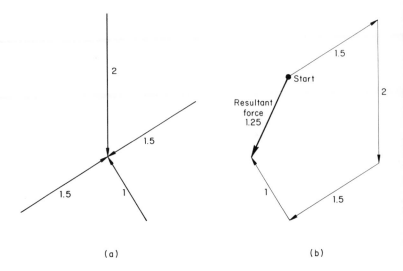

(a)   (b)

Fig. 3.9.
Resultant force acting on a joint *not* held in equilibrium by four forces. (a) The four forces which do not hold the joint in equilibrium. (b) The resultant is the closing side of the (five-sided) polygon of forces; it acts from the start of the first vector to the end (marked by an arrow) of the last vector.

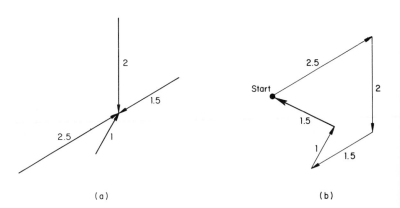

(a)   (b)

Fig. 3.10.
Determination of an unknown force acting on a joint in equilibrium. (a) Four of the forces acting at the joint; the fifth force is unknown in magnitude and direction. (b) The unknown force is the closing side of the polygon; it acts from the end of the fourth known vector to the start of the vector diagram.

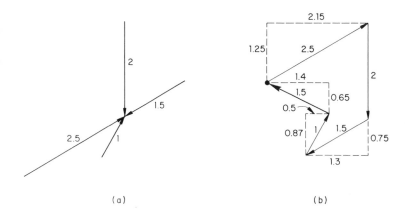

Fig. 3.11.

Determination of the horizontal and vertical components of an unknown force acting on a joint in equilibrium. (a) The four unknown forces acting on the joint. (b) The unknown force is the closing side of the polygon, and its horizontal and vertical components are the horizontal and vertical projections of that line.

fundamental laws of equilibrium:

$$\sum F_x = 0$$

and

$$\sum F_y = 0$$

or *the sum of all the forces in the x-direction, and the sum of all the forces in the y-direction are zero for a structure in equilibrium.*

The magnitude of the horizontal and vertical components of the unknown force can be determined graphically, as in Fig. 3.11, or by trigonometry.

If the directions of *two* forces of unknown magnitude are given, the problem is also soluble by statics (see Section 4.2). We can draw the polygon of forces if we know the directions of two forces, and the magnitude *and* direction of the remainder; we can then measure the length of the two unknown sides on the polygon (see Fig. 4.4b).

3.3. The Lever Principle

The triangle of forces can be used for the composition of any two forces (that is, to determine their resultant), provided they are not parallel. The point of intersection of two almost parallel forces may be inconveniently remote for a graphical solution (Fig. 3.12a), but if we use trigonometry for an arithmetic solution, as in Eq. (3.2), this

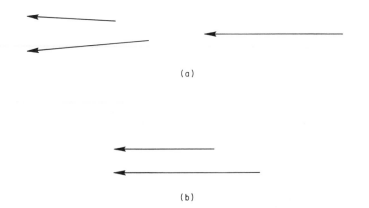

(a)

(b)

Fig. 3.12.
(a) Two nonparallel lines have a finite point of intersection, even though the resultant (on the right-hand side of the diagram) may be inconveniently remote if the angle is small. (b) Two parallel lines intersect only at infinity.

presents no practical problem. Only one case is insoluble with the triangle of forces—namely the composition of two lines which are parallel but not in line (that is, nonconcurrent, as shown in Fig. 3.12b)

For the composition of two parallel lines, we use the lever principle, which was discovered by Archimedes of Syracuse circa 250 B.C.

Let us balance a ruler on a pivot, and hang from it two weights as shown in Fig. 3.13. Let us then hang a weight W from a point on the other side until it balances the other two weights. There is no unique solution to this problem; the farther we move the weight from the pivot, the smaller we can make it. This accords with the common experience that heavy weights are easier to shift with long levers, and Archimedes used this principle extensively in devising machines for throwing missiles in the defence of Syracuse when it was attacked by the Romans.

In dealing with parallel forces we must therefore consider not only their magnitude, but also their distance from a pivot. We call the product

$$\text{force} \times \text{distance} = \text{moment}$$

and we solve problems containing parallel forces in terms of moments. If the force is measured in pounds and the distance in inches, we measure moments in pound-inches (lb in). We indicate direction by considering those which tend to rotate the lever *clockwise* about the pivot to be of opposite sign to those which tend to rotate it *anticlockwise* (counterclockwise). Two clockwise moments reinforce one another, and are therefore added. An anticlockwise moment is subtracted from a clockwise moment, or vice versa. The terms

Statics

clockwise and *anticlockwise* therefore correspond to *positive* and *negative*. They are opposing terms like *up* and *down* (for vertical components), or *left* and *right* (for horizontal components of forces.)

From the experiment illustrated in Fig. 3.13 it is evident that the resultant moment $W \times x$ balances the component moments $W_1 \times x_1$ and $W_2 \times x_2$. We may therefore conclude that *a simple lever, capable of turning about a given hinge or pivot, is in equilibrium if the moments of all the weights hanging from it, taken about the pivot, add up to zero* (or if the clockwise moments equal the anticlockwise moments). This is the second fundamental law of equilibrium. We can also express it in mathematical terms as

$$\sum M = 0$$

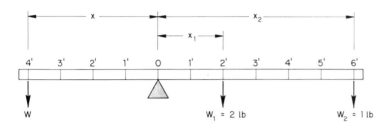

Fig. 3.13.
Experimental demonstration of the lever principle. The two loads W_1 and W_2 can be balanced by a large or a small force W, depending on its distance from the pivot of the lever; the smaller the distance, the greater the load required. For the case shown

$$W \times 4 \text{ in} = 2 \times 2 \text{ in} + 1 \times 6 \text{ in}$$

which gives $W = 2\frac{1}{2}$ lb. The argument is equally true if the distances are measured in metres and the forces in newtons.

Example 3.2. *Determine the reactions of the simply supported beam shown in Fig. 3.14.*

Let us take moments about the left-hand support. Since $\sum M = 0$, the sum of the moments of the individual loads about the left-hand support equals the moment of the right-hand reaction about the left-hand support.

$$200 \text{ lb} \times 3 \text{ ft} + 400 \text{ lb} \times (5 + 3) \text{ ft} + 500 \text{ lb} \times (6 + 5 + 3) \text{ ft}$$

$$+ 300 \text{ lb} (4 + 6 + 5 + 3) \text{ ft} = R_R \times 20 \text{ ft}$$

Solving for the right-hand reaction:

$$R_R = \frac{600 \text{ lb ft} + 3{,}200 \text{ lb ft} + 7{,}000 \text{ lb ft} + 5{,}400 \text{ lb ft}}{20 \text{ ft}} = 810 \text{ lb}$$

Let us now take moments about the right-hand support:

$$R_L \times 20 \text{ ft} = 200 \text{ lb} \times 17 \text{ ft} + 400 \text{ lb} \times 12 \text{ ft} + 500 \text{ lb}$$

$$\times 6 \text{ ft} + 300 \text{ lb} \times 2 \text{ ft}$$

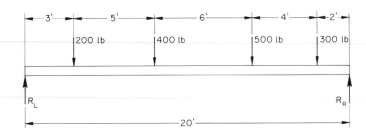

Fig. 3.14.
Determination of reactions (see Example 3.2).

Consequently the left-hand reaction $R_L = 590$ lb.

For vertical equilibrium, $\Sigma F_y = 0$, or the sum-total of the loads must equal the sum of the reactions. We may thus check the values determined for R_L and R_R:

$$200 \text{ lb} + 400 \text{ lb} + 500 \text{ lb} + 300 \text{ lb} = 590 \text{ lb} + 810 \text{ lb}$$

If it should turn out that the check equation does not balance, the mistake will be found in the arithmetic rather than the statics.

It is advisable for the student to carry the units throughout the computations, as shown in Example 3.2, until he becomes thoroughly familiar with the units of measurement, since confusion of units is a common source of error in structural calculations. However, the units are normally omitted in the intermediate stages of structural calculations, and this is done in subsequent examples.

Example 3.3. *Determine the moment reaction required at the end of the cantilever shown in Fig.* 3.15.

The anticlockwise restraining moment must equal the clockwise moments about the support produced by the forces due to the loads acting on the cantilever

$$M = 200 \times 3 + 400 \times 8 + 500 \times 14 + 300 \times 18 = 16{,}200 \text{ lb ft}$$

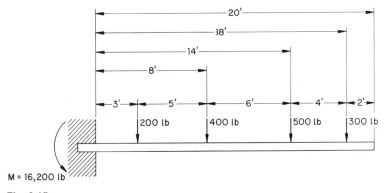

Fig. 3.15.
Determination of moment reaction (see Example 3.3).

Statics

Example 3.4. *Determine the moment reaction required at the end of a cantilever, 20 ft long, carrying a uniformly distributed load of 1,400 lb.*

The load equals 70 lb per foot run of beam. If only the restraining moment is required, it can be replaced by a single load of 1,400 lb halfway along the beam (Fig. 3.16). Consequently

$$M = 1,400 \times 10 = 14,000 \text{ lb ft}$$

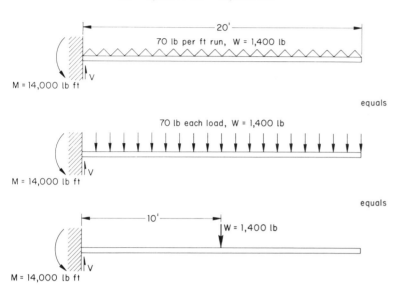

Fig. 3.16.
Determination of moment reaction (see Example 3.4).

3.4. The Conditions for Static Equilibrium

A substantial part of structural design, including most of the simpler problems, is performed with the aid of statics alone. Structures that can be solved purely by applying the conditions of static equilibrium are called *statically determinate or isostatic*. Structures that cannot be solved by statics alone are called *statically indeterminate*, or *hyperstatic*. Statically indeterminate structures are by no means indeterminate, but they require the application of some physical laws in addition to statics, and the solutions are generally harder to obtain. In this and the next two chapters we shall confine our attention to statically determinate problems.

We have seen in Section 3.2 that forces can be resolved into two components at right angles (Eq. 3.3), $T \cos \theta$ and $T \sin \theta$. In practice it is convenient to choose the horizontal and the vertical as the two directions. Verticality is, by definition, the direction of gravitational forces, and the weight of the structure and the superimposed loads act vertically. Thus, the majority of the forces acting on the structure act vertically, and the remainder mostly act horizontally—e.g., wind and earthquake forces.

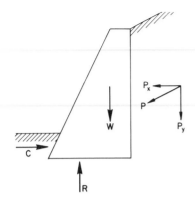

Fig. 3.17.
Conditions for static equilibrium:

$$\sum F_x = C - P_x = 0$$

and $$\sum F_y = R - W - P_y = 0$$

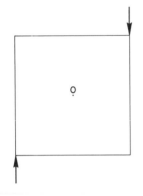

Fig. 3.18.
The forces on the structure balance horizontally and vertically, but cause it to rotate about 0.

Let us consider a vertical force W. Its horizontal component is $W \cos 90°$, which is zero. Consequently we cannot learn anything about the equilibrium of a structure by considering only the horizontal direction if most of the loads are vertical. Similarly we obtain no information on the equilibrium of horizontal forces by considering vertical equilibrium. We must resolve both horizontally *and* vertically to determine whether the structure in Fig. 3.17 is in equilibrium. Each force has either a horizontal or a vertical component, or both. If these balance both horizontally and vertically, the structure does not move in any direction. If they do not balance, then the structure will move bodily in the direction of the resultant force; the problem then ceases to be one of statics and becomes one of dynamics.

Horizontal and vertical equilibrium is sufficient to ensure that the structure does not move bodily; but it does not tell us whether the structure might rotate under the action of the imposed loads. Thus the structure in Fig. 3.18 satisfies the conditions of both horizontal and vertical equilibrium, but it will obviously rotate about a pivot 0 under the action of the two loads. To ensure full equilibrium, we must also take moments about some convenient point and check whether they balance. The point may be chosen at random, or to suit the particular problem. Thus if we wish to know the magnitude of the force T in the bottom chord (frequently called the "tie") of the roof truss shown in Fig. 3.19, due to a known load W and reaction R, we would do well to take moments about the point A. Three of the forces (W, C, and P) pass through this point, and their moments about A are zero. We thus obtain a single, simple equation; but the answer would be exactly the same if we chose instead some other, less convenient point, such as B. We would have to do more work to obtain the result, as three equations would be required for the simultaneous solution of three unknown forces.

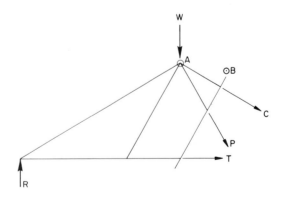

Fig. 3.19.
It is easier to determine the force in the bottom chord of this roof truss by taking moments about A, than by taking moments about a point B outside the truss, but the answer is the same.

Statics

As previously stated, there are two fundamental conditions for the equilibrium of a plane (or two-dimensional) structure. First, the sum of all the forces in the plane must be zero (see Section 3.2); this is tested by resolving *both* horizontally and vertically (Eq. 3.5a and 3.5b). Second, the sum of the moments must be zero (see Section 3.3); this is tested by taking moments about any one point in the plane (Eq. 3.5c).

The two conditions of equilibrium therefore require the solution of three equations:

(i) *The sum of all the horizontal forces is zero.* (3.5a)

(ii) *The sum of all the vertical forces is zero.* (3.5b)

and (iii) *The sum of all the moments about any one point is zero.* (3.5c)

Or in mathematical terms:

$$\sum F_x = 0 \qquad\qquad (3.5a)$$

$$\sum F_y = 0 \qquad\qquad (3.5b)$$

$$\sum M = 0 \qquad\qquad (3.5c)$$

Equation (3.5) is of fundamental importance for statically determinate as well as statically indeterminate structures. All buildings and all parts of buildings must be in statical equilibrium.

If the first condition (Eq. 3.5a) were not satisfied, there would be an unbalanced horizontal force, and the structure would behave like a vehicle moving along a horizontal track with constant acceleration as long as the unbalanced horizontal force was acting on it. If the second condition (Eq. 3.5b) were not satisfied, the structure would become a rocket, accelerating upward as long as the unbalanced vertical force was acting on it. If the third condition were not satisfied, the structure would behave like a rotary engine as long as the unbalanced moment was acting on it.

The conditions of statical equilibrium can be used to determine the reactions of all statically determinate and some statically indeterminate structures. We will now solve some typical problems.

Example 3.5. *Determine the reactions of the roof truss shown in Fig. 3.20.*

The truss carries vertical dead loads at all panel points, and in addition there is wind pressure on the windward face and wind suction on the leeward face. We will assume that the columns supporting the truss are sufficiently flexible to allow slight horizontal expansion of the truss under load; but one horizontal reaction is needed for equilibrium to keep the entire building from being pushed over by the wind. In addition, there are vertical reactions at both columns.

Resolving horizontally:

$$(100 + 200 + 100)\cos 45° + (150 + 300 + 150)\cos 45° = R_H$$

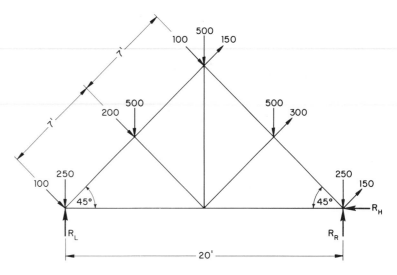

Fig. 3.20.
Determination of reactions (see Example 3.5).

This gives $R_H = 707$ lb.

Resolving vertically:

$$(3 \times 500 + 2 \times 250) + 400 \sin 45° - 600 \sin 45° = R_L + R_R$$

This gives $R_L + R_R = 2,000 - 200 \times 0.707 = 1,859$ lb.

Taking moments about the right-hand support:

$$250 \times 20 + 500 \times 15 + 500 \times 10 + 500 \times 5 + 100 \times 14$$
$$+ 200 \times 7 - 150 \times 14 - 300 \times 7 = R_L \times 20$$

This gives $R_L = 930$ lb.

Since $R_L + R_R = 1,859$ lb, $R_R = R_L$; however, this result is not obvious.

Examples 3.6. *Determine the vertical and horizontal reactions of the retaining wall shown in Fig. 3.21.*

The wall retains soil, which exerts a horizontal pressure of 50,000 lb at two-thirds of the depth. The wall itself weighs 150,000 lb. Let us summarize the reactions as a horizontal force R_H, which must be provided by friction or mechanical anchorage under the wall, and a vertical reaction R_V, which must be provided by the upward pressure of the soil under the wall.

Resolving horizontally: $R_H = 50,000$ lb.

Resolving vertically: $R_V = 150,000$ lb.

Since the horizontal reaction acts at the base of the wall, the foundation pressure is unlikely to be uniform; and to determine its distribution, we must determine its line of action, defined by the distance a from the vertical wall face.

Taking moments about the bottom of the vertical face, we obtain

$$150,000 \times 2.8 + 50,000 \times 4 = R_H \times 0 + R_V \times a$$

which gives $a = 620,000/150,000 = 4.13$ ft.

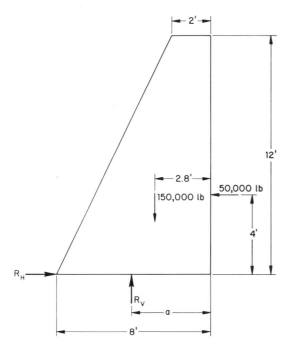

Fig. 3.21.
Determination of reactions (see Example 3.6).

In order to determine the reactions of a space frame, we must resolve along all three axes. The y-axis is customarily vertical, but the location of the x and z axes follows no fixed rule. A three-dimensional object can rotate about any of three axes while being in rotational equilibrium about the other two. Consequently there are three moment-equilibrium conditions, about the x-, y-, and z-axes. The space frame equilibrium conditions therefore become

$$\sum F_x = 0 \qquad (3.6a)$$

$$\sum F_y = 0 \qquad (3.6b)$$

$$\sum F_z = 0 \qquad (3.6c)$$

$$\sum M_x = 0 \qquad (3.6d)$$

$$\sum M_y = 0 \qquad (3.6e)$$

$$\sum M_z = 0 \qquad (3.6f)$$

Fortunately most space frames are symmetrical, so that many of the equations become equal to zero on both sides. If a symmetrical space frame is symmetrically loaded with vertical loads only, then the reactions can be determined simply by dividing the total load by the number of (vertical) reactions.

3.5. Centers of Gravity

The center of gravity of a body is that point in space through which the weight of the body acts. In practice we usually confine ourselves to determining centers of gravity of plane shapes, since most statical problems can be reduced to two dimensions.

Let us first consider a *rectangle* (Fig. 3.22a), and take a thin horizontal element. In accordance with the lever principle, the center of gravity of this element is in the middle. By considering further elements, it becomes evident that the center of gravity lies on the line that connects the midpoints of the two sides. Turning the rectangle now through 90°, we consider elements of the rectangle in its new position, and evidently the center of gravity again lies on the line connecting the midpoint of the two sides. Since these two lines intersect halfway up the height of the rectangle, whichever way it is placed, the center of gravity of a rectangle is at half-height.

Let us next consider a *triangle* with one of its sides horizontal

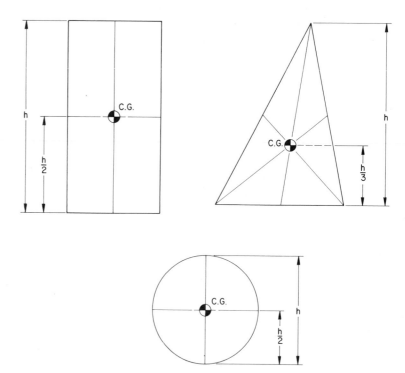

Fig. 3.22.
The center of gravity (C.G.) of the rectangle, the triangle, and the circle.

Statics

(Fig. 3.22b). The center of gravity of a thin horizontal element is, in accordance with the lever principle, in the middle, and the center of gravity therefore lies on the line connecting the middle of the side to the opposite apex. By turning the triangle with one of the other sides horizontal, we find similarly that the center of gravity is on the line connecting the middle of that side to the opposite apex. The center of gravity therefore lies at the intersection of the three lines connecting the middle of the sides to the opposite apexes. These three lines intersect at one-third of the height. (This statement is proved in most elementary geometry books.)

The center of gravity of a *circle* is evidently at its center (Fig. 3.22c).

Example 3.7. *Determine the center of gravity of the retaining wall shown in Fig. 3.23.*

If the wall is made of the same material throughout (say, concrete) the weight of any part of the section is proportional to the area of that section. For convenience, we therefore divide the wall into a rectangle whose center of gravity lies at half-height and half-width, and a triangle whose center of gravity lies at one-third height and one-third width.

We are concerned with determining the location of the center of gravity through which the weight of the combined wall acts, and since this is a vertical force, we are interested only in the location of the center of gravity along the x axis. The component rectangle has a weight proportional to the area of 2×12, and its center of gravity (C.G.$_3$) is 1 ft from the right-hand (vertical) face. Its moment about that face is $2 \times 12 \times 1$. The triangle has an area of $\frac{1}{2} \times 6 \times 12$, and its center of gravity (C.G.$_1$) is $2 + 2$ ft from the vertical face. The moment about it is $\frac{1}{2} \times 6 \times 12 \times 4$.

If the distance of the center of gravity of the entire wall (C.G.$_2$) from the vertical face is $\bar{x}$, then taking moment about the vertical face

$$\left(2 \times 12 + \tfrac{1}{2} \times 6 \times 12\right)\bar{x} = 2 \times 12 \times 1 + \tfrac{1}{2} \times 6 \times 12 \times 4$$

so that $\bar{x} = 168/60 = 2.8$ ft.

Example 3.8. *Determine the location of the center of gravity of the prestressed concrete section shown in Fig. 3.24.*

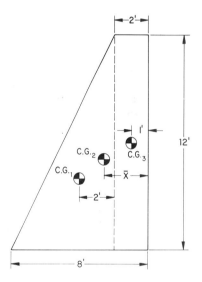

Fig. 3.23.
Determination of center of gravity (C.G.) (see Example 3.7).

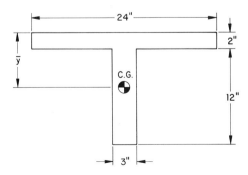

Fig. 3.24.
Determination of center of gravity (see Example 3.8).

Since the section is vertically symmetrical, the center of gravity is evidently situated centrally, 12 in from either side. To determine its vertical location, let us take moments about the top face

$$(24 \times 2 + 12 \times 3)\,\bar{y} = 24 \times 2 \times 1 + 12 \times 3 \times (2 + 6)$$

$$\bar{y} = \frac{336}{84} = 4.00 \text{ in}$$

Steel sections are rolled to standard profiles of I, T, or angle shape, because it is economical to concentrate the expensive metal in the most-highly stressed regions (see Section 6.4). The center of gravity of these sections is not readily obtainable by calculation, but a graphical method, based on the principles already discussed in this section, can be used. In practice this need not be done by the designer, since the properties of standard sections are readily available in section tables, which are obtainable, free of charge or at a small fee, from manufacturers. Readers interested in the graphical method are referred to one of the older textbooks on the strength of materials (e.g., Arthur Morley, *Strength of Materials*, Longmans, London, 1934, p. 136). However, the designer must still work out the center of gravity of composite sections.

Example 3.9. *Determine the depth of the center of gravity of a composite steel section consisting of an American Standard 10-in I-beam (35 lb per ft) and a 10-in $\times$ 1-in plate, as shown in Fig. 3.25.*

From section tables, the cross-sectional area of the I-beam is 10.22 sq in. Its center of gravity is symmetrically placed, i.e., 5 in from the edge of the flange. The plate has an area of 10 sq in.

Taking moments about the top face of the compound section

$$(10.22 + 10)\,\bar{y} = 10 \times 0.5 + 10.22(5 + 1)$$

$$\bar{y} = \frac{66.32}{20.22} = 3.28 \text{ in}$$

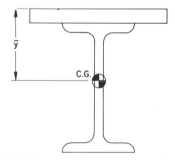

Fig. 3.25.
Determination of center of gravity (see Example 3.9).

3.6. Forces on Simple Foundations and Retaining Walls

Foundations serve to distribute the concentrated loads of columns to the ground. While special consideration must be given to weak and fine-grained (clay) soils, it is sufficient on strong coarse-grained (sandy) soil and on rock to assume the foundation pressure to be uniformly distributed (Fig. 3.26).

Let us consider a footing for a wall which carries a load P per foot run. If we make the footing B ft wide, we distribute this load for each foot run of wall over an area B ft wide and 1 ft long, and the resulting pressure on the foundation is w lb per sq ft. This must be less than the permissible bearing pressure specified in the local

Statics

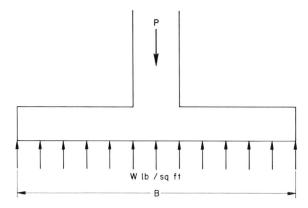

Fig. 3.26.
Foundation pressure under wall footing (see Example 3.10).

building regulations. In addition the effect of settlement should be considered, where local soils are known to contain soft clay, peat, or uncompacted fill.

Example 3.10.   Determine the width of the foundation required for a wall carrying a load of 10,000 lb per foot run (including the weight of the wall and the footing) if the maximum permissible bearing pressure of the soil is 3,500 lb per sq ft.

The width required is

$$\frac{10,000}{3,500} = 2.86 \text{ ft (Say 3 ft 0 in.)}$$

Example 3.11.   Determine the size of the footing required for a reinforced concrete column, 2 ft square, carrying a load of 400,000 lb, if the maximum permissible bearing pressure is 3,500 lb per sq ft.

Let us provide the footing with a reinforced concrete slab, 2 ft thick, and assume the weight of the concrete as 150 lb per cu ft. The footing slab therefore weighs 300 lb per sq ft, and we must subtract that from the maximum permissible bearing pressure, which leaves

$$3,500 - 300 = 2,200 \text{ lb per sq ft}$$

Let us assume a square footing, of area B^2 (Fig. 3.27) so that

$$B = \sqrt{\frac{400,000}{3,200}} = 11.18 \text{ ft (Say 12 ft square)}$$

Retaining walls are required to retain water, or more commonly soil, whose pressure increases with depth. The hydrostatic pressure of liquids is the same in all directions, and its increase is directly proportional to the depth.

Since the weight of water is 62.4 lb per cu ft (1,000 kg per cu m), its pressure at a depth of h feet equals 62.4 h lb per sq ft.

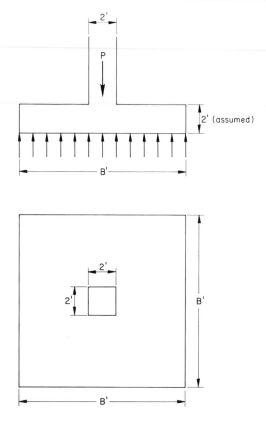

Fig. 3.27.
Foundation pressure under column footing (see Example 3.11).

Example 3.12. *Determine the water pressure at the bottom of a water tower, if the depth of water is 6 ft.*

The water pressure is $6 \times 62.4 = 374.4$ lb per sq ft.

Most soils weigh far more than water; however, most soils exert a smaller horizontal pressure than water, because they are granular and not fluid. It is convenient to specify the *equivalent fluid weight* of soil so that the pressure exerted by the soil can be treated as a hydraulic pressure.

While we must accept the foundation soil on the building site unless we are willing to incur very heavy costs of excavation and transport, we can at reasonable expense backfill a retaining wall with a different soil, at least immediately behind its face, to ensure proper drainage and prevent building up of high pore-water pressure in the retained soil.

Since the pressure builds up uniformly with depth, and the resultant pressure acts at the center of gravity of the triangular pressure diagram (Fig. 3.28), the horizontal force H acts at two-thirds of the depth of the retaining wall.

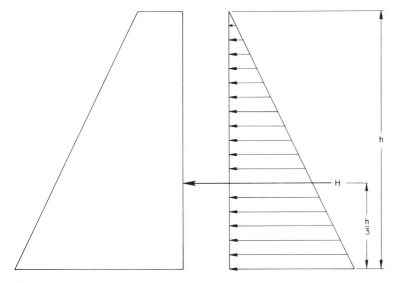

Fig. 3.28.
Pressure behind retaining wall (see Example 3.13).

Example 3.13. *Determine the resultant horizontal force on a wall, retaining soil 12 ft deep. The equivalent fluid weight of the soil is 30 lb per cu ft.*

The horizontal pressure of the soil varies from zero at the surface to 12 ft depth × 30 lb per cu ft = 360 lb per sq ft at the bottom of the wall. The average soil pressure is $\frac{1}{2}$ × 360 lb per sq ft. The area on which this average pressure acts is a depth of 12 ft and a unit length of 1 ft. The resultant horizontal pressure per foot run of wall is

$$H = \tfrac{1}{2} \times 360 \times 12 \times 1 = 2{,}160 \text{ lb per ft run of wall}$$

This resultant H acts at $\frac{1}{3}h = 4$ ft above the base of the wall (Fig. 3.28).

3.7. The Middle-Third Rule

A mass retaining wall, consisting of concrete or masonry, is held in equilibrium by the weight of the wall, W, the horizontal pressure of the soil, H, and the reaction of the foundation under the wall, R (Fig. 3.29).

The horizontal force H, acting at the center of gravity of the horizontal pressure diagram (one-third of the height of the wall) and the weight of the wall, acting vertically at the center of gravity of the wall, combine to form a resultant R. This can be done numerically or by drawing the triangle of forces. The resultant R then continues in a straight line until it intersects the base of the retaining wall. The horizontal component R_H is there resisted by friction between the base of the wall and the soil, the vertical component R_V by the upward pressure of the soil. Since the vertical

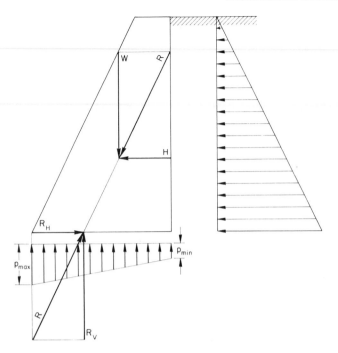

Fig. 3.29.
Transmission of forces acting on a mass retaining wall to the foundation.

force R_V is off center, the foundation pressure must vary from a maximum pressure p_{max} to a minimum pressure p_{min}, so that the center of gravity of the trapezoidal pressure diagram coincides with the line of action of R_V (which is determined by the magnitude and location of W and H, and by the geometry of the wall).

If p_{min} changes sign and becomes tensile, the wall lifts off the foundation, because the wall cannot suck up the soil and the development of a significant amount of tension between the wall and the soil is impossible. Although uplift does not immediately lead to failure, it is clearly undesirable, and mass retaining walls are designed to ensure compression in the soil below the entire length of the retaining wall.

The useful limit is therefore reached when $p_{min} = 0$, and the pressure diagram below the footing is triangular (Fig. 3.30). Since the center of gravity of a triangle is at one-third of its depth, the line of the resultant pressure must not approach the edge of the retaining wall closer than one-third of the base length, i.e., *it must lie within the middle third.*

The middle-third rule is easily demonstrated by compressing a porous block eccentrically (Fig. 3.31). It applies to eccentrically loaded foundations, and also to eccentrically loaded masonry structures, whether they be built of bricks or blocks of stone joined with a weak mortar. In reinforced concrete and reinforced brickwork,

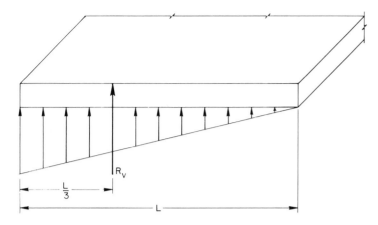

Fig. 3.30.
The middle-third rule. The foundation pressure is wholly compressive, provided the resultant force falls within the middle third.

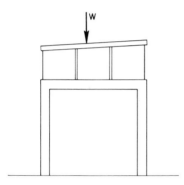

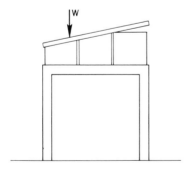

Fig. 3.31.
Experimental demonstration of the middle-third rule. A sponge or block of foamed plastic is visibly in compression along its entire length if loaded within its middle third. If loaded outside the middle third, the joint opens up.

tensile stresses are permitted because they are taken by the reinforcement.

A famous application of the middle-third rule to traditional construction is the modern design of the Gothic buttress (Fig. 3.32). The thrust of the flying buttress is turned downward by the weight of the pinnacle, and turned farther downward by the weights of masonry in the buttress. If tension in the joints is to be avoided, the buttress must be made wide enough to keep the resultant within the middle third.

The original Gothic builders did not know the triangle of forces, which was discovered two centuries after the building of Chartres Cathedral. It was, however, used in the neo-Gothic revival of the nineteenth century. The middle-third rule is conservative, because a buttress does not immediately collapse if tension develops in the masonry joints. So long as there are insufficient hinges to turn the masonry structure into a mechanism (see Fig. 2.1), it will not fall down. The determination of the collapse loads of masonry structures with weak mortar joints is beyond the scope of this book. Moreover, it has little relevance to modern architectural design, since a simple remedy lies in the use of tension reinforcement.

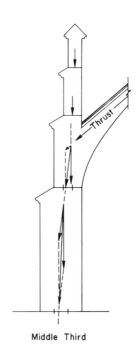

Middle Third

Fig. 3.32.
Gothic buttress designed by the middle-third rule. The thrust of the flying buttress is turned downward by the weight of the pinnacle and further turned down by the weight of masonry in the buttress. If tension in the joint is to be avoided, the buttress must be made wide enough to keep the resultant within the middle third.

Suggestions for Further Reading

J. L. MERIAM: *Statics*. Wiley, New York, 1966. Chapters 1 and 2, pp. 1–74.

W. MORGAN and D. T. WILLIAMS: *Structural Mechanics*. Pitman, London, 1963. Chapters 1–6, pp. 1–118.

G. A. NEVILL: *Programmed Principles of Statics*. Wiley, New York 1969. Chapters 1–6, pp. 1–92.

T. J. REYNOLDS and L. E. KENT: *Introduction to Structural Mechanics*. English Universities Press, London, 1973. Chapters 1–7, pp. 1–130.

Problems

[†]**3.1.** Determine the magnitude of the force needed to restore equilibrium in the system of forces shown in Fig. 3.33.

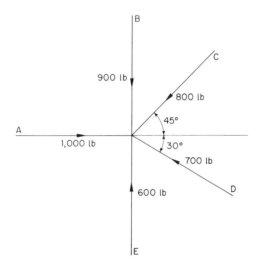

Fig. 3.33.
Problem 3.1.

3.2. Find the resultant of all the forces acting on the bracket shown in Fig. 3.34. Calculate the shortest distance from point A to the line of action of the resultant.

3.3. Determine the reactions of a cantilevered beam, shown in Fig. 3.35, which carries a total load of 1,300 lb.

3.4. The cantilever shown in Fig. 3.36 carries a uniformly distributed load of 500 lb per foot run, as well as concentrated loads totaling 3,000 lb. Determine the vertical and moment reactions at the support.

[†]Similar problems in metric units are given in Appendix G.

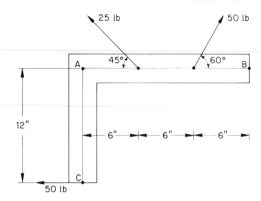

Fig. 3.34.
Problem 3.2.

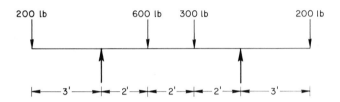

Fig. 3.35.
Problem 3.3.

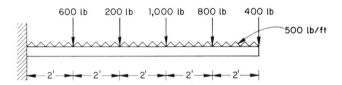

Fig. 3.36.
Problem 3.4.

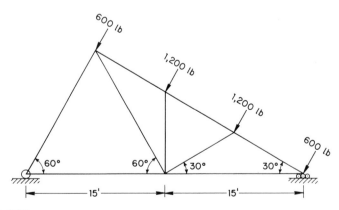

Fig. 3.37.
Problem 3.5.

Statics

3.5. The roof truss shown in Fig. 3.37 is pinned at the left-hand support and able to slide at the right-hand support. It carries a wind load of 3,600 lb. Determine the magnitude of the support reactions.

3.6. Figure 3.38 shows a steel plate with two holes in it. Determine the distance of its center of gravity from the bottom left corner.

3.7. Figure 3.39 shows a large rectangular steel plate with a circular hole in it. The plate is freely suspended by a cable at a point B. There is a weight hanging from one corner, point E. The plate weighs 10 lb per sq ft. If the magnitude of the weight W is such that the plate hangs with the side AC exactly perpendicular to the cable BD, calculate the force in the cable.

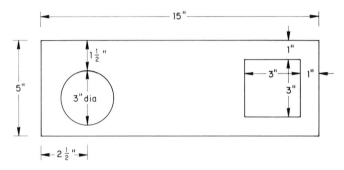

Fig. 3.38.
Problem 3.6.

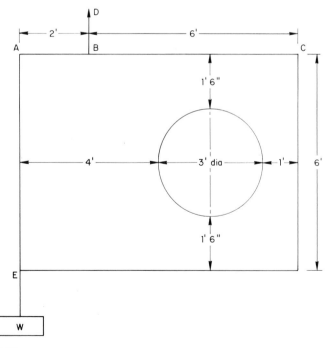

Fig. 3.39.
Problem 3.7.

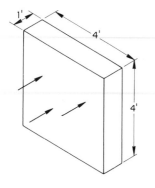

Wind Pressure = 40 lb/sq ft

Fig. 3.40.
Problem 3.8.

3.8. A concrete block, of density 150 lb per cu ft, measures 4 ft by 4 ft by 1 ft. There is a horizontal pressure of 40 lb per sq ft on one side of the block as shown in Fig. 3.40. Calculate whether the block will overturn.

3.9. A tall, narrow building has the following dimensions: plan 100 ft by 30 ft, height 330 ft. If it weighs approximately 7 lb per cu ft, determine the horizontal wind pressure required to overturn the building. Assume the wind velocity profile to be uniform.

3.10. A masonry wall is 5 ft high and 1 ft thick. There is a horizontal pressure of 10 lb per sq ft, acting uniformly on one side of the wall. Calculate the minimum average density of the material from which the wall is constructed, if no tension develops in any part of the structure.

Forces in Statically Determinate Trusses

4

*It is no paradox to say that in our most theoretical
moods we may be nearest to our practical applications.*
Alfred North Whitehead

In this chapter, we examine the criteria that make plane trusses and space frames statically determinate, or isostatic. We then explain the principal methods available for their solution. Plane trusses can be solved easily by graphic statics, by trigonometry, or by algebra; for large trusses, the simultaneous equations may be expressed in matrix algebra and evaluated by computer. The design of wind bracing and of purlins is discussed in Section 8.8.

Space frames are most conveniently solved by algebra because of the complexity of the problem. Space frames need not be too complicated to be included in the architect's repertoire, but they are never simple.

4.1. Statically Determinate Trusses, Statically Indeterminate Trusses, and Mechanisms

The term *pin joint* originated in the days of cast iron, when flexible joints were in fact made by pushing a pin through holes cast into the ends of the members to be joined. It was relatively simple to cast such holes into the material, and the cast sections were not so flexible as to permit rotation of the joints. In structural steel and aluminum trusses, joints are invariably bolted, rivetted, or welded; but it is possible to make the members sufficiently flexible to ensure that moments are not transmitted through them to any significant extent. We call such joints *pin jointed*, even though actual pins are reserved today for bridge bearings and other special uses.

Although trusses with rigid joints are used in buildings, for example, the Vierendeel truss (see Sections 1.6 and 8.6), they are excluded from consideration in this chapter, which is confined to pin-jointed trusses.

At this stage we will furthermore confine ourselves to those trusses that can be solved by statics alone. These are called *statically determinate*, or *isostatic*, and they form the boundary between *mechanisms*, which can be moved by only nominal forces, and *statically indeterminate* or *hyperstatic* trusses, which have too many members for a static solution (Fig. 4.1). For any given form there is only one statically determinate truss, but there is a whole range of mechanisms and statically indeterminate trusses. Thus the frame in Fig. 4.1a can be pushed sideways by a small horizontal force, and it has one degree of freedom. We can take a further member away to give it two degrees of freedom. We can also add a diagonal; this will just turn it into a truss (Fig. 4.1b) which is statically determinate. Further members can be added, and each will form one redundancy. Thus the frame in Fig. 4.1c is a statically indeterminate frame with one redundancy.

Evidently, the simplest statically determinate truss has three pin joints and three members (Fig. 4.2a). If we add one joint, we require two additional members (Fig. 4.2b), and so on (Fig. 4.2c). A stati-

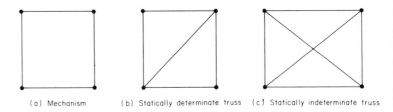

(a) Mechanism   (b) Statically determinate truss   (c) Statically indeterminate truss

Fig. 4.1.
Distinction between a mechanism, a statically determinate truss, and a statically indeterminate truss.

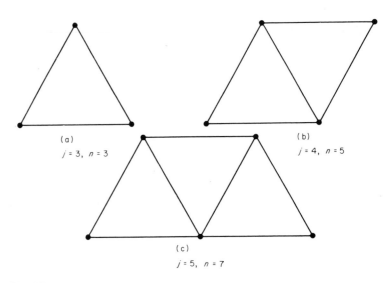

(a)
$j = 3$, $n = 3$

(b)
$j = 4$, $n = 5$

(c)
$j = 5$, $n = 7$

Fig. 4.2.
The simplest statically determinate truss has three members and three pin joints. Each additional pin joint requires two additional members.

Forces in Statistically Determinate Trusses

cally determinate truss with j joints therefore requires

$$n = 2j - 3 \qquad (4.1)$$

members (Fig. 4.3).

We can generalize Eq. (4.1) by writing it as

$$2j = n + r \qquad (4.2)$$

where r is the number of reactions needed under *any* condition of loading. A plane truss requires three reactions: two vertical reactions at the two supports, and a horizontal reaction at one of the supports to prevent the truss from being pushed off the supports by a horizontal load (see Fig. 3.20). If we substitute $r = 3$, we obtain Eq. (4.1).

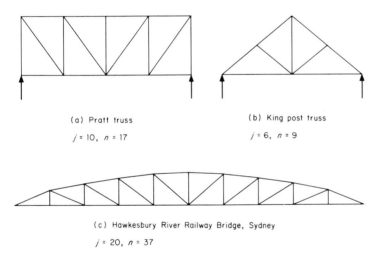

(a) Pratt truss

$j = 10, \; n = 17$

(b) King post truss

$j = 6, \; n = 9$

(c) Hawkesbury River Railway Bridge, Sydney

$j = 20, \; n = 37$

Fig. 4.3.
Statically determinate trusses with j pin joints require $n = 2j - 3$ members. If there are more, the truss is statically indeterminate. If there are fewer, the truss becomes a mechanism.

4.2. Bow's Notation, and Stress Diagrams for Trusses

Let us consider the truss shown in Fig. 4.4a, and specifically the joint at the extreme left. Since the joint, like the rest of the building, does not move through space with ever-increasing velocity, it is in equilibrium under the forces acting on it; these are the support reaction, the internal force in the tie, and the internal force in the rafter. We can therefore draw a triangle of forces for this joint. We know the magnitude of this reaction; by symmetry it is half the total load acting on the truss (or we can work it out by the methods described in Section 3.4). We can therefore draw it to a suitable

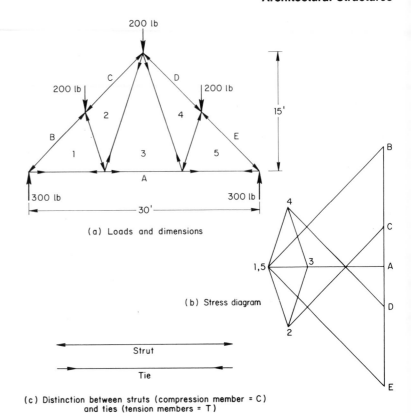

(a) Loads and dimensions

(b) Stress diagram

(c) Distinction between struts (compression member = C)
and ties (tension members = T)

Member	Force (lb)	Strut or Tie
A1 A5	300	T
A3	200	T
B1 ·E5	424	C
C2 D4	363	C
12 45	163	C
23 34	163	T

(d) Magnitude of forces scaled from stress diagram

Fig. 4.4.
Stress diagram for timber roof truss.

load scale such as 100 lb to 1 in. (The solution for a similar truss in metric units is given in Appendix G.)

We know that the force in the rafter must be parallel to the rafter, and the force in the tie horizontal. Consequently, we can draw lines in these directions, and they intersect at the third point of the triangle. We can now scale off the lengths of the other two sides of the triangle, and obtain the internal forces in the two members intersecting at the left-hand support. We can then proceed in the same way with all other joints.

Forces in Statistically Determinate Trusses

This method was proposed by Clerk Maxwell, Professor of Physics at Kings College, London, in 1864. The stress diagram is greatly simplified by *Bow's notation*, which labels the *spaces* between the members and external forces.*

Using Bow's notation, the support reaction of 300 lb is drawn upward as AB, A1 is drawn horizontally, and B1 parallel to the rafter; the intersection of the two lines gives the point 1, and AB1 is the triangle of forces for equilibrium at the left-hand support (joint AB1). We now proceed to the equilibrium condition of the joint BC12, and we draw the load of 300 lb down from B, giving us C. We already have B1, and we draw lines parallel to C2 through C and to 12 through 1; we thus obtain point 2, and BC21 is the polygon of forces for equilibrium at the joint BC12. We then proceed to the joint A123. We already have A, 1, and 2, and we draw a line parallel to 23 through 2, which intersects the horizontal line through A at 3. The remainder of the diagram is symmetrical, and not strictly necessary. However, we draw it here for completeness.

We next measure the lengths of the various lines according to our chosen load scale and tabulate the internal forces in the members.

It then only remains to distinguish between ties and struts (or tension and compression members). We do this by taking each triangle or polygon of forces in turn, and marking the arrows on the members. Thus taking the joint AB1, we know that the reaction AB is up, and therefore we proceed around the triangle in the direction A–B–1–A. We mark these arrows on the ends of the members B1 and A1 near the joint. Taking the next joint, we know that BC acts down, and therefore we proceed B–C–2–1–B. Again we mark the arrows on the truss. The arrows denote the effect of the forces on the joints, and the internal forces in the members are equal and opposite. Consequently two arrows pushing out denote a strut and two arrows pulling in denote a tie (Fig. 4.4c).

The solution is shown in the table (Fig. 4.4d).

†Example 4.1. *By graphic statics determine the forces in the king-post truss shown in Fig. 4.5a.*

The stress diagram is drawn in Fig. 4.5b, and the arrows are marked on Fig. 4.5a. The results measured off the stress diagram are tabulated in Fig. 4.5c. The force BC goes straight to the reaction AB and does not contribute

---

*Although conventions vary, it is common practice to letter the spaces external to the truss clockwise. If the truss is supported at both ends, the load line should be drawn as far as possible to the right, since the stress diagram falls entirely to the left of the load line. The stress diagram of a cantilevered truss falls opposite to that of a truss supported at both ends.

It is recommended that letters be used for external spaces and numbers for internal spaces. This means that a line in the stress diagram such as AB denotes an external force, a line such as A2 denotes an outside (or principal) member, and a line such as 23 denotes an internal (or short) member of the truss.

†This and all subsequent examples are worked in metric units in Appendix G.

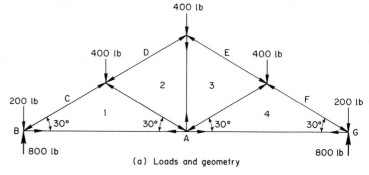

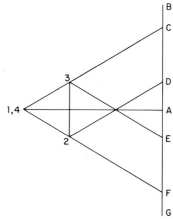

(a) Loads and geometry

(b) Stress diagram

Member	Force (lb)	Strut or Tie
Al A4	1,040	T
Cl F4	1,200	C
D2 E3	800	C
12 34	400	C
23	400	T

(c) Magnitude of forces scaled from stress diagram

Fig. 4.5.
Stress diagram for king post truss (Example 4.1)

to the stress diagram. Note that the king post is not in compression, but in tension.

Example 4.2. *By graphic statics determine the forces in the parallel chord truss shown in Fig. 4.6a.*

The stress diagram is drawn in Fig. 4.6b. The arrows have been placed on Fig. 4.6a to distinguish between struts and ties, and the results are tabulated in Fig. 4.6c.

It may be noted that the reactions AB and GA are transmitted along the verticals B1 and G8, so that the bottom chords A1 and A8 have no internal

Forces in Statistically Determinate Trusses

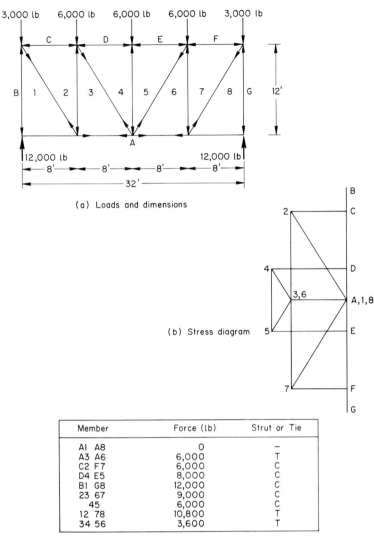

(a) Loads and dimensions

(b) Stress diagram

Member	Force (lb)	Strut or Tie
A1 A8	0	—
A3 A6	6,000	T
C2 F7	6,000	C
D4 E5	8,000	C
B1 G8	12,000	C
23 67	9,000	C
45	6,000	C
12 78	10,800	T
34 56	3,600	T

(c) Magnitude of forces scaled from stress diagram

Fig. 4.6.
Stress diagram for Pratt truss (Example 4.2)

forces and, theoretically, they could be removed. As a matter of common sense, these members are required nevertheless; they become important as soon as the smallest horizontal force is placed against the truss. Unstressed members do occur under certain conditions of loading, particularly in space frames, where even a statically determinate frame has a large number of members.

Roof trusses are loaded vertically because of dead load and snow load, and are also loaded at an angle because of wind pressure and wind suction (see Fig. 3.20). If there are nonvertical loads, it is best to draw a separate stress diagram for the vertical loads and the appropriate reactions, and another for the inclined loads and their reactions. The internal forces due to the two can then be added (or subtracted) arithmetically. This is less work than a combined graphical solution.

If more than two unknown members meet at a joint, the stress diagram becomes insoluble, since the triangle or polygon of forces can determine the internal forces only in two members for which the direction of the forces is known. (It is actually a graphical solution to the horizontal and vertical resolution of forces.) An example is the trussed rafter with four panels (Fig. 4.7), which is called a "Fink truss" in the United States and a "French truss" in England. It is of some practical importance, because it is suitable for factory roof spans of 50 to 60 ft, which is a common span for industrial buildings. We can solve the stress diagram, starting at the left-hand reaction AB, for the joints AB1, BC21, and A123. Thereafter the graphical method breaks down. If we move to joint CD4532, we encounter three unknowns, which is one too many. If we proceed instead to joint A357 we encounter the same problem. We could now start at the right-hand end, but we are stopped in the same way by the symmetrically corresponding joints.

There are a number of classical solutions to this problem, which may be found in the older textbooks on structural design (e.g., H. Parker and C. M. Gay: *Materials and Methods of Architectural Construction*, Wiley, New York, 1958, p. 549, *First Edition 1932*; T. J. Reynolds and L. E. Kent: *Introduction to Structural Mechanics*, English Universities Press, London, 1973, p. 143, *First Edition 1944*). The simplest of these is to calculate the member A7 by the method of sections (see Section 4.3), and draw it to scale. The stress

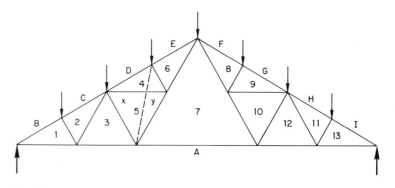

Fig. 4.7.
The Fink or French Truss cannot be solved by graphic statics alone because, on either side, joints are reached where the magnitude of more than two internal forces is unknown. However, a graphical solution can be obtained by substituting an imaginary member x-y for the actual member 4-5. Having reached the joint ED46 with this imaginary member x-y, we can then go back with actual diagonal 4-5 to complete the stress diagram.

Forces in Statistically Determinate Trusses

diagram may then be completed by graphical means.

Because of the ease with which digital computers handle routine calculations and even detail design previously done by means of working drawings, the trend is away from graphical methods and more toward computation techniques. It is therefore doubtful whether the arithmetic evaluation of a single member in an otherwise graphical solution is still warranted when the entire solution can be done by arithmetic.

4.3. The Method of Sections

In this method, first proposed by A. Ritter in 1862, we make a fictitious section along a convenient line, replace the internal forces by equivalent external forces, and take moments about some convenient point. The frame on either side of the section is kept in equilibrium by the external forces, including the "equivalent external forces" acting across the section.

Let us consider the Fink or French truss of Fig. 4.7 and make a section as shown in Fig. 4.8. If we wish to obtain the force in the tie BC, we can do so by the method of sections without solving the entire truss. Let us take moments about the point A. The internal forces in AC and AD have no moment about A, and we obtain:

$$BC \times 14 = 1,400 \times 28 - 400 \times 21 - 400 \times 14 - 400 \times 7$$

which gives BC = $1,400 \times 2 - 400(1.5 + 1 + 0.5) = 1,600$ lb.

This particular example of the method of sections has become well-known, because the Fink truss cannot be solved by graphic statics alone (see Section 4.2) and the method of sections may be used to supplement the stress diagram.

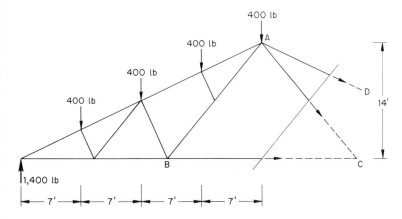

Fig. 4.8.
The method of sections. The force in the tie BC is obtained by making a cut, and taking moments about A. (The horizontal distance between panel points is 7 ft).

As we have pointed out, the numerical evaluation of a single member is hardly warranted in conjunction with a solution by graphic statics, when the solution of trusses by calculation has become much simpler by modern techniques. The method of sections has, however, still a useful function in determining the forces in critical members, without having to solve the entire truss.

The critical members are not always obvious in gabled trusses; the largest tension and compression under uniform loading usually occurs immediately adjacent to the support (i.e., in members A1 and B1 of Fig. 4.7); but this depends on the slope of the roof.

The method of sections is of greater value with parallel-chord trusses, where the location of the largest internal forces is generally obvious.

Example 4.3. *Determine the magnitude of the greatest tensile and compressive forces in the horizontal chords of the truss shown in Fig. 4.6.*

Since the truss is simply supported at its ends, the bending moment is a maximum at the center of the truss (see Sections 1.1 and 5.4). Since the truss has parallel chords, the resistance arm of the moment is the same throughout, and the highest tensile and compressive forces therefore occur in the panels nearest to the center.

Let us therefore cut the truss as shown in Fig. 4.9, and replace the internal forces in AD, AC, and BC by imaginary external forces.

AD and AC have no moments about A, and the moments of the remaining "external" forces are:

$$BC \times 12 = 12,000 \times 8 - 3,000 \times 8$$

which gives BC = 6,000 lb.

Since we have drawn the force as tensile and the answer is positive, the force is, in fact, tensile.

Let us now take moments about C:

$$AD \times 12 = -12,000 \times 16 + 3,000 \times 16 + 6,000 \times 8$$

which gives AD = $-8,000$ lb.

Since we have drawn the force as tensile and the answer is negative, the force is in fact compressive.

These answers may be compared with those obtained in Example 4.2. Evidently we got our solutions more expeditiously by the methods of sections, but we calculated only the maximum forces in the horizontal chords. To work out all members by the method of sections would be tedious and inefficient.

We can now determine the approximate size of the two major members, namely the top and bottom chords. Although some of the verticals and diagonals may have larger forces in them, these are short members of secondary importance. If we wish merely to satisfy ourselves, at the preliminary design stage, that the truss is a feasible and reasonably economical structure, this is a useful method.

Forces in Statistically Determinate Trusses

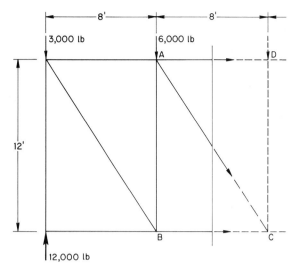

Fig. 4.9.
Approximate design of Pratt truss by the method of sections (Example 4.3). The internal forces in the horizontal chords are obtained by making a suitable cut. The other members are of lesser importance, and they need not be considered at the preliminary design stage.

4.4. Resolving at the Joints

The method of resolution at the joints, proposed by D. J. Jourawski in 1850, lost popularity after the solution by stress diagram established itself. The result is obtained in the form of a series of simultaneous equations, which for large frames are tedious to solve without mechanical aids. However, it is the method which lends itself best to mechanized computation, and it is therefore now coming back into favor. It also provides the simplest approach to the solution of space-frame problems.

We consider each joint in turn. Since the joint, like the entire structure, is at rest, we can write down two equations *for each joint*:

$$\sum F_x = 0$$

$$\sum F_y = 0$$

We thus obtain twice as many equations as there are joints. Since there are only $2j - 3$ members in the truss, we are left with three equations. These yield us no new information. If we insert the results already obtained, they must come to zero on both sides, and are therefore called the *check equations*. If the equations do not check, this indicates a mistake in the arithmetic. The method is best illustrated by an example.

Example 4.4. *Determine the forces in the king-post truss shown in Fig. 4.10.*

Bow's notation is unsuitable for resolving at the joints, and we letter the joints instead. Since the truss is obviously symmetrical, we can save ourselves some equations by lettering statically identical joints with the same letter.

In writing down our equations, we will assume that all internal forces in the truss are tensile. If the result comes out negative, the internal force is compressive.

Taking joint A first and resolving horizontally (see Fig. 3.6 and Fig. 4.10b)

$$T_{AB} \cos 30° + T_{AD} = 0 \qquad \text{(i)}$$

Resolving vertically at joint A:

$$200 - T_{AB} \sin 30° - 800 = 0 \qquad \text{(ii)}$$

Equation (ii) gives

$$T_{AB} = \frac{200 - 800}{\sin 30°} = \frac{-600}{0.5} = -1,200 \text{ lb (compression)}$$

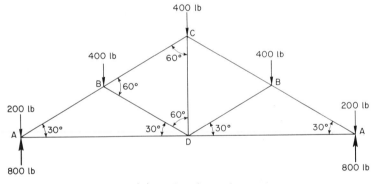

(a) Loads and geometry

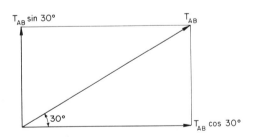

(b) Horizontal and vertical components of internal force T_{AB}

Fig. 4.10.
The method of resolving at the joints (Example 4.4).

Forces in Statistically Determinate Trusses

We can now substitute this value of T_{AB} into Eq. (i):

$$-1{,}200 \times 0.866 + T_{AD} = 0$$

which gives

$$T_{AD} = +1{,}040 \text{ lb (tension)}$$

Taking next joint B and resolving horizontally:

$$-T_{AB} \cos 30° + T_{BC} \cos 30° + T_{BD} \cos 30° = 0$$

which simplifies to

$$T_{BC} + T_{BD} = -1{,}200 \qquad\qquad\text{(iii)}$$

Resolving vertically at B:

$$400 - T_{BC} \sin 30° + T_{AB} \sin 30° + T_{BD} \sin 30° = 0$$

which simplifies to

$$400 - T_{BC} \times 0.5 - 1{,}200 \times 0.5 + T_{BD} \times 0.5 = 0$$

$$T_{BC} - T_{BD} = -400 \qquad\qquad\text{(iv)}$$

Adding Eq. (iii) to Eq. (iv)

$$2T_{BC} = -1{,}600 \quad\text{and}\quad T_{BC} = -800 \text{ lb (compression)}$$

From Eq. (iii) $T_{BD} = -400$ lb (compression).
Resolving horizontally at C we obtain:

$$T_{BC} = T_{BC}$$

which is one of our check equations.
Resolving vertically at C:

$$400 + T_{BC} \sin 30° + T_{BC} \sin 30° + T_{CD} = 0$$

This gives

$$T_{CD} = -400 + 800 \times 0.5 + 800 \times 0.5 = +400 \text{ lb (tension)}$$

Resolving horizontally at D we obtain another obvious check equation:

$$T_{AD} + T_{BD} \cos 30° = T_{BD} \cos 30° + T_{AD}$$

Resolving vertically at D, we obtain a more useful check on our calculation:

$$T_{CD} + 2T_{BD} \sin 30° = 0$$

Substituting numbers:

$$400 - 2 \times 400 \times 0.5 = 0$$

which checks.

The solution may be compared with the work done in Example 4.1. Choice of method depends on the designer's special abilities and on his facilities. Graphic statics requires a good drawing board, and some practice in precision drafting. Resolving at the joints requires some skill in getting the signs in the equations correct.

It is only proper to point out that Example 4.4 has been specially selected for its simplicity. In practice, trusses are specified by span and rise, and the individual joints are specified as fractions of the span and rise. It is actually simpler to set out the horizontal and vertical coordinates of a truss with a scale than to set out the angles with a protractor.

Resolving at the joints, however, is complicated if coordinates are given, instead of angles. We must first calculate the angles from the coordinates (which are the tangent and cotangent of the angle), and then look up the sines and cosines.

*4.5. Tension Coefficients

The remainder of this chapter involves the use of a substantial amount of mathematics, and we suggest that readers omit Sections 4.5, 4.6, 4.7 and 4.8 on first reading, and proceed to Chapter 5. These section heads are preceded by an asterisk.

We can simplify Jourawski's method for resolving at the joints by replacing the sines and cosines by Cartesian coordinates (x and y). Let us consider a member AB inclined at an angle θ to the horizontal (Fig. 4.11). Its length is L_{AB}, its horizontal projection x_{AB}, and its vertical projection y_{AB}. Then

$$\cos \theta = \frac{x_{AB}}{L_{AB}} \qquad \text{and} \qquad \sin \theta = \frac{y_{AB}}{L_{AB}}$$

Consequently the horizontal and vertical components of a force T_{AB} are

$$T_{AB} \cos \theta = \frac{T_{AB} x_{AB}}{L_{AB}}$$

and

$$T_{AB} \sin \theta = \frac{T_{AB} y_{AB}}{L_{AB}}$$

In 1920 Southwell proposed a simplified notation by defining a *tension coefficient*

$$t_{AB} = \frac{T_{AB}}{L_{AB}}$$

The horizontal and vertical components of the force T_{AB} then become

$$t_{AB} x_{AB} \qquad \text{and} \qquad t_{AB} y_{AB}$$

Forces in Statistically Determinate Trusses

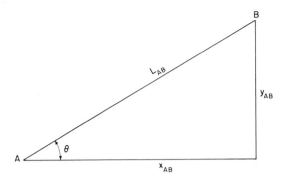

Fig. 4.11.
The method of tension coefficients.

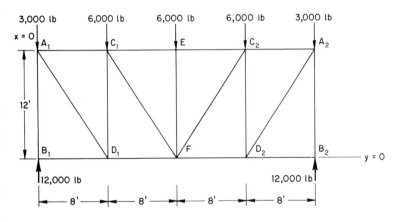

(a) Loads and dimensions

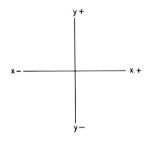

(b) Sign convention. Distances *and* loads positive up and right,
negative down and left.

Fig. 4.12.
Solution of Pratt truss by tension coefficients (Example 4.5).

It should be emphasized that the tension coefficient has no physical meaning as such. It is the force in the member, divided by its length; its usefulness lies in the simplicity of the notation. We can now write down our equations in terms of the vertical and horizontal coordinates of the joints and the external forces acting on the truss. We can then calculate the tension coefficients from the simultaneous equations, and from them obtain the forces by multiplying each tension coefficient by the length of the member.

It is best to carry out the computation in tabular form, particularly if the truss has many members.

Example 4.5. *Using tension coefficients, determine the forces in the truss shown in Fig. 4.12.*

The frame being symmetrical, A, B, C, and D are repeated with the suffix 2 on the right-hand side.

Since the solution is largely mechanical, errors in sign are not so easily discovered as in the previous methods, where they are often evident by inspection. The sign convention must therefore be strictly observed.

As with ordinary Cartesian coordinates, x-distances are positive to the right and negative to the left, and y-distances are positive up and negative down. If we take joint C_1, the x-distances of E and F are $+8$ and the x-distances of A and B are -8; the x-distance of D is zero. The y-distances of A and E are zero, and the y-distances of B, D, and F are -12.

Similarly the loads at A, C, and E are negative, because they are downward; and the reaction at A is positive, because it acts upward.

We will now write down the equations for each joint in turn, starting with the condition of horizontal equilibrium at the joint (x), and followed by the condition of vertical equilibrium (y):

Joint	Direction	Equation	Eq. No.
A	x	$0t_{AB} + 8t_{AC} + 8t_{AD} = 0$	(i)
	y	$-12t_{AB} + 0t_{AC} - 12t_{AD} - 3{,}000 = 0$	(ii)
B	x	$+8t_{BD} = 0$	(iii)
	y	$+12t_{AB} + 12{,}000 = 0$	(iv)
C	x	$-8t_{AC} + 8t_{CE} + 8t_{CF} = 0$	(v)
	y	$-12t_{CD} - 12t_{CF} - 6{,}000 = 0$	(vi)
D	x	$-8t_{BD} - 8t_{AD} + 8t_{DF} = 0$	(vii)
	y	$+12t_{AD} + 12t_{CD} = 0$	(viii)
E	x	$-8t_{CE} + 8t_{CE} = 0$	(ix)
	y	$-12t_{EF} - 6{,}000 = 0$	(x)
F	x	$-8t_{DF} + 8t_{DF} - 8t_{CF} + 8t_{CF} = 0$	(xi)
	y	$+12t_{CF} + 12t_{CF} + 12t_{EF} = 0$	(xii)

Most of these equations are very easily solved:

(ix) and (xi) are check equations only and give $0 = 0$.

(iii) $t_{BD} = 0$

(iv) $t_{AB} = -1{,}000$

(x) $t_{EF} = -500$

(ii) $t_{AD} = -t_{AB} - 250 = +750$

Forces in Statistically Determinate Trusses

(i) $t_{AC} = -t_{AD} = -750$

(vii) $t_{DF} = t_{AD} + t_{BD} = +750$

(viii) $t_{CD} = -t_{AD} = -750$

(xii) $t_{CF} = -\frac{1}{2}t_{EF} = +250$

(v) $t_{CE} = t_{AC} - t_{CF} = -1,000$

This leaves one equation as a numerical check

(vi) $-12(-750) - 12(+250) - 6,000 = 0$, which is correct.

We now tabulate the tension coefficients and the lengths of the members, and thus obtain the forces. Positive answers denote tension, and negative answers compression.

Member	Tension Coefficient (lb/ft)	Length (ft)	Force (lb)
AB	− 1,000	12	12,000 C
AC	− 750	8	6,000 C
AD	+ 750	14.4	10,800 T
BD	0	8	0
CD	− 750	12	9,000 C
CE	− 1,000	8	8,000 C
CF	+ 250	14.4	3,600 T
DF	+ 750	8	6,000 T
EF	− 500	12	6,000 C

This solution may be compared with that of Example 4.2.

*4.6.   Solving Trusses by Matrices and Computers

Although the solutions of structural problems by matrix algebra and by computers is beyond the scope of this book, it may be helpful to point out that the method discussed in the last section is the one that lends itself most to mechanized evaluation.

Let us consider the truss of Fig. 4.12. At joint A_1 we have a y-equation

$$y_{AB}t_{AB} + y_{AC}t_{AC} + y_{AD}t_{AD} + Y_A = 0$$

where y_{AB} is the vertical distance of B from A, and y_{AC} and y_{AD} have similar meaning; Y_A is the vertical force (or the vertical component of an inclined force) acting on A. The sign is positive up and negative down.

Similarly we have an x-equation at A_1

$$x_{AB}t_{AB} + x_{AC}t_{AC} + x_{AC}t_{AD} + X_A = 0$$

even though X_A, the horizontal force acting on joint A, is zero in this instance. This is usually so in frames which carry only vertical loads.

Evidently we could adopt a form of shorthand by calling all joints adjacent to A by the letter k, and giving this in turn all possible values; in this instance k would in turn become B, C, and D. This then gives us

$$x_{Ak}t_{Ak} + X_A = 0$$

and

$$y_{Ak}t_{Ak} + Y_A = 0$$

We can now go around all the other joints and obtain an x- and a y-equation for each. We can further cut down on the writing by considering that the letter i can stand for any joint, so that the equations required for the entire truss become

$$x_{ik}t_{ik} + X_i = 0 \tag{4.3}$$

and

$$y_{ik}t_{ik} + Y_i = 0 \tag{4.4}$$

in which i and k can have any value corresponding to a member in the truss.

We can solve these equations by putting all appropriate values of i and k into Eqs. (4.3) and (4.4) in turn, and then solving them individually, as in Section 4.5. Alternatively, we can consider x and y as *matrices* of the horizontal and vertical coordinates of the joints in the frame, and X and Y as matrices of the horizontal and vertical forces acting externally on the structure. We can then evaluate all the simultaneous equations in one operation by matrix algebra, which is a convenient shorthand for solving a large number of simultaneous equations. We thus obtain the matrix for t, which then gives us the internal forces in the members of the frame in tabular form.

The details are beyond the scope of this book; but it may be appropriate to point out that matrix algebra is warranted only if the truss is sufficiently large. In the same way, it is not worthwhile to obtain the services of a shorthand writer to record and transcribe a short note; it is less trouble to write it out by conventional means.

Matrices are particularly suitable for evaluation by electronic digital computer. A general instruction, or *program*, setting out the method of solution of isostatic trusses by matrices is punched on computer cards or tape, and is then available whenever required for the solution of a particular truss. The matrices for the x- and y-coordinates and the X- and Y-forces acting on a particular truss are also punched on cards or tape and the computer produces, from the combination of the general program and the specific data for a particular truss, the matrix for the tension coefficients. By an extension of the program, the internal forces (which are obtained by multiplying each tension coefficient by the length of the member) can be printed out directly.

Forces in Statistically Determinate Trusses

Once a computer has been employed in the solution, the data already stored in it may be used further. They can, for example, be used for choosing suitable sections for the members of the truss, and they can then be used further to work out the exact length of each member, the cutting required, and the details of the connections. The record can also be used for issuing instructions to the steel fabricator, making out labels for the members to identify them during fabrication and making out an invoice for the cost of the truss.

On this basis, the use of a computer may be worthwhile even for medium-sized and small trusses, provided a sufficient number are similar, but not quite identical, so that the calculations must be repeated many times. If all are exactly the same, the calculations need to be done only once, and the computer offers no advantage.

Since the design of statically determinate plane trusses is neither complicated nor difficult, the use of computers cannot be justified for the architect's preliminary design.

*4.7. Space Frames

The use of matrix algebra and of computerized solutions is of much greater significance for space frames. It is difficult to draw a stress diagram in space. Although a model could be constructed, say from balsawood, this would be neither an accurate nor an economical method for solving space frames. The method of resolution at the joints and the method of sections are both awkward to use, because computation of angles in space is time-consuming. However, Cartesian coordinates function easily in three dimensions, and we therefore employ the method of tension coefficients described in Sections 4.5 and 4.6.

Using the notation introduced in Section 4.6, it is easy to see that the use of a third dimension z, in addition to x and y, adds a third equation to (4.3) and (4.4):

$$x_{ik}t_{ik} + X_i = 0 \qquad (4.3)$$

$$y_{ik}t_{ik} + Y_i = 0 \qquad (4.4)$$

$$z_{ik}t_{ik} + Z_i = 0 \qquad (4.5)$$

These equations can be solved by putting all appropriate values of i and k into Eqs. (4.3) to (4.5), and solving them individually, as in Section 4.5. However, because of the larger number of equations and the larger number of members of a space frame, the solution by matrix algebra, particularly with the aid of a computer, is far more attractive than in the case of the plane frame.

We have already discussed the determination of the reactions for a space frame (see Section 3.4). The simplest structure which is

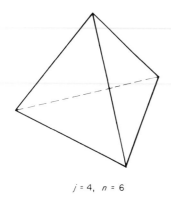

$j = 4, \; n = 6$

Fig. 4.13.
The simplest statically determinate space frame has six members and four joints.

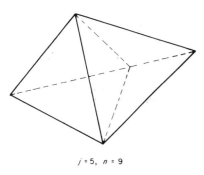

$j = 5, \; n = 9$

Fig. 4.14.
A statically determinate space frame with pin joints requires $n = 3j - 6$ members. If there are more, the truss is statically indeterminate. In this truss $j = 5$, and $n = 9$.

completely stable in space under a system of loading is the tetrahedron (Fig. 4.13). This has four joints and six members. If we want to add a joint (Fig. 4.14), we must add three members. A statically determinate space frame with j joints therefore requires

$$n = 3j - 6 \qquad (4.6)$$

members.

We noted in Section 4.1 that in a plane frame $n + r$ is twice the number of joints. Similarly for a space frame

$$3j = n + r \qquad (4.7)$$

This means that a true space frame requires six reactions, of which two must be mutually perpendicular horizontal reactions.

The determination of the forces even in simple space frames is laborious, because of the large number of simultaneous equations which have to be solved.

Example 4.6. *Determine the forces in the bracket shown in Fig. 4.15. The bracket is fixed to a pin in the wall at A, and it is sliding against the wall at C_1 and C_2.*

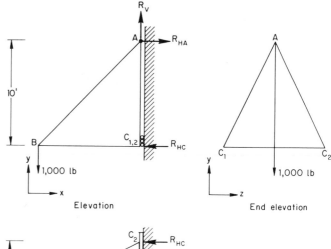

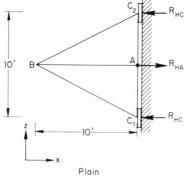

Fig. 4.15.
Solution of space frame by tension coefficients (Example 4.6).

Forces in Statistically Determinate Trusses

The frame being symmetrical, C_1 and C_2 are identical. There are four joints and six members, so that the frame is statically determinate. We obtain the reactions from Eq. (3.6).

From $\sum F_x = 0$ we obtain $R_v = 1,000$ lb.

From $\sum F_y = 0$ we obtain $R_{HA} = 2R_{HC}$.

From $\sum M_{xy} = 0$ we obtain, by taking moments about A

$$2R_{HC} \times 10 = 1000 \times 10$$

which gives $R_{HC} = 500$ lb and $R_{HA} = 1,000$ lb.

The remaining equilibrium conditions give $0 = 0$.

We will now write down the equations for each joint in turn, as in Example 4.5.

Joint	Direction	Equation	Eq. No.
A	x	$-10\,t_{AB} + 0\,t_{AC} + 1,000 = 0$	(i)
	y	$-10\,t_{AB} - 10\,t_{AC} - 10\,t_{AC} + 1,000 = 0$	(ii)
	z	$0\,t_{AB} - 5\,t_{AC} + 5\,t_{AC} = 0$	(iii)
B	x	$+10\,t_{AB} + 10\,t_{BC} + 10\,t_{BC} = 0$	(iv)
	y	$+10\,t_{AB} + 0\,t_{BC} - 1,000 = 0$	(v)
	z	$0\,t_{AB} - 5\,t_{BC} + 5\,t_{BC} = 0$	(vi)
C_1	x	$0\,t_{AC} - 10\,t_{BC} + 0\,t_{CC} - 500 = 0$	(vii)
	y	$+10\,t_{AC} + 0\,t_{BC} + 0\,t_{CC} = 0$	(viii)
	z	$+5\,t_{AC} + 5\,t_{BC} + 10\,t_{CC} = 0$	(ix)

As in Example 4.5, most of these equations solve very easily:

(i) $t_{AB} = 100$

(viii) $t_{AC} = 0$

(iv) $t_{BC} = -\frac{1}{2}t_{AB} = -50$

(ix) $t_{CC} = -\frac{1}{2}t_{BC} = +25$

The remainder are check equations only:

(ii) $-1,000 + 0 + 1,000 = 0$

(iii) $0 = 0$

(v) $1,000 - 1,000 = 0$

(vi) $0 = 0$

(vii) $+500 - 500 = 0$

It may seem at first sight that most of this operation is a waste of time if so many equations are superfluous; however, it is not always obvious which equations are redundant until they are written out.

We now tabulate the tension coefficients and the lengths of the members, and thus obtain the forces.

Member	Tension Coefficient (lb per ft)	Length (ft)	Force (lb)
AB	$+100$	14.2	1,420 T
AC	0	11.2	0
BC	-50	11.2	560 C
CC	$+25$	10.0	250 T

*4.8.   Triangulated Frames for Roofing Curved Surfaces

Statically determinate plane frames are formed by triangulation, so that the number of members required to connect j joints (see Section 4.1) is

$$n = 2j - 3 \qquad (4.1)$$

The same applies to statically determinate frames formed into a curved surface.

It is evidently possible for a structure to be a surface structure without being plane. If we do not insist on it being also statically determinate, then we have a wide range to choose from. Thus every shell form (see section 1.11) can be copied as a triangulated surface structure; but some are stable only if the joints are made stiff (instead of being pin-jointed), so that the structure becomes statically indeterminate.

As we shall show in Section 9.3, the membrane shell, in which the forces are purely within the surface of the shell, is statically determinate, so that the membrane forces can be worked out by the equations of static equilibrium alone (see Section 3.4). However, this is dependent on suitable boundary conditions; this means that the internal forces within the shell must be compatible with the external forces at the supports. If they are not, bending stresses are set up in the shell, and these are statically indeterminate, so that the shell can no longer be solved by statics alone.

It is relatively easy to achieve suitable boundary conditions in hemispherical domes, hexagonal domes (Fig. 1.54), and square domes; the *geodesic dome* is therefore the best known example of a statically determinate curved-surface structure.

Geodesy is the science of mapping the earth's surface, and *geodesic lines* are the shortest lines possible between two points on the earth's surface. If we place a string on the surface of a globe and pull it tight, it forms a geodesic line, or *great circle*. The great circles mapped on a terrestrial globe are the meridians of longitude and the equator. The circles of latitude other than the equator are "small circles," and a distance along a small circle is not the shortest distance between two points.

A geodesic dome is thus formed by three sets of interesecting great circles, forming a series of equilateral spherical triangles with pin joints. In practice, geodesic domes of small span can be made very light. Snow cannot easily accumulate on a small-diameter sphere, and it offers little resistance to wind. It therefore carries mainly its own weight, and this is made initially small for a membrane structure.

The geodesic dome can also be considered as a space frame, using Eq. (4.7). This shows that the number of reactions

$$r = 3j - n = b + 3 \qquad (4.8)$$

where b is the number of joints on the boundary of the geodesic dome. We must therefore provide a reaction at each boundary joint. This can be done with a stiff boundary member, or the entire boundary of the dome can be fixed directly to the foundation. Analysis can be made by the statically determinate space frame Eqs. (4.3) to (4.5), or it may be considered as a membrane shell with a large number of openings in it (see Section 9.3).

Masonry and concrete domes have declined in popularity because of the difficulty of making the formwork in timber (see Section 1.11). This problem does not apply to the geodesic dome, which is formed from identical great circles; these are easily made in metal, plastic, or laminated timber. If the surface is divided into a sufficiently large number of triangles, it is admissible to flatten the curves and build up the dome from plane triangles.

Suggestions for Further Reading

H. S. Howard: *Structural Analysis—An Architect's Approach.* McGraw-Hill, New York, 1966. 297 pp.

Z. Makowski: *Steel Space Structures.* Michael Joseph, London, 1964. 214 pp.

C. H. Norris and J. B. Wilbur: *Elementary Structural Analysis.* McGraw-Hill, New York, 1960. Chapters 4, 5, 7, and 9; pp. 115–188, 221–240, 256–275.

H. Parker, C. M. Gay, and J. W. MacGuire: *Materials and Methods of Architectural Construction.* Wiley, New York, 1958. Chapter 21, pp. 528–565.

Problems

4.1. List and write detailed notes on the assumptions which must be made in order that actual structures can be analyzed by structural mechanics. Explain the specific assumptions which are made about trusses in order that they may be analyzed as pin-jointed. When are these assumptions no longer valid?

4.2. Discuss the various factors which influence the choice of a roof-truss system for an industrial roof, and illustrate your answer by means of typical examples.

4.3. How many members do you require for (a) a plane frame, (b) a three-dimensional surface structure, and (c) a space frame for a tower, if the solution is to be statically determinate?

What would be the effect of (i) the removal, and (ii) the addition of structural members?

With the aid of sketches, describe three architecturally successful examples of the use of three-dimensional frames, and give reasons for your choice in each case.

†**4.4.** Draw the stress diagram for the truss shown in Fig. 4.16, and find the magnitude of the forces in the members D4, 34, and A3. State whether the forces are tensile or compressive.

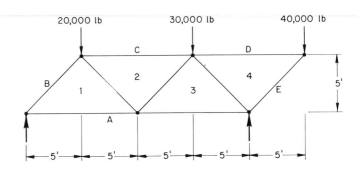

Fig. 4.16.
Problem 4.4.

4.5. Draw the stress diagram, and determine the forces in all the members of the Pratt truss shown in Fig. 4.17, indicating whether it is tensile or compressive.

4.6. Check the forces in the two members next to the left-hand support of the truss in Fig. 4.17 by the method of resolving at the joints.

4.7. Check the force in the vertical member nearest to the left-hand support of the truss shown in Fig. 4.17 by the method of sections.

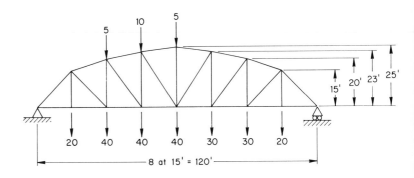

Fig. 4.17.
Problems 4.5, 4.6, and 4.7. Loads are in kip (or 1,000 lb) units.

4.8. Figure 4.18 shows a truss of 24-ft span and 18-in depth, carrying 5 loads of 1,000 lb each. Calculate the forces in the members AO, OB, CD, and ML.

†Similar problems in metric units are given in Appendix G.

Forces in Statistically Determinate Trusses

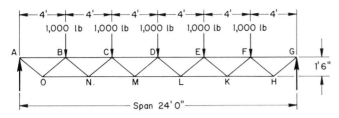

Fig. 4.18.
Problem 4.8.

4.9. Using the method of sections, determine the magnitude of the forces in the members BD, CD, CE, DE, and DF of the truss shown in Fig. 4.19. State whether the forces are tensile or compressive.

4.10. If the load at point D in Fig. 4.19, is increased by 3,000 lb, find the change in the magnitude of the force in the member AB by drawing the triangle of forces.

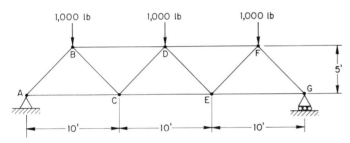

Fig. 4.19.
Problems 4.9 and 4.10.

4.11. Figure 4.20 shows a bracket on a pole carrying electric cables. Assuming that the joints are free to rotate and the self-weight of the bracket is negligible, calculate:

(i)   The force in the member XY; state whether it is compressive or tensile.

(ii)   The angle which the reaction Z makes with the vertical center line of the pole; what is the horizontal component of that reaction?

4.12. An advertising sign weighing 600 lb is suspended from a shop front by two oblique ties and a horizontal strut as shown in Fig. 4.21. Using the method of tension coefficients, determine the forces in each of the three members, and state whether they are in tension or compression. (It is only necessary to solve the equations for the joint B.)

4.13. Determine the forces in each of the members of the crane shown in Fig. 4.22, and also the reactions at A, B, and C. Distinguish between tension and compression members. (It is advisable to start at point E.).

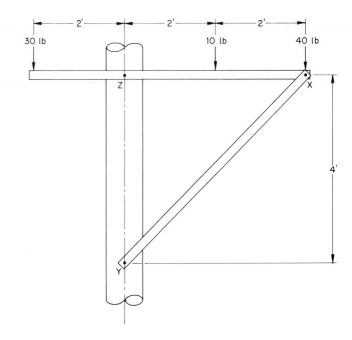

Fig. 4.20.
Problem 4.11.

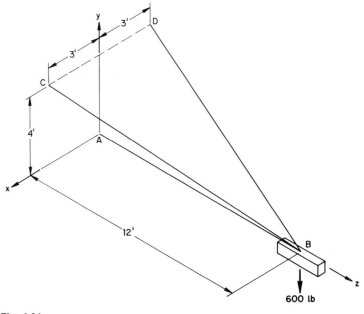

Fig. 4.21.
Problem 4.12.

Forces in Statistically Determinate Trusses

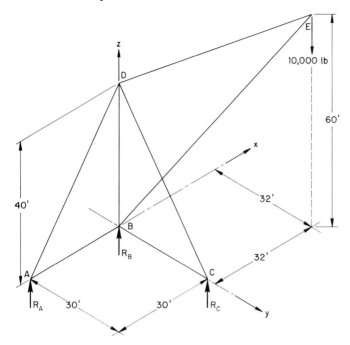

Fig. 4.22.
Problem 4.13.

Bending Moments and Shear Forces in Statically Determinate Beams and Frames

Give me the structure, and I will find you naturally
the forms which should result from it; but if you
change the structure, I will have to change the forms.

Viollet-le-Duc

In this chapter we examine the criteria which make beams, arches, and frames with rigid joints statically determinate, or isostatic. We then explain the methods used for determining bending moments and shear forces in each case. The stresses due to the bending moments and shear forces are discussed in Chapter 6.

5.1. Statically Determinate Beams, Statically Indeterminate Beams, and Mechanisms

If we know the loads and the support reactions of a beam, we can always determine the bending moment and shear force at any section. A beam without pin joints (see Section 4.1) is therefore statically determinate, or isostatic, if the number of support reactions does not exceed the number of equilibrium equations (see Section 3.4).

Let us consider a simply supported beam (Fig. 5.1a) carrying five vertical loads. The three equations of static equilibrium are

$$\sum F_x = 0 \tag{3.5a}$$

$$\sum F_y = 0 \tag{3.5b}$$

$$\sum M = 0 \tag{3.5c}$$

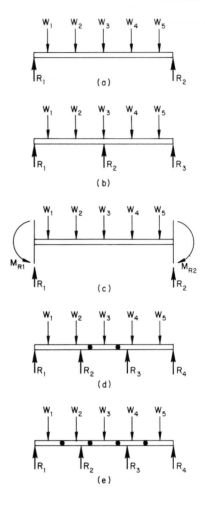

Fig. 5.1.
Statically determinate beams, statically indeterminate beams, and mechanisms.

(a) A beam with two supports is statically determinate.

(b) The third reaction cannot be determined by statics, and the beam is statically indeterminate.

(c) A beam built-in at the two ends is statically indeterminate, because the restraining moments due to the building-in cannot be determined by statics alone.

(d) A beam with two additional supports can be rendered statically determinate by inserting two pin joints at which the bending moment is, by definition, zero. These give two additional equations of static equilibrium, which allow the magnitude of the two extra reactions to be determined. This is a Gerber beam (see Section 5.6).

(e) If the number of pin joints exceeds the number of additional reactions, the beam ceases to be a structure and becomes a mechanism.

Bending Moments and Shear Forces

The condition of horizontal equilibrium $\Sigma F_x = 0$ yields $0 = 0$, which is correct, but uninformative. The condition of vertical equilibrium $\Sigma F_y = 0$ tells us that the sum of the reactions $R_1 + R_2$ is equal to the sum of the loads $W_1 + W_2 + W_3 + W_4 + W_5$. The condition of moment equilibrium $\Sigma M = 0$ allows us to separate the two reactions by taking moments about either support (see Examples 3.2 to 3.4). Knowing the reactions, we can determine the internal moments and forces in the beam, which is thus statically determinate.

However, if we add a third support (Fig. 5.1b), the problem becomes statically indeterminate. We can still determine the sum of the reactions from $\Sigma F_y = 0$, but the only remaining equation does not allow us to separate the three reactions. The problem is therefore not soluble by statics alone. It can be solved by the combined use of statics and elasticity (see Section 8.3), but this introduces additional complications. The beam is called *continuous* over an intermediate support, or over *two spans*. Each intermediate support adds a statically indeterminate reaction, so that a four-span continuous beam has three statically undeterminate reactions, and we require three elastic equations for its solution, in addition to the conditions of static equilibrium. In this chapter we consider only statically determinate beams; statically indeterminate beams are briefly discussed in Sections 8.3 and 8.4.

We also obtain statically indeterminate reactions if we build a single-span beam firmly into the supporting walls (Fig. 5.1c). Since the beam is built-in at the ends, it cannot rotate, and consequently it has restraining moments M_{R1} and M_{R2} in addition to the vertical reactions R_1 and R_2. Since we can determine only two reactions from the conditions of static equilibrium, the built-in beam is statically indeterminate, and we shall not consider it further in this chapter (see Section 8.4).

While an additional reaction makes a beam more statically indeterminate, a pin joint has the opposite effect. A pin joint is defined as a joint which allows free rotation. It therefore does not transmit moment across the joint, but it has the capacity to transmit both axial and shear forces.

In the nineteenth century, connections were frequently made by pushing a large pin through holes cast into iron beams (Fig. 5.2a), and this guarantees free rotation. It is still the method used for large spans, such as occur in long bridges.

Pins were an economical method of jointing before the invention of automatic screw-cutting machines. For modern architectural structures, simpler "pin joints" are satisfactory. In structural steel, a good pin joint can be made by cutting the beam and rejoining it with a web cleat (Fig. 5.2b); this transmits vertical forces, but very little moment because it does not touch the flanges. Reinforced concrete is heavy, and it is generally sufficient to seat a small inset girder on the supporting girder with a halved joint (Fig. 5.2c); this

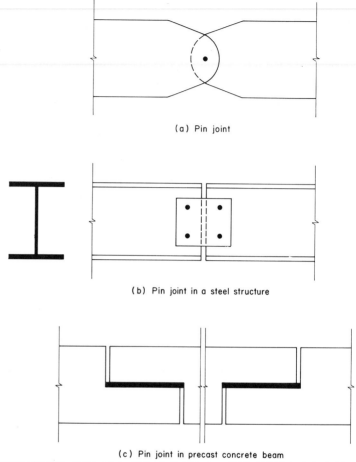

(a) Pin joint

(b) Pin joint in a steel structure

(c) Pin joint in precast concrete beam

Fig. 5.2.
Pin joints. (a) The original pin joint was formed by a pin passing through holes in the members to be fitted. (b) A pin joint in a steel structure is formed by a web connection which offers little resistance to bending, but is strong enough to transmit shear. (c) The halved joints are made strong enough (by additional reinforcement if need be) to transmit shear, and the joint faces right or left according to the sign of the shear force.

allows rotation, and also transmits the vertical force (provided the halved joint is adequately reinforced).

Since the moment at a pin joint is zero, we obtain an additional $\Sigma M = 0$ equation for each pin joint. Consequently the beam in Fig. 5.1d is statically determinate, because the two pin joints cancel out the two statically indeterminate reactions. A statically determinate beam with short insets simply supported from cantilever overhangs is called a *Gerber beam* (see Section 5.6). It was patented in 1866 by a Bavarian engineer of that name, who was concerned about the effect of foundation settlement on continuous girder bridges. Evidently if the two middle supports of a four-span continuous girder bridge settled into the mud of the river, the span would be increased three times and the bridge fail. Gerber beams are useful in buildings whose foundations are liable to settlement (see Section 8.9), and in buildings which are to be assembled rapidly by dry construction methods from prefabricated components. They are evidently less rigid than continuous beams.

If we introduce more pin joints than redundant reactions, the structure becomes a mechanism, and falls down (Fig. 5.1e).

The only other group of statically determinate structures are the cantilevers (Fig. 5.3). If the beam is built into the wall at one end, it does not need to be supported at all at the other. This is particularly useful for balconies and canopies, where columns may form an obstruction, and for light curtain walls which look better without columns. Evidently cantilevers have to be deeper because of the absence of supports at one end (see Section 5.2.).

The equation for horizontal equilibrium, $\Sigma F_x = 0$, gives us $0 = 0$; the equation for vertical equilibrium, $\Sigma F_y = 0$, gives us R; and the equation for moment equilibrium, $\Sigma M = 0$, gives us the support moment M. A cantilever is thus statically determinate.

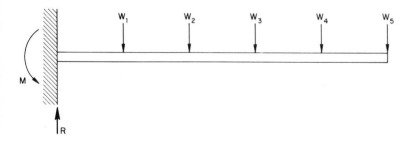

Fig. 5.3.
A cantilever also has two reactions, both at the same end, and it is consequently statically determinate.

5.2. Bending Moment and Shear Force Diagrams for Cantilevers

As we pointed out at the beginning of this book (see Section 1.1), vertical loads acting on a structure bridging a horizontal gap set up bending moments and shear forces, and their magnitude is independent of the form of the structure. We will now consider the magnitude of these moments and forces in detail, because in beams these are resisted by internal moments and shear forces that are exactly equal and opposite. The existence of the internal moment and shear force is easily demonstrated with a model (Fig. 5.4).

The beam must be sufficiently strong to resist both the bending moment and the shear force (see Sections 6.4 and 6.6); in most architectural structures, however, the bending moment is the more important consideration, because it is dependent on the magnitude

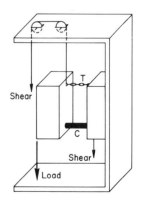

Fig. 5.4.
Model illustrating the internal bending moment and internal shear force in a beam. The internal moment is represented by a tensile force (the chain T) and a compressive force (the strut C) separated by a lever arm. The internal shear force is represented by the two weights, one tending to slide the section upward, and the other downward.

113

of the loads as well as the span, and its importance increases as the span increases. The design of short-span structures carrying heavy loads (for example, a pin connecting two engine parts) is dominated by the shear force, which is dependent only on the magnitude of the loads. The beams of a building are normally *designed* for the bending moment and then *checked* for the shear force (and modified if, as occasionally happens, they are not big enough). The bending moment is therefore the primary design factor for beams, and the only one that need normally be considered in a preliminary design, while the shear force is a secondary design factor, to be considered only for detail design.

Let us first consider a cantilever carrying a single concentrated load W at its end (Fig. 5.5), for example, a brick wall at the end of a balcony.

The vertical support reaction $R = W$, and the moment reaction $M_R = -WL$. The vertical force tending to shear (or cut) through the beam is constant along its entire length, so that the shear force

$$V = W$$

The shear force (S.F.) diagram is therefore a rectangle (Fig. 5.5b).

Taking moments about a section $X–X$ at a distance x from the free end of the beam, the bending moment

$$M = -W \times x$$

and this reaches its maximum at the support where

$$M = -WL$$

The bending moment (B.M.) diagram is therefore a triangle.

[†]Example 5.1. *Draw the bending moment and shear force diagrams for a cantilever, spanning 10 ft, carrying a load of 1,000 lb (a) at the end, and (b) 8ft from the support*

The solution is shown in Fig. 5.6.

Let us next consider a cantilever carrying a uniformly distributed load (Fig. 5.7), for example, a reinforced concrete balcony slab carrying its own weight. If we are concerned only with the maximum bending moment and shear force, as we normally are, the problem can be solved easily by replacing the uniformly distributed load by a concentrated load of the same magnitude, acting at its center of gravity, i.e., a load W at a distance $\frac{1}{2}L$ from the support. We then obtain the support reactions as

$$R = W \quad \text{and} \quad M_R = \tfrac{1}{2}WL$$

By static equilibrium these are also the maximum values of the

[†]This and all subsequent examples are worked in metric units in Appendix G.

Bending Moments and Shear Forces

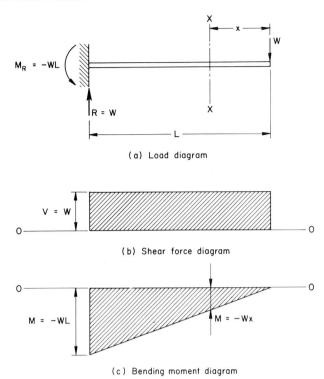

(a) Load diagram

(b) Shear force diagram

(c) Bending moment diagram

Fig. 5.5.
Cantilever carrying a concentrated load.

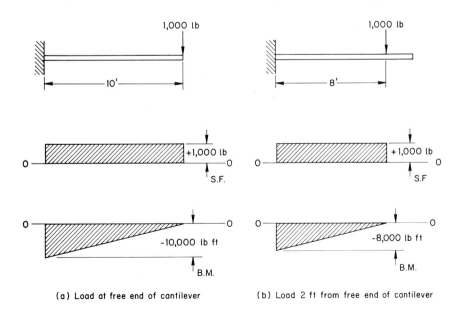

(a) Load at free end of cantilever

(b) Load 2 ft from free end of cantilever

Fig. 5.6.
Cantilever carrying a concentrated load (Example 5.1).

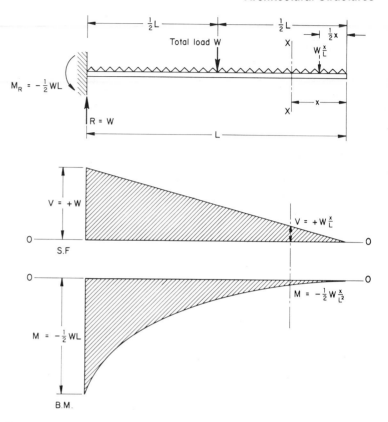

Fig. 5.7.
Cantilever carrying a uniformly distributed load.

bending moment and the shear force:

$$V = W \quad \text{and} \quad M = -\tfrac{1}{2}WL$$

It is only rarely necessary to determine the variation of bending moment and shear force along the beam in architectural structures, since it is usually most economic to use a beam of constant section, designed for the maximum moment. However, the complete diagram is easily determined.

Let us consider a section, X–X, at a distance x from the free end. The total load on the length x is Wx/L, and this is evidently the force tending to shear through the beam (Fig. 5.7). The shear force diagram is therefore triangular, and V increases uniformly from zero to W.

The uniformly distributed load Wx/L can be replaced by a concentrated load of the same magnitude, acting at $\tfrac{1}{2}x$ from the section X–X. Its moment about X–X is

$$M = -\frac{Wx}{L} \times \tfrac{1}{2}x = -\tfrac{1}{2}\frac{Wx^2}{L}$$

Bending Moments and Shear Forces

which becomes $-\frac{1}{2}WL$ when $x = L$, so that the bending moment M increases parabolically from 0 to $-\frac{1}{2}WL$ (Fig. 5.7).

Example 5.2. *Determine the maximum bending moment and shear force for a cantilever, 10 ft long, carrying a load of 1,000 lb (a) if it is uniformly distributed over the whole of its length, and (b) if it is uniformly distributed over a length of 8 ft from the support.*

The solution is shown in Fig. 5.8.

The bending moments and shear forces acting on a beam carrying a multiple load are best determined by the *principle of superposition*. Because the stresses in the beams are proportional to their deformation (see Section 6.4), the effect of combined loading is simply the sum of the separate effects. We can therefore work each simple case out separately and then add them together (or superimpose one case upon the other). This is much simpler and quicker than considering the combined case. The maximum bending moments and shear forces in cantilevers always occur at the fixed end.

If we are interested only in the maxima, we need merely add the separate effects together. If we are interested in the variation along the beam, we work out the separate effects at, say, 6 to 8 points x along the beam, add them together, and plot the curve, which is naturally a compound curve.

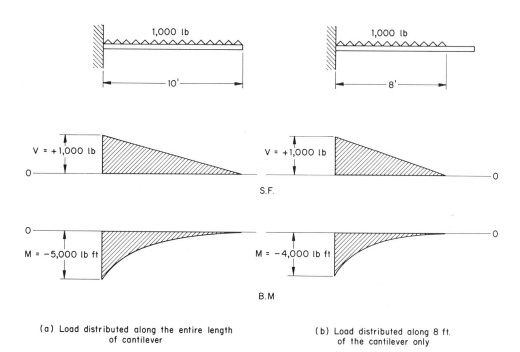

(a) Load distributed along the entire length of cantilever

(b) Load distributed along 8 ft. of the cantilever only

Fig. 5.8.
Cantilever carrying a uniformly distributed load (Example 5.2).

117

Example 5.3. *A balcony slab, cantilevering 10 ft, carries at the free end a concentrated load of 500 lb (due to a dwarf brick wall), a load of 1,000 lb uniformly distributed over the whole span (due to its own weight), and a load of 400 lb uniformly distributed over a length of 8 ft from the support (due to the live load, caused by people and furniture on the balcony). Determine the maximum bending moment and the maximum shear force.*

The maximum shear force

$$V = 500 + 1,000 + 400 = 1,900 \text{ lb}$$

The maximum bending moment

$$M = -500 \times 10 - \tfrac{1}{2} \times 1,000 \times 10 - \tfrac{1}{2} \times 400 \times 8 = -11,600 \text{ lb ft}$$

5.3. Sign Convention for Bending Moment and Shear Force

Evidently the absolute sign convention described in Section 4.5 is inadequate for bending moments; otherwise the bending moment would be of opposite sign, depending on whether the cantilever was pointing right or left.

There is no universal agreement on the sign convention for bending moment and shear force. We shall use the convention commonly employed in textbooks on architectural structures; this defines a positive bending moment as one which causes a beam to sag and produce a concave curve, and a negative moment as one

(a) Positive bending moment

(b) Negative bending moment

Fig. 5.9.
Sign convention for bending moments. (a) A positive bending moment causes a beam to sag, and it produces a concave curvature. (b) A negative bending moment causes a beam to hog or hump, and it produces a convex curvature.

Bending Moments and Shear Forces

which causes a beam to hog and produce a convex curve (Fig. 5.9). Consequently, cantilever bending moments are negative, and simply supported beam bending moments are positive (Fig. 5.10).

By the same convention a positive shear force causes the left-hand portion of a beam to shear upward relative to the right-hand portion, while a negative shear force does the opposite (Fig. 5.11).

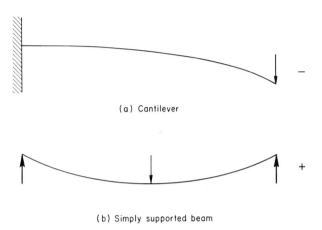

(a) Cantilever

(b) Simply supported beam

Fig. 5.10.
Sign convention for bending moments. (a) The deflected shape of a cantilever is convex, and the bending moment is negative. (b) The deflected shape of a simply supported beam is concave, and the bending moment is positive.

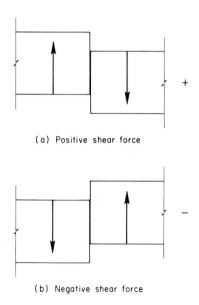

(a) Positive shear force

(b) Negative shear force

Fig. 5.11.
Sign convention for shear forces.

5.4. Bending Moment and Shear Force Diagrams for Simply Supported Beams

Let us next consider a simply supported beam carrying a central concentrated load (Fig. 5.12). From the condition of vertical equilibrium the two support reactions

$$R = \tfrac{1}{2}W$$

and the force tending to shear through the beam

$$V = R = \tfrac{1}{2}W$$

until we reach the central concentrated load, when it changes to

$$V = R - W = -\tfrac{1}{2}W$$

If the load was indeed perfectly concentrated, the change from $+\tfrac{1}{2}W$ to $-\tfrac{1}{2}W$ would be sudden. In practice even a concentrated load must rest on a short length of beam, and the change from positive to negative shear force is distributed over the supporting length of the load.

The bending moment at a distance x from the left-hand support is

$$M = Rx = \tfrac{1}{2}Wx$$

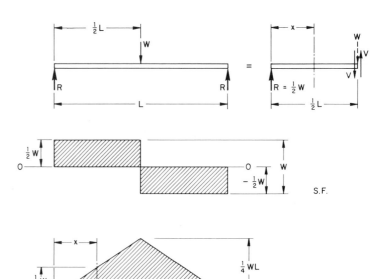

Fig. 5.12.
Simply supported beam carrying a central concentrated load.

Bending Moments and Shear Forces

The maximum occurs under the load when $x = \frac{1}{2}L$ and

$$M = \frac{1}{2}W \times \frac{1}{2}L = \frac{1}{4}WL$$

Example 5.4. *Determine the maximum bending moment and shear force due to a central concentrated load of 1,000 lb., carried on a simply supported span of 10 ft.*

The solution is $V = \pm 500$ lb and $M = 2,500$ lb ft.

A comparison with Example 5.1 shows the structural superiority of the simply supported beam over the cantilever. The shear force is half that in a cantilever carrying the same load on the same span, and the maximum bending moment (with the load in the worst location in each case) is a quarter that of the cantilever.

Let us now consider a single concentrated load located unsymmetrically (Fig. 5.13). For vertical equilibrium

$$W = R_1 + R_2$$

Taking moments about R_2

$$R_1 = \frac{WL_2}{L}$$

so that

$$R_2 = \frac{WL_1}{L}$$

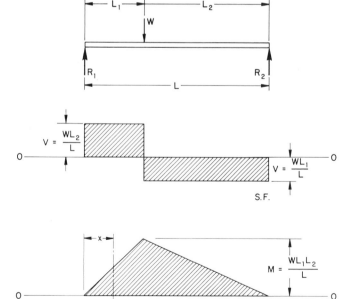

Fig. 5.13.
Simply supported beam carrying a concentrated load off-center.

121

Consequently the shear force from R_1 to W is

$$V = R_1 = \frac{WL_2}{L}$$

and from W to R_2

$$V = R_1 - W = \frac{WL_1}{L} = -R_2$$

The bending moment at a distance x from R_1

$$M = R_1 x = \frac{WL_2 x}{L}$$

until we reach the load W at $x = L_1$, when

$$M = R_1 x - W(x - L_1)$$

which reduces to 0 at the right-hand support.

The maximum bending moment, which occurs immediately under the load

$$M = \frac{WL_1 L_2}{L}$$

If $L_1 = L_2 = \frac{1}{2}L$, this reduces to the previously considered case of concentric loading:

$$M = \tfrac{1}{4}WL$$

Example 5.5. *Determine the bending moment and shear force due to a concentrated load of 1,000 lb on a simply supported span of 10 ft if the load is located (a) at $\frac{1}{10}$ the span, (b) at $\frac{1}{4}$ span, and (c) at mid-span.*

For $\frac{1}{10}$ span, $L_1 = 1$ ft and $L_2 = 9$ ft, and $L_2/L = 0.9$.
For $\frac{1}{4}$ span, $L_1 = 2.5$ ft. and $L_2 = 7.5$ ft, and $L_2/L = 0.75$.
For mid-span, $L_1 = L_2 = 5$ ft., and $L_1/L = L_2/L = 0.5$.
Consequently the maximum shear force is

(i) $V = 0.9 \times 1,000 = 900$ lb

(ii) $V = 750$ lb

(iii) $V = 500$ lb

The maximum bending moment, which occurs directly under the load is

(i) $M = 1,000 \times 1 \times 9/10 = 900$ lb ft

(ii) $M = 1,000 \times 2.5 \times 7.5/10 = 1,875$ lb ft

(iii) $M = 1,000 \times 5 \times 5/10 = 2,500$ lb ft

Bending Moments and Shear Forces

The maximum bending moment increases as the load moves toward the center of the span; the maximum shear force increases as the load moves toward either support.

Assuming that we wished to determine the design condition for a heavy movable load, for example, a steel safe or a computing machine, we should consider (i) mid-span which gives the highest bending moment $M = \frac{1}{4} WL$, and (ii) a location just off either support which gives the highest shear force, namely $V = W$. Since the bending moment is usually critical for the design, it may be sufficient to consider the mid-span location only.

Let us next consider two symmetrically located concentrated loads, i.e., two loads at $\frac{1}{3}$ span. This is a common problem when the floor slab is supported on secondary beams, which in turn are supported on primary girders (Fig. 5.14b). A subdivision into two panels is rarely used because the secondary beam is then carried by the primary beams at mid-span, where it causes a high bending moment. A four-panel subdivision is also uncommon, because it produces either an uneconomically short span for the floor slab, or a long span (and correspondingly great depth) for the primary girder.

If we take the total load of the beam as W, and each concentrated load as $\frac{1}{2} W$ (Fig. 5.15), the end reactions are each $\frac{1}{2} W$, and the shear force is $\pm \frac{1}{2} W$ in the outer thirds. In the middle third $V = \frac{1}{2} W - \frac{1}{2} W = 0$.

The bending moment in the outer thirds is

$$M = Rx_1 = \tfrac{1}{2} Wx_1$$

which reaches the maximum value

$$M = \tfrac{1}{2} W \times \tfrac{1}{3} L = \tfrac{1}{6} WL$$

at $x_1 = \frac{1}{3} L$. In the middle third

$$M = \tfrac{1}{2} W \times x_2 - \tfrac{1}{2} W \left(x_2 - \tfrac{1}{3} L \right) = \tfrac{1}{2} W \times \tfrac{1}{3} L = \tfrac{1}{6} WL$$

The bending moment is therefore constant in the middle third, i.e., in the region where the shear force is zero. Although we cannot in an elementary textbook prove this, we may mention that a region of constant bending moment always coincides with a region of zero shear.

If we compare the two-panel subdivisions of Fig. 5.14a with the three-panel subdivision of Fig. 5.14b, we observe that the bending moment is reduced from $\frac{1}{4} WL$ to $\frac{1}{6} WL$.

Example 5.6. *A primary girder spanning 30 ft carries two secondary beams at the third points of the span, each imposing a reaction of 100,000 lb. Determine the resulting maximum bending moment in the primary girder.*

$$M = \tfrac{1}{6} WL = \tfrac{1}{6} \times 200{,}000 \times 30 = 1{,}000{,}000 \text{ lb ft}$$

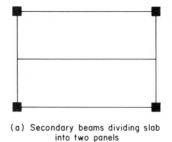

(a) Secondary beams dividing slab
into two panels

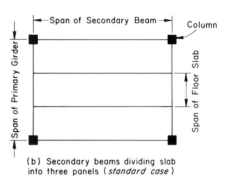

(b) Secondary beams dividing slab
into three panels (*standard case*)

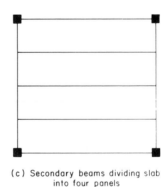

(c) Secondary beams dividing slab
into four panels

Fig. 5.14.
Floor slab supported on primary girders and secondary beams.

The principle of superposition is not as useful for simply supported beams as it is for cantilevers, where the maximum bending moment always occurs at the support. In a simply supported beam carrying several unsymmetrically arranged concentrated loads the maximum bending moment occurs under one of the loads near mid-span, but the exact location is not immediately obvious, and it is simpler to draw the entire bending moment diagram.

Bending Moments and Shear Forces

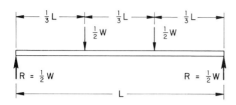

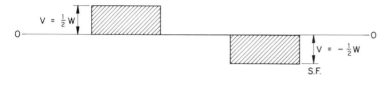

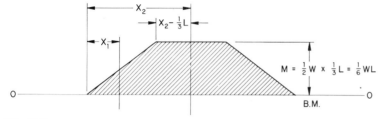

Fig. 5.15.
Simply supported beam carrying two symmetrically placed concentrated loads (e.g., secondary beams dividing a slab into three panels).

Example 5.7. *Determine the distribution of shear force and bending moment for the beam shown in Fig. 5.16.*

The beam carries a total load of 1,000 lb so that

$$R_1 + R_2 = 1,000 \text{ lb}$$

Taking moments about R_2

$$R_1 = \frac{300 \times 9 + 100 \times 6 + 200 \times 4 + 400 \times 3}{10} = 530 \text{ lb}$$

which gives $R_2 = 470$ lb.

The shear force at the left-hand support is therefore 530 lb. This is reduced by 300 lb at the first concentrated load, by 100 lb at the next, etc. (Fig. 5.16b)

The bending moment at the first concentrated load is $530 \times 1 = 530$ lb ft; and at the second concentrated load it is

$$M = 530 \times 4 - 300 \times 3 = 1,220 \text{ lb ft}$$

The bending moment at the fourth concentrated load, proceding from the right-hand support, is $470 \times 3 = 1,410$ lb ft; and at the third concentrated load it is

$$M = 470 \times 4 - 400 \times 1 = 1,480 \text{ lb ft}$$

The maximum bending moment is 1,480 lb ft, 6 ft from the left-hand and 4 ft from the right-hand support (Fig. 5.16).

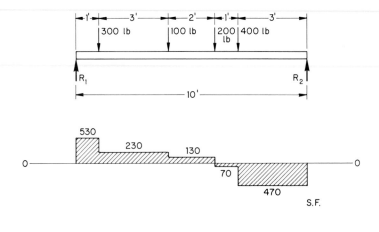

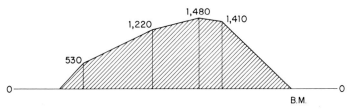

Fig. 5.16.
Simply supported beam carrying several concentrated loads (Example 5.7).

Finally, we consider the simply supported beam carrying a uniformly distributed load—for example, its own weight (Fig. 5.17). If the beam carries a total load W, the reactions on either side are $\frac{1}{2} W$, and this is the shear force at the supports.

Let us consider a section at a distance x from the left-hand support. The shear force is reduced by the load carried on the length of beam x, which is Wx/L, so that

$$V = \tfrac{1}{2} W - \frac{Wx}{L}$$

and the shear force diagram varies linearly from $+\frac{1}{2} W$ at the left-hand support, through zero at the center, to $-\frac{1}{2} W$ at the right-hand support.

The bending moment at a distance x from the left-hand support (Fig. 5.17)

$$M = \tfrac{1}{2} Wx - W\frac{x}{L} \times \tfrac{1}{2} x$$

The bending moment diagram is therefore a parabolic curve (Fig. 5.17) and the maximum bending moment, which occurs at mid-span

Bending Moments and Shear Forces

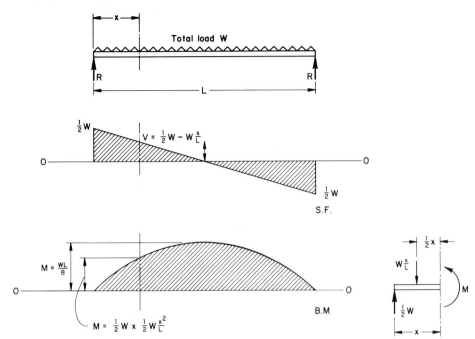

Fig. 5.17.
Simply supported beam carrying a uniformly distributed load.

(where $x = \frac{1}{2} L$):

$$M = \tfrac{1}{2} W \times \tfrac{1}{2} L - W \times \frac{\frac{1}{2} L}{L} \times \tfrac{1}{2} \times \tfrac{1}{2} L = \tfrac{1}{4} WL - \tfrac{1}{8} WL = \tfrac{1}{8} WL$$

Example 5.8. *Determine the maximum shear force and bending moment for a beam carrying a uniformly distributed load of 1,000 lb over a simply supported span of 10 ft.*

The answer is

$$V = \tfrac{1}{2} \times 1,000 = 500 \text{ lb}$$

and

$$M = \tfrac{1}{8} \times 1,000 \times 10 = 1,250 \text{ lb ft}$$

We may compare this with the single concentrated load (Example 5.4), the two symmetrically arranged concentrated loads, and the series of irregularly arranged concentrated loads (Example 5.7), all of which amount to 1,000 lb, over a simply supported span of 10 ft.

For symmetrical arrangement, the shear force is always the same, namely 500 lb. The unsymmetrical arrangement of Example 5.7 makes only a slight difference to the maximum shear force, which is increased to 530 lb.

127

The bending moment, however, varies considerably. For a single concentrated load it is 2,500 lb ft. For two loads at third points, $M = \frac{1}{6} WL$ = 1,670 lb ft. For the unsymmetrical arrangement in Example 5.7 it is 1,480 lb ft. For the uniformly distributed load it is 1,250 lb ft.

*5.5 Beams with Cantilever Overhangs

The remainder of this chapter involves the use of a substantial amount of mathematics, and we suggest that readers omit Sections 5.5, 5.6, 5.7, 5.8 and 5.9, on first reading, and proceed to Chapter 6. Headings for these sections will be preceded by an asterisk.

A simply supported beam with one or two cantilever overhangs is statically determinate: there are two reactions, and two equations for determining them.

Let us first consider a beam carrying a uniformly distributed load on the cantilever overhangs only (Fig. 5.18); we will neglect the weight of the beam in the center span. The shear force increases uniformly from zero at the ends to W_1 at the supports. Since the support reactions are each equal to W_1, the shear force in the center span

$$V = W_1 - W_1 = 0$$

The bending moment increases parabolically (see Section 5.2) to $-\frac{1}{2} W_1 L_1$ and then remains uniform in the center span where (Fig.

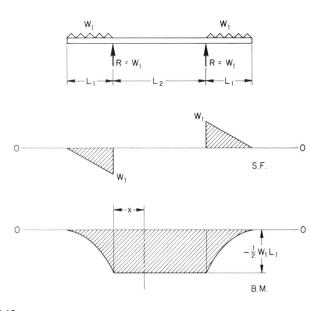

Fig. 5.18.
Simply supported beam carrying a uniformly distributed load on two identical cantilever overhangs only.

Bending Moments and Shear Forces

5.18) at a distance x from one of the supports

$$M = - W_1\left(\tfrac{1}{2}L_1 + x\right) + W_1 x = -\tfrac{1}{2}W_1 L_1$$

irrespective of the value of x.

Let us next consider the same beam with a uniformly distributed load on the center span only (Fig. 5.19). We will neglect the weight of the cantilevers, which thus carry no load. Since they make no contribution to the reactions, the shear force and the bending moment, and the diagrams are the same as for a simply supported beam (see Fig. 5.17).

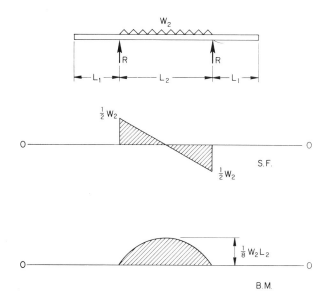

Fig. 5.19.
Cantilevered beam carrying a uniformly distributed load between supports, and none on the overhangs.

Let us next combine these two diagrams to form a beam with cantilever overhangs, carrying a load uniformly distributed along its entire length (Fig. 5.20). By the principle of superposition we can add the two separate effects. For the shear force (Fig. 5.20b) the result is obvious, since two equal cantilever overhangs produce zero shear force in the center span, and there is no shear force in the cantilevers when they are unloaded. For the bending moment we must subtract the negative from the positive moment in the center span (Fig. 5.20c) to obtain the net bending moment diagram (Fig. 5.20d).

Example 5.9. A beam, 30 ft long, carries a uniformly distributed load of 200 lb per foot run. It is simply supported on two columns, which can be placed anywhere under the beam. Determine the location of the columns which equalizes the maximum positive and negative bending moments.

Let us call the cantilever overhangs L. Then, using the notation of Fig. 5.20:

$$L_1 = L, \qquad L_2 = 30 - 2L$$

$$W_1 = 200L, \qquad W_2 = 200(30 - 2L)$$

Equalizing the maximum positive and negative bending moments (Fig. 5.20d) requires that

$$\tfrac{1}{8} W_2 L_2 - \tfrac{1}{2} W_1 L_1 = \tfrac{1}{2} W_1 L_1$$

$$\tfrac{1}{8} \times 200(30 - 2L)(30 - 2L) = \left(\tfrac{1}{2} + \tfrac{1}{2} \right) \times 200L^2$$

$$\tfrac{1}{2}(15 - L)(15 - L) = L^2$$

$$225 - 30L + L^2 = 2L^2$$

$$L^2 + 30L - 225 = 0$$

$$L = -15 + \sqrt{225 + 225} = 15(\sqrt{2} - 1) = 6.213 \text{ ft}$$

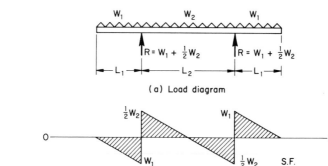

(a) Load diagram

(b) Shear force diagram

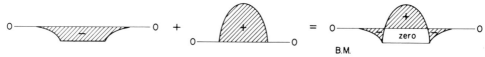

(c) Components of the bending moment diagram

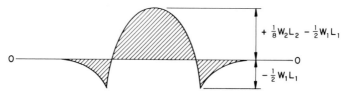

(d) Combined bending moment diagram

Fig. 5.20.
Simply supported beam with two cantilever overhangs, carrying a uniformly distributed load along its entire length.

Bending Moments and Shear Forces

We thus obtain $L_1 = 6.21$ ft, $L_2 = 17.57$ ft, $W_1 = 1,243$ lb, $W_2 = 3,515$ lb.

The negative moment is $-\frac{1}{2} \times 1,243 \times 6.21 = -3,860$ lb ft, and the positive moment is $+\frac{1}{8} \times 3,515 \times 17.57 - 3,860 = +3,860$ lb ft.

If we wish to use a beam of uniform section, and bending moment is the design criterion (see Section 6.4), as it usually is, then this is the best location of the columns, because it gives us the lowest possible maximum bending moment. If, on the other hand, the shear force is the design criterion, which may happen in foundation beams (see Section 6.6), we require

$$W_1 = \tfrac{1}{2} W_2$$

which evidently means that the columns should be $\frac{1}{4} \times 30$ ft, or 7.5 ft from the end. This makes $L_1 = 7.5$ ft and $L_2 = 15$ ft.

Let us next consider a beam simply supported on two columns, which has one cantilever overhang (Fig. 5.21a). The reactions are now unequal. Taking moments about the left-hand support

$$R_2 L_2 = W_2 \times \tfrac{1}{2} L_2 - W_1 \times \tfrac{1}{2} L_1$$

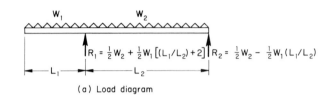

(a) Load diagram

(b) Shear force diagram

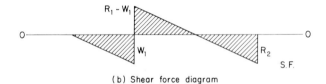

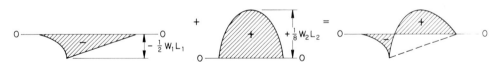

(c) Components of the bending moment diagram

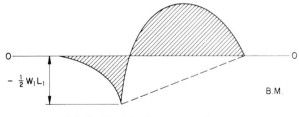

(d) Combined bending moment diagram

Fig. 5.21.
Simply supported beam with one cantilever overhang.

which gives

$$R_2 = \tfrac{1}{2}\,W_2 - \tfrac{1}{2}\,W_1 \times \frac{L_1}{L_2}$$

and

$$R_1 = W_1 + W_2 - R_1 = \tfrac{1}{2}\,W_2 + \tfrac{1}{2}\,W_1\!\left(2 + \frac{L_1}{L_2}\right)$$

The shear force in the cantilever increases uniformly to $-W_1$. At the first support it changes by R_1, so that just inside the support it is $\tfrac{1}{2}\,W_2 + \tfrac{1}{2}\,W_1 L_1 / L_2$. It then varies uniformly to R_2, but zero shear is not at the center of the span L_2.

The bending moment due to the load W_1 increases parabolically in the cantilever to reach a maximum of $-\tfrac{1}{2}\,W_1 L_1$, and this then decreases linearly to 0 at the other support. The bending moment due to the load on the span L_2 varies parabolically to a maximum $+\tfrac{1}{8}\,W_2 L_2$. If we combine these, we obtain the net bending moment diagram for the entire beam (Figs. 5.21c and d).

Although the maximum bending moment is not precisely in the middle of the span L_2, we can obtain an approximate maximum value by considering the bending moment at the center of the span L_2; this is $\tfrac{1}{8}\,W_2 L_2 - \tfrac{1}{4}\,W_1 L_1$. Then exact maximum positive bending moment occurs at the section where the shear force is zero, which is easily determined from the shear force diagram. We can then obtain the maximum bending moment by taking moments about that section.

Example 5.10. *Sketch the shear force and bending moment diagrams for the beam shown in Fig. 5.22.*

Taking moments about the right-hand support

$$R_1 = \frac{200 \times 30 \times \tfrac{1}{2} \times 30}{20} = 4{,}500\ \text{lb}$$

which gives $R_2 = 200 \times 30 - R_1 = 1{,}500$ lb.
The shear force at the inside end of the cantilever

$$V = -200 \times 10 = 2{,}000\ \text{lb}$$

and shear force in the span L_2, just beyond the cantilever is

$$V = -2{,}000 + R_1 = +2{,}500\ \text{lb}$$

The shear force at the other support is 1,500 lb, and the point of zero shear occurs at $20 \times 1{,}500/4{,}000 = 7.5$ ft from the right-hand support.
The maximum cantilever bending moment

$$M = -\tfrac{1}{2}\,W_1 L_1 = -\tfrac{1}{2} \times 200 \times 10^2 = -10{,}000\ \text{lb ft}$$

Bending Moments and Shear Forces

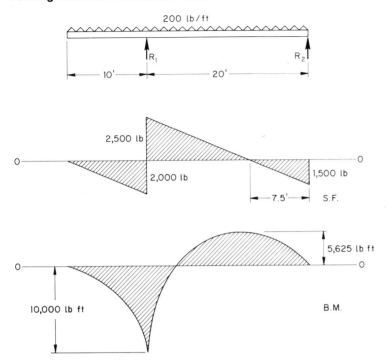

Fig. 5.22.
Simply supported beam with one cantilever overhang (Example 5.10).

The moment $\frac{1}{8} W_2 L_2 = \frac{1}{8} \times 200 \times 20^2 = +10,000$ lb ft, which occurs 10 ft from the right-hand support. Consequently the moment at the center of span L_2

$$M = \frac{1}{8} W_2 L_2 - \frac{1}{4} W_1 L_1 = 10,000 - 5,000 = 5,000 \text{ lb ft}$$

The maximum bending moment occurs 7.5 ft from the right-hand support:

$$M = R_2 \times 7.5 - \frac{1}{2} \times 200 \times 7.5^2 = 11,250 - 5,625 = 5,625 \text{ lb ft}$$

This is 12.5% more than the moment at the center of the span L_2.

*5.6. Gerber Beams

W. Gerber in 1866 patented a form of construction in which continuous (and therefore statically indeterminate) beams are rendered statically determinate by inserting a sufficient number of pin joints. The patent expired long ago, and with improved soil exploration methods we have fewer cases of uneven foundation settlement (which Gerber originally aimed to counter).

133

Gerber beams are of theoretical interest because they demonstrate that not all statically determinate beams are simple, and that beams can be continuous over several supports and at the same time statically determinate. They also have a practical architectural application in precast concrete construction (Fig. 5.23a), when short inset beams are used alternately with cantilevered spans. The system forms a number of continuous spans L_1 (Fig. 5.23b), in which alternate spans have two pin joints (see Fig. 5.2c).

The inset spans carry a load W_3 over a simply supported span L_3 (Fig. 5.23c), which produces a maximum bending moment $\frac{1}{8} W_3 L_3$. The reactions are transmitted to the ends of the cantilevers as concentrated loads $\frac{1}{2} W_3$, which are additional to the uniformly distributed loads W_1 and W_2 (see Fig. 5.20).

Evidently the maximum negative moment in the cantilevered beams is (Fig. 5.23d)

$$M = -\tfrac{1}{2}(W_2 + W_3)L_2$$

and the maximum positive moment in the cantilevered beams is

$$M = \tfrac{1}{8} W_1 L_1 - \tfrac{1}{2}(W_2 + W_3)L_2$$

The maximum positive bending moment in the inset beams is

$$M = \tfrac{1}{8} W_3 L_3$$

The maximum positive moments are the same in the cantilevered beam and the inset beam, irrespective of the location of the pin joints, provided that the column spacing and the load are uniform. Let us consider a uniformly distributed load w lb per foot run, so that $W_1 = wL_1$, $W_2 = wL_2$, and $W_3 = wL_3$. Let us further consider equal column spacing so that $L_2 = \frac{1}{2}(L_1 - L_3)$. The maximum positive moment in the cantilevered beams

$$M = \tfrac{1}{8} W_1 L_1 - \tfrac{1}{2}(W_2 + W_3)L_2$$

$$= \tfrac{1}{8} wL_1^2 - \tfrac{1}{2} w(L_2 + L_3)L_2$$

$$= \tfrac{1}{8} wL_1^2 - \tfrac{1}{8} w(L_1 - L_3 + 2L_3) \times (L_1 - L_3)$$

$$= \tfrac{1}{8} wL_1^2 - \tfrac{1}{8} w(L_1^2 - L_3^2)$$

$$= \tfrac{1}{8} wL_3^2$$

Bending Moments and Shear Forces

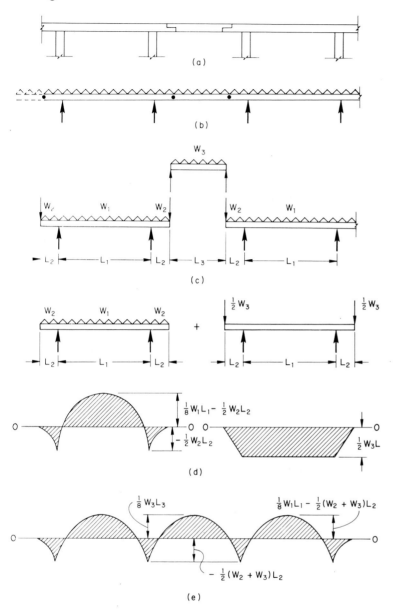

Fig. 5.23.

Gerber beam.

(a) Arrangement of precast concrete pin joints.

(b) Load diagram.

(c) Separation of Gerber beam into short simply supported beams carried on longer cantilevered beams. The reactions of the short beams are transmitted to the cantilevers as concentrated loads.

(d) The bending moment diagram of the cantilevered beams is composed of two parts, one for the uniformly distributed load, and the other for the reactions from the short beams.

(e) Bending moment diagram for the entire Gerber beam.

which is also the maximum moment in the inset span.

The location of the pins is, however, critical, if we wish to make the absolute maximum moment a minimum. The shorter the inset span, the greater the maximum negative moment; the longer the inset span, the greater the maximum positive moment.

Example 5.11. A Gerber beam system in precast concrete carries a load of 400 lb per foot run over continuous spans of 30 ft each, with two pin joints in alternate spans. Determine the location of the pin joints which produce the lowest maximum bending moments.

Let us call the length of the inset span L, so that the length of the cantilever is $\frac{1}{2}(30 - L) = 15 - \frac{1}{2}L$. This gives $L_1 = 30$, $L_2 = 15 - \frac{1}{2}L$, and $L_3 = L$.

The maximum positive moment, for equal column spacing and uniformly distributed loading is

$$M_+ = \tfrac{1}{8}wL^2$$

The maximum negative moment is

$$M_- = -\tfrac{1}{2}w(L_2 + L_3)L_2 = -\tfrac{1}{2}w\left(15 - \tfrac{1}{2}L + L\right) \times \left(15 - \tfrac{1}{2}L\right)$$

$$= -\tfrac{1}{8}w(30^2 - L^2) = -\tfrac{1}{8}w \times 30^2 + \tfrac{1}{8}wL^2$$

Equating positive and negative moments

$$\tfrac{1}{8}w \times 30^2 = \tfrac{1}{8}wL^2 + \tfrac{1}{8}wL^2$$

which gives the inset span $L = 30/\sqrt{2} = 21.21$ ft.

The cantilever span $L_2 = \frac{1}{2}(30 - 21.21) = 4.40$ ft.

Consequently the maximum positive moment is

$$M_+ = \tfrac{1}{8} \times 400 \times 21.21^2 = 22{,}500 \text{ lb ft}$$

and the maximum negative moment is

$$M_- = \tfrac{1}{2} \times 400(4.40 + 21.21) \times 4.40 = 22{,}500 \text{ lb ft}$$

The inset span is a little over $\frac{2}{3}$ and the cantilever overhangs are a little less than $\frac{1}{6}$ of the column spacing. If the pin joints divided the span exactly in the ratio of 1:4:1, which would be easier to set out, the maximum moments would not be significantly altered.

*5.7. Rectangular Three-Pinned Portals

When a portal frame is loaded vertically, its foundations tend to spread, due to the elastic deformation of the structure. The elastic deformation of a simply supported beam rarely presents a problem; it either slides on the supports to the small extent necessary, or the

Bending Moments and Shear Forces

columns are sufficiently flexible to allow the beam to bend. Portal frames, however, include the columns, and it is difficult in practice to provide for the necessary freedom of movement. A roller joint, such as is shown in Fig. 5.24a, is not a practical solution for architectural structures, because the cost of maintaining freedom of movement of the rollers is out of all proportion to the magnitude of the problem. Portals are therefore either hinged, or built in at the base.

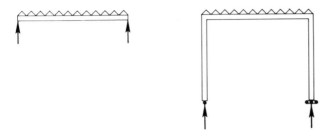

(a) A simply supported beam corresponds to a portal with one sliding joint
(*statically determinate*).

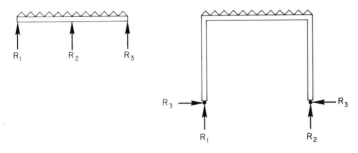

(b) A two-pin portal corresponds to a two-span continuous beam
(*statically indeterminate*).

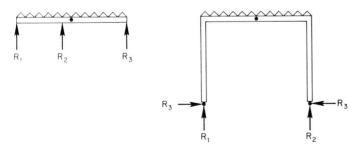

(c) A three-pin portal corresponds to a Gerber beam
(*statically determinate*).

Fig. 5.24.
Comparison of beams and corresponding portal frames.

The built-in portal is evidently statically indeterminate; but the hinged portal is also statically indeterminate, because the horizontal restraint imposes an additional reaction R_3 (Fig. 5.24b). Resolving horizontally, we merely obtain that the two horizontal reactions are equal

$$R_3 = R_3$$

which is correct, but obvious. We cannot determine R_3 by statics, just as we cannot determine the third reaction in a two-span continuous beam. The frame can, however, be made statically determinate by introducing a third pin, and the three-pin portal thus corresponds to the Gerber beam (Fig. 5.24c).

Let us consider a rectangular portal frame, height H, and span L, carrying a uniformly distributed load W on the beam. To render it statically determinate, it has two pins at the supports and a third at mid-span (Fig. 5.25a). The vertical reactions equal the vertical load for equilibrium, and by symmetry

$$R_V = \tfrac{1}{2} W$$

The horizontal reactions are equal, by horizontal equilibrium; but we can determine them statically only by taking moments about the third pin. Let us therefore replace the internal forces at the top pin by equivalent external forces (Fig. 5.25b). By horizontal equilibrium, the horizontal reaction at the top pin is also R_H, there being no other horizontal force. There is no vertical reaction at the top pin in this particular frame (although there might be in others), since $R_V = \tfrac{1}{2} W$ balances. Let us now take moments about one of the pins, say the top pin. The bending moment at a pin is zero by definition, and we therefore obtain

$$R_V \times \tfrac{1}{2} L - \tfrac{1}{2} W \times \tfrac{1}{4} L - R_H \times H = 0$$

which gives the horizontal reactions

$$R_H = \frac{\tfrac{1}{4} WL - \tfrac{1}{8} WL}{H} = \frac{\tfrac{1}{8} WL}{H}$$

We can now draw the bending moment and shear force diagrams for the frame. The bending moment builds up uniformly in the column, from zero at the pin joint to $-R_H \times H$ at the knee, and this moment is transferred around the corner into the beam. The negative bending moment is reduced parabolically by the positive moment acting on the "simply supported beam" (see Fig. 5.17) until it reaches zero at the center pin joint. Let us consider a section

Bending Moments and Shear Forces

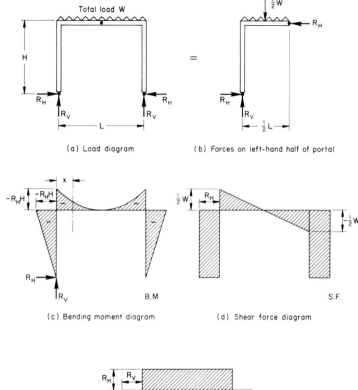

(a) Load diagram

(b) Forces on left-hand half of portal

(c) Bending moment diagram

(d) Shear force diagram

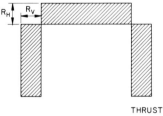

THRUST

(e) Thrust diagram

Fig. 5.25.
Rectangular three-pin portal carrying a vertical uniformly distributed load. In part (b), the vertical reaction at the top pin joint is zero by symmetry.

across the beam at a distance x from the support (Fig. 5.25c):

$$M = -R_H \times H + R_V x - W \frac{x}{L} \times \tfrac{1}{2}x$$

$$= -\tfrac{1}{8}WL + \tfrac{1}{2}Wx - \frac{\tfrac{1}{2}Wx^2}{L}$$

At $x = \tfrac{1}{2}L$ this becomes zero. The bending moment is therefore negative throughout. It varies linearly in the column, and parabolically in the beam; however, the parabola is concave, not convex as in a simply supported beam.

The shear force in the column is constant (Fig. 5.25d) and equal to R_H. Being a force (and therefore a directional quantity), it is not transmitted through a right-angle joint, and the shear force in the beam is precisely the same as in a simply supported beam of the same span carrying the same load (see Fig. 5.17).

In addition to bending moments and the shear force, the portal frame is subject also to direct compressive forces, or *thrusts* (*see* Section 7.3). For vertical equilibrium, the thrust in the columns is $R_V = \frac{1}{2} W$, and the thrust in the beam, for horizontal equilibrium, is $R_H = \frac{1}{8} WL/H$ (Fig. 5.25e).

The design of portal frames is usually controlled by the magnitude of the bending moments; the shear forces and thrusts are only of secondary importance.

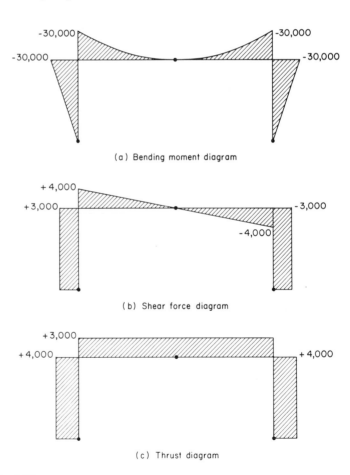

(a) Bending moment diagram

(b) Shear force diagram

(c) Thrust diagram

Fig. 5.26.
Three-pin portal (Example 5.12).

Bending Moments and Shear Forces

Example 5.12. *A rectangular three-pinned portal frame, 10 ft high, carries a load of 8,000 lb over a span of 30 ft. Sketch the diagrams for bending moment, shear force, and thrust.*

The horizontal reactions are

$$\frac{\frac{1}{8} \times 8,000 \times 30}{10} = 3,000 \text{ lb}$$

and the maximum bending moment (at the knee of the portal) is

$$3,000 \times 10 = 30,000 \text{ lb ft}$$

The diagrams for bending moment, shear force, and thrust are shown in Fig. 5.26.

*5.8. Gabled Frames with Three Pins

A gabled frame incorporates the sloping roof of the building. It can be symmetrical, be north- (or south-) lit, or have monitors, and it can have one or several bays (see Figs. 1.42 and 1.43). However, in this chapter we will confine ourselves to single-bay symmetrical frames.

Let us consider a frame carrying a uniformly distributed vertical load W (Fig. 5.27). By symmetry the vertical reactions

$$R_V = \tfrac{1}{2} W$$

The horizontal reaction is obtained by taking moment about the third pin (Fig. 5.26).

$$R_H = \frac{R_V \times \frac{1}{2} L - \frac{1}{2} W \times \frac{1}{4} L}{H} = \frac{\frac{1}{8} WL}{H}$$

as for the rectangular portal.

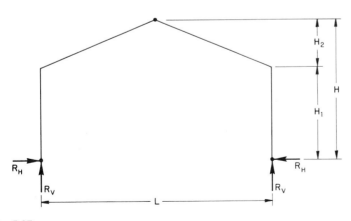

Fig. 5.27.
Gabled frame with three pins.

However, the bending moment in the inclined gable members is different, since it is affected by the horizontal reaction. The shear force and thrust are similarly altered.

Example 5.13. A gabled three-pinned portal frame, with dimensions as shown in Fig. 5.28, carries a load of 8,000 lb over a span of 30 ft. Sketch the diagrams for bending moment, shear force and thrust.

The vertical reactions are

$$R_V = \tfrac{1}{2} \times 8,000 = 4,000 \text{lb}$$

and the horizontal reactions are

$$R_H = \frac{\tfrac{1}{8} \times 8,000 \times 30}{25} = 1,200 \text{ lb}$$

The bending moment increases uniformly to the knee of the portal, where it is

$$M = -1,200 \times 10 = -12,000 \text{ lb ft}$$

The bending moment in the gable members, at a distance x from the gable, is

$$M = R_H \times y - \left(\frac{W}{L} \right) x \times \tfrac{1}{2} x$$

Since for a 45° gable $x = y$

$$M = 1,200x - \tfrac{1}{2} \left(\frac{8,000}{30} \right) x^2 = (3,600 - 400x) \frac{x}{3}$$

For an approximate bending moment diagram (Fig. 5.28 b) it is sufficient to plot two intermediate values:

For $x = 0$, $M = 0$
For $x = 5$, $M = +2,667$ lb ft
For $x = 10$, $M = -1,333$ lb ft
For $x = 15$, $M = -12,000$ lb ft

The shear force in the column is constant and equal to the horizontal reaction, $R_H = 1,200$ lb. In the gable the shear force is

$$V = R_H \cos 45° - \frac{W}{L} x \sin 45°$$

which varies uniformly from $1,200/\sqrt{2} = +848$ lb at the top pin to $(1,200 - 4,000)/\sqrt{2} = -1,980$ lb at the knee (Fig. 5.28c, left-hand side).

The thrust in the column is constant and equal to the vertical reaction $R_V = 4,000$ lb. At the gable the thrust is

$$P = R_H \sin 45° + \frac{W}{L} x \cos 45°$$

which varies uniformly from $1,200/\sqrt{2} = 848$ lb at the top pin to $(1,200 + 4,000)/\sqrt{2} = 3,680$ lb at the knee (Fig. 5.28c, right-hand side).

Bending Moments and Shear Forces

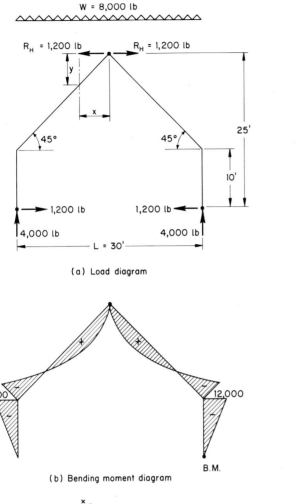

(a) Load diagram

(b) Bending moment diagram

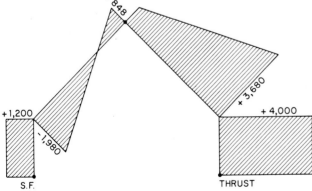

(c) Shear force (left-hand) and thrust (right-hand) diagram

Fig. 5.28.
Three-pin gabled frame carrying a uniformly distributed vertical load (Example 5.13).
In part (a), the vertical reaction at the top joint is zero by symmetry.

143

It should be emphasized that the bending moment diagram is the one which mainly determines the dimensions of the structure, and in a preliminary design it would be sufficient to determine the maximum bending moment, particularly if the frame is to be formed from a single standard section. The maximum bending moment in a three-pin portal almost invariably occurs at the knee, even in this rather extreme example, where the ratio of the span to the height of the knee is 3:1.

The bending moment at the knee is simply

$$M = R_H H_1$$

and the basic design is therefore simple, even though a complete determination of all the forces and moments takes time.

*5.9. Three-pin Arches

In arches, there is a gradual transition from beam to column, and the bending moment, shear force, and thrust throughout the arches are affected by both the horizontal and the vertical reactions. It is possible to design an arch in which the moments due to the horizontal forces are precisely out by the moments due to vertical forces, so that the arch, for a particular kind of loading, is entirely free from bending. This makes it possible to bridge a long span with a comparatively small amount of material (see Section 1.4). The long architectural history of arches is, however, not due merely to their structural efficiency. The beauty of many ancient and modern buildings is due to the choice of the curve adopted for the load-bearing arches. Although some structurally correct arches are widely accepted as having aesthetic merit, the two criteria do not necessarily agree. The Romans almost invariably used circular arches because they regarded the circle as the perfect curve, and this practice was revived in the Renaissance. Many fine buildings have been produced with circular arches, but the bending moment cannot be eliminated in circular arches, as in parabolic or catenary arches, and over a long period of history the accepted aesthetic criterion was at variance with the principle of structural economy.

Let us first consider a semicircular arch, because its constant radius of curvature makes the solution relatively simple (Fig. 5.29). If the span, or diameter of the semicircle, is L, and the rise of the arch, or radius of the semicircle, is H, then

$$H = \tfrac{1}{2} L$$

Let us assume that the arch carries a total load W, uniformly distributed along the span of the arch (not along the length of the arch, which would give a higher load toward the ends where the arch is steeper). The vertical reactions are

$$R_V = \tfrac{1}{2} W$$

Bending Moments and Shear Forces

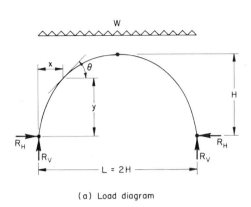

(a) Load diagram

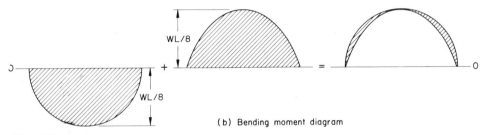

(b) Bending moment diagram

Fig. 5.29.
Three-pin semicircular arch carrying a uniformly distributed vertical load.

and the horizontal reactions are obtained by taking moments about the top pin:

$$R_H \times H = R_V \times \tfrac{1}{2}L - \tfrac{1}{2}W \times \tfrac{1}{4}L = \tfrac{1}{2}W\left(\tfrac{1}{2}L - \tfrac{1}{4}L\right) = \tfrac{1}{8}WL$$

Since $H = \tfrac{1}{2}L$

$$R_H = \tfrac{1}{4}W = \tfrac{1}{2}R_V$$

The bending moment at a point along the arch with coordinates x and y from the support is

$$M = -R_H \times y + R_V \times x - \left(\frac{W}{L}\right)x \times \tfrac{1}{2}x$$

The first part of this equation $-R_H \times y$ is a diagram which has the same shape as the arch, namely a semicircle. Its maximum value occurs at mid-span, where $y = H = \tfrac{1}{2}L$, and $R_H \times y = \tfrac{1}{4}W \times \tfrac{1}{2}L = \tfrac{1}{8}WL$.

The second part of this equation is the ordinary bending moment diagram for a simply supported beam carrying a uniformly distributed load (see Fig. 5.17). If we combine these two figures we get the net bending moment diagram for the arch (Fig. 5.29). This is a negative moment consisting of the difference between the semi-circular diagram for the first, negative, part and the parabolic diagram for the second, positive, part. The bending moment is a maximum approximately at quarter span, and it is zero at the three pins.

The shear force at a point along the arch whose tangent makes an angle θ with the horizontal is

$$V = R_{\mathrm{H}} \sin \theta - R_{\mathrm{V}} \cos \theta + \left(\frac{W}{L} \right) x \cos \theta$$

The thrust at a point along the arch is similarly

$$P = R_{\mathrm{V}} \sin \theta - \left(\frac{W}{L} \right) x \sin \theta + R_{\mathrm{H}} \cos \theta$$

Example 5.14. A semicircular three-pin arch, spanning 30 ft, carries a load of 8,000 lb uniformly distributed in plan. Sketch the bending moment diagram.

The solution is shown in Fig. 5.30. The bending moment along the entire length of the arch is negative; it consists of the difference between the semicircular and the parabolic curves of height $\frac{1}{8} WL = 30,000$ lb ft. The maximum can be scaled off the curve; it is the longest *vertical* line.

We can also calculate it, although this may be beyond the mathematical competence of some readers. The bending moment

$$M = -R_{\mathrm{H}} \times y + R_{\mathrm{V}} \times x - \left(\frac{W}{L} \right) x \times \tfrac{1}{2} x \tag{i}$$

As we have seen, $R_{\mathrm{H}} = \frac{1}{4} W$ and $R_{\mathrm{V}} = \frac{1}{2} W$. Furthermore, we may calculate the y-coordinate in terms of the radius R, which is equal to the height of the arch, or half the span, so that $R = \frac{1}{2} L$ and

$$R^2 = (R - x)^2 + y^2$$

which gives

$$y = (2Rx - x^2)^{1/2}$$

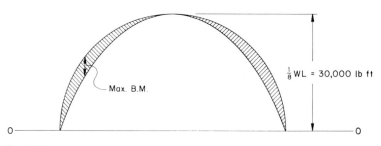

Fig. 5.30.
Bending moment diagram for semicircular three-pin arch (Example 5.14).

Bending Moments and Shear Forces

The bending moment therefore becomes

$$M = -\tfrac{1}{4} W (2Rx - x^2)^{1/2} + \tfrac{1}{2} Wx - \frac{\tfrac{1}{4} Wx^2}{R}$$

The maximum bending moment occurs at the value of x for which the differential coefficient equals zero.

$$\frac{dM}{dx} = 0 = -\tfrac{1}{4} W \times \tfrac{1}{2}(2Rx - x^2)^{-1/2} \times (2R - 2x) + \tfrac{1}{2} W - \frac{\tfrac{1}{2} Wx}{R}$$

This gives

$$-\tfrac{1}{2}(2Rx - x^2)^{-1/2} \times (R - x) + 1 - \frac{x}{R} = 0$$

$$-\tfrac{1}{2}(2Rx - x^2)^{-1/2} + \frac{R - x}{R(R - x)} = 0$$

$$2Rx - x^2 = \tfrac{1}{4} R^2$$

$$x = R - \left(R^2 - \tfrac{1}{4}R^2\right)^{\frac{1}{2}} = R(1 - 0.866) = 0.134R = 2.01 \text{ ft}$$

From geometry, the corresponding value of y is

$$y = \left[R^2 - (R - x)^2\right]^{\frac{1}{2}} = (15^2 - 12.99^2)^{\frac{1}{2}} = 7.50 \text{ ft}$$

Substituting these values of x and y, we obtain the maximum bending moment from Eq. (i):

$$M = -\tfrac{1}{4} \times 8000 \times 7.50 + \tfrac{1}{2} \times 8{,}000 \times 2.01 - \frac{\tfrac{1}{2} \times 8000 \times 2.01^2}{30}$$

$$= -7{,}500 \text{ lb ft}$$

If we wish to eliminate the bending moment in the arch altogether, we must make the shape of the arch equal to the bending moment diagram of the load which it carries (see Section 1.4). In the case of a load uniformly distributed along the span, the bending moment diagram is parabolic. Consequently we can eliminate the bending moment altogether if we make the arch *parabolic*.

We then also eliminate the shear force, because the shear due to the vertical forces is exactly canceled out by the shear due to the horizontal forces. We do, however, retain the thrust, which becomes a maximum at the springings (abutments, or base) of the arch.

Thus the three-pin arches in Fig. 5.31 are all free of bending moment. The first carries a concentrated load, and the bending moment diagram for a simply supported beam carring a central concentrated load is a triangle. The second is the parabolic arch. The third is a catenary, which is the shape assumed by a cable hanging under its own weight (see Section 1.3), and this is the moment-free shape for an arch of uniform cross-section carrying its

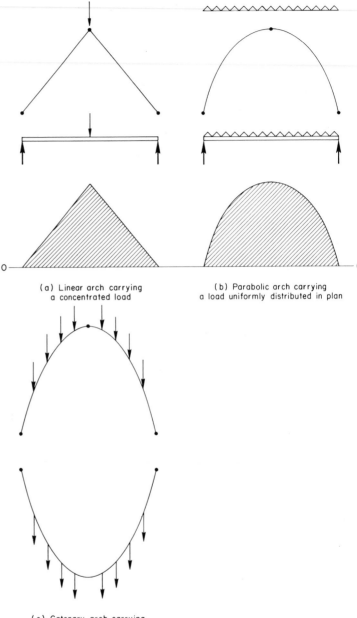

(a) Linear arch carrying
a concentrated load

(b) Parabolic arch carrying
a load uniformly distributed in plan

(c) Catenary arch carrying
a load uniformly distributed
along the length of the arch

Fig. 5.31.
Moment-free arches.

own weight only. There is, however, little difference in practice between the shape of a catenary and a parabola of the same height and span, and one can be substituted for the other with little error.

Example 5.15. *Determine the maximum forces and moments in a parabolic three-pin arch, carrying a load of 8,000 lb uniformly distributed in plan. The arch has a span of 30 ft and a rise of 15 ft.*

Bending Moments and Shear Forces

Since the arch is parabolic, the bending moment and the shear force are zero along the length of the arch. The maximum thrust occurs at the springings.

The vertical reactions are (Fig. 5.32)

$$R_V = \tfrac{1}{2} \times 8{,}000 = 4{,}000 \text{ lb}$$

and the horizontal reactions are

$$R_H = \frac{R_V \times \tfrac{1}{2}L - \tfrac{1}{2}W \times \tfrac{1}{4}L}{H} = \frac{\tfrac{1}{8}WL}{H} = \tfrac{1}{4}W = 2{,}000 \text{ lb}$$

The tangent of the angle θ of a parabola (Fig. 5.33)

$$\tan \theta = \frac{2H}{\tfrac{1}{2}L} = \frac{4H}{L}$$

(This result may be found in a geometry textbook which includes conic sections, or in a structural handbook which includes a mathematical section, e.g., I. E. Morris: *Handbook of Structural Design*, Reinhold, New York, 1963, p. 100).

Consequently

$$\cos \theta = 1/(1 + \tan^2 \theta)^{1/2} = 1/(1 + 16H^2/L^2)^{1/2} = 1/\sqrt{5}\ .$$

From Fig. 5.32 it is evident that the thrust P at the crown is horizontal and equal to R_H for horizontal equilibrium. Some distance down the arch the thrust has increased; but its horizontal component remains

$$P_x \cos \theta_x = R_H$$

At the springings P reaches its maximum and $\cos \theta$ its minimum value, and the horizontal component of the maximum thrust in the arch is still equal to the horizontal reaction:

$$P \cos \theta = R_H$$

Consequently the maximum thrust

$$P = \frac{R_H}{\cos \theta} = 2{,}000\sqrt{5} = 4{,}472 \text{ lb}$$

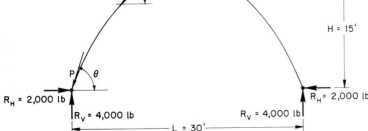

Fig. 5.32.
Load and thrust diagram for parabolic three-pin arch. The vertical reaction at the top pin joint is zero by symmetry (Example 5.15).

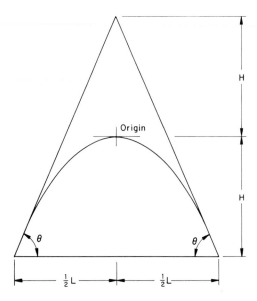

Fig. 5.33.
It is a geometric property of the parabola that tangents drawn at a distance H below the origin intersect at a distance H above the origin.

Suggestions for Further Reading

S. W. CRAWLEY and D. M. DILLON: *Steel Buildings.* Wiley, New York, 1970. Chapter 2, pp. 9–31.

J. C. McCORMAC: *Structural Analysis.* International Text Book Company, Scranton, Pennsylvania, 1967. Chapter 4, pp. 37–49

H. PARKER: *Simplified Design of Structural Steel.* Wiley, New York, 1965. Chapter 4, pp. 42–78.

M. SALVADORI and M. LEVY: *Structural Design in Architecture.* Prentice-Hall, New York, 1967. Chapters 4 and 10, pp. 44–95 and 204–221.

Problems

[†]**5.1.** Determine the maximum shear force and the maximum bending moment in the cantilever shown in Fig. 3.36, p. 78.

5.2. Draw the bending moment and shear force diagrams for the cantilever shown in Fig. 5.34, which carries a uniformly distributed load of 500 lb per foot run over the 4 ft nearest to the support, and reactions from two secondary beams and a safety rail, as shown. What are the maximum values?

[†]Similar problems in metric units are given in Appendix G.

Bending Moments and Shear Forces

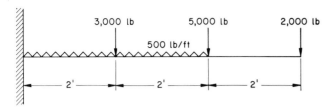

Fig. 5.34.
Problem 5.2.

5.3. Draw the bending moment and shear force diagrams for the beam carrying the concentrated loads shown in Fig. 5.35, and determine the maximum values.

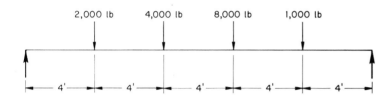

Fig. 5.35.
Problem 5.3.

5.4. A horizontal beam AB supports three equal loads of N lb each at quarter points. Determine the equation for the bending moment along the beam.

5.5. A simply supported beam carries a uniformly distributed load of 2,000 lb per foot run over part of its length, and one of 1,000 lb per foot run over the remainder, as shown in Fig. 5.36. Sketch the bending moment and shear force diagrams, and determine the maximum shear force and bending moment.

5.6. A beam carries a single central load (due to a computer), and uniformly distributed loads on two cantilevers, as shown in Fig. 5.37. Sketch the bending moment and shear force diagrams, and determine the maximum values.

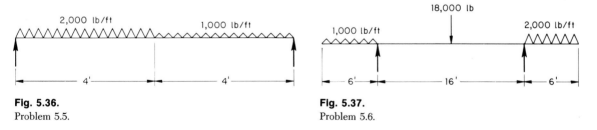

Fig. 5.36.
Problem 5.5.

Fig. 5.37.
Problem 5.6.

5.7. A carport roof is to be supported by edge beams 20 ft long, each beam being supported on two columns located at a certain distance in from the ends. Calculate the positions of these columns

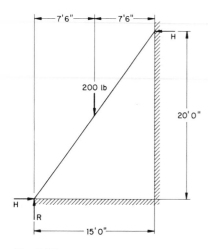

Fig. 5.38.
Problem 5.8.

to give the smallest maximum bending moment under uniformly distributed loading, neglecting any effect due to nonuniform or partial loading.

5.8. Figure 5.38 shows a ladder resting against a smooth wall, with a 200-lb man at the center. Calculate the reactions, and the maximum bending moment in the ladder.

5.9. A beam, 9 ft in length and weighing 10 lb per foot run is suspended horizontally by three cables, as shown in Fig. 5.39. There is a weight of 100 lb hanging from a position 3 ft from one end of the beam. Find the force in the cable AE.

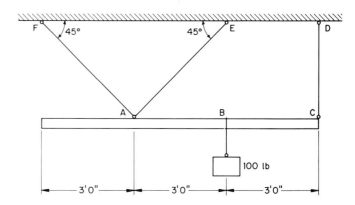

Fig. 5.39.
Problem 5.9.

5.10. A Gerber beam is continuous over 12 spans of 20 ft, with a uniformly distributed load of 1,000 lb per foot run. Two pins are inserted in alternate spans at a distance of 3 ft from the supports to make it statically determinate, as shown in Fig. 5.40. Calculate the bending moment and shear force diagrams for two adjacent interior spans.

Fig. 5.40.
Problem 5.10.

5.11. Figure 5.41 shows a three-hinged frame with a uniformly distributed load of 20,000 lb on the top member. Calculate the reactions and the maximum bending moment, and draw the bending moment diagram.

Bending Moments and Shear Forces

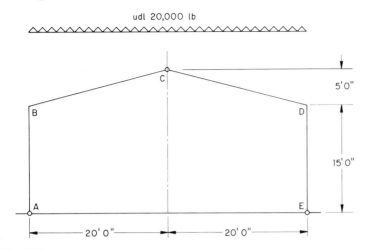

Fig. 5.41.
Problem 5.11.

5.12. A three-pin frame for a series of carports, shown in Fig. 5.42, has a rigid joint at the top of the rear column, and the front column is hinged at both ends. Calculate the reactions, and draw the bending moment and shear force diagrams for a wind load of 1,000 lb uniformly distributed over the rear column. What are the maximum values of the bending moment and the shear force?

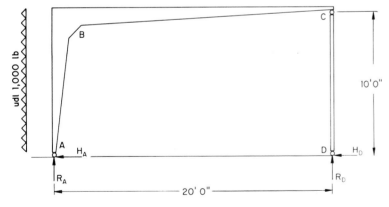

Fig. 5.42.
Problem 5.12.

5.13. The symmetrical monitor frame shown in Fig. 5.43 is pin-jointed at the supports and at the center; it carries a single concentrated load of 30,000 lb in the position indicated. Neglecting the weight of the frame, sketch the bending moment diagram, and determine the magnitude and position of the maximum bending moment. Indicate the location of the points of zero bending moment.

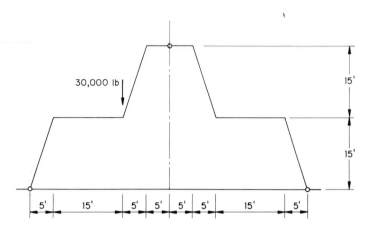

Fig. 5.43.
Problem 5.13.

5.14. A semicircular arch, spanning 80 ft, supports a superstructure which transmits to it two reactions of 50,000 lb, as shown in Fig. 5.44. Draw the bending moment diagram for the arch, neglecting its own weight.

A steel cable is to be used to provide the necessary horizontal reaction. If the cable is stressed to 100,000 p.s.i., determine the area of cable required.

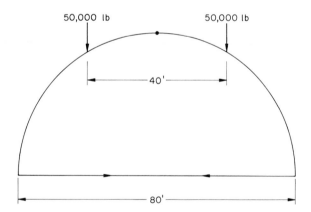

Fig. 5.44.
Problem 5.14.

Elasticity and Stress   6

Nature and nature's law lay hid in night:
God said, "Let Newton be!", and all was light.
Alexander Pope

This could not last; the devil howling, "Ho!
Let Einstein be!", restored the status quo.
Anon.

We derived the forces and moments in simple structures in Chapters 4 and 5. This information, however, does not enable us to say whether a structure is safe or unsafe until we have determined the stresses caused by these forces and moments. If the stresses caused by the loads are greater than the permissible stresses, the structure must be redesigned to bring them within the permissible range. If the maximum stresses in the structure are much below the permissible values, we must also redesign because we are wasting material. In this chapter, then, we will determine the stresses caused by forces and moments. We shall also discuss the calculation of deformations caused by loads.

6.1. The Behavior of Structural Materials Under Load

All solid materials deform under load. Until a critical load is reached, the deformation is *elastic*; i.e., the structure resumes its original shape when the load is removed, and the deformation is fully recovered. The critical load, or *elastic limit*, is very low for some solids, and these do not make suitable structural materials.

Deformation is an essential condition for structural equilibrium, since the loads imposed on the materials must be resisted by internal forces in the crystals. So long as the forces within a crystal are not sufficient to break the interatomic bonds, the condition of equilibrium requires that the material recover its shape when the load is removed.

When the elastic limit is exceeded, a few of the interatomic bonds are broken. Some materials then crack or break into two separate parts; i.e., the interatomic bonds are broken and do not re-form; these are called *brittle materials*. In others the atoms jump one position, and the interatomic bonds then re-form, so that material deforms permanently, but remains in one piece without significant loss of strength; these are called *ductile materials*.

Among the major structural materials, steel is typically ductile, and concrete is typically brittle under normal circumstances. Brittleness and ductility are properties dependent on crystal structure, temperature, and pressure; the same material may be both ductile and brittle under different conditions. Thus, a number of welded ships broke in half during World War II, because the very cold winter weather in the Arctic Ocean made the ductile material brittle. On the other hand, rocks, which are brittle under normal conditions can be turned ductile by high temperature and pressure, as our geological records show. We are not concerned in an elementary text with designing structures for exceptional heat or cold, nor with pressure vessels, and we may thus classify the structural metals, such as steel and aluminum, as ductile; and natural and artificial stone, such as rock, brick, and concrete, as brittle. Timber is a complex organic material that has different properties along and across the grain. However, it suffices at this stage to consider it as ductile.

It is evidently much easier for a brittle material to fail in tension than in compression. A crack, once formed, soon spreads in tension. Far more force is needed to cause a compresssion failure. Brittle materials are, therefore, much stronger in compression than in tension (see Section 6.8). Concrete, the most important brittle material, has a compressive strength 10 times as great as its tensile strength. Its weakness in tension is overcome by reinforcement (see Section 7.11) or by prestressing (see Section 7.13).

In ductile materials the atomic bonds re-form, and their tensile and compressive strength is therefore about the same (see Section 6.8). This is the reason why steel is the normal structural material for resisting tension, even in concrete, where it is used as reinforcement.

The typical load-deformation diagram of steel (Fig. 6.1a) has an elastic stage AB, during which the deformation is fully recovered when the load is removed. The steel then *yields*, as shown in BC. Its strength is not impaired, but the inelastic deformation is not recovered when the load is removed. The yield-stage provides a useful safety margin, since in case of an overload, the structure does not immediately collapse. However, permanently increasing deformation is evidently unacceptable under normal working (service) loads, which must consequently always be within the range AB of the curve. During the stage CD the crystal lattice of steel is more and more distorted, and eventually cracks begin to form. The material finally breaks at D.

The typical load-deformation curve of concrete (Fig. 6.1b) also has an elastic stage AB during which the deformation is recoverable. Cracks then begin to form; at first these are invisible or barely visible (*microcracks*), later they become prominent, and eventually they separate the material into two or more parts, at D.

Elasticity and Stress

Structural design may be based on elastic deformation or ultimate strength (see Section 2.2). In this elementary text we will confine ourselves mainly to elastic design; however, ultimate strength design is now normal procedure for reinforced concrete sections, and this is discussed in Sections 7.9 to 7.12.

The elastic deformation of every structural material is directly proportional to the load it carries. This law, enunciatied by Robert Hooke in 1678, is the basis of elastic design. The linear relationship contained in *Hooke's Law* is fortunate, and structural design would be much more complicated if the relationship were nonlinear.

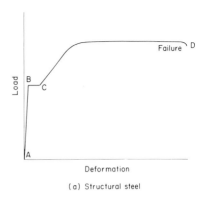

(a) Structural steel

6.2. The Concepts of Stress and Strain

Elastic design is thus concerned only with the elastic range of the load-deformation diagram AB (Fig. 6.1a and b). To ensure the safety of the structure, we fix a limiting load or force for the most highly stressed part of the material which allows us an adequate *factor of safety* (see Section 2.2). Since the internal force may vary throughout a piece of the material (as it always does in beams, which are the most common structural members), we now introduce the concept of *stress*. This is defined as the *force per unit area*, and the unit is taken very small when the stress varies across the section. Furthermore, we introduce the concept of *strain*, which is defined as the deformation per unit length, taking the same very small unit.

Since structural design is mainly concerned with the determination of the maximum stresses in the structure, to ensure that they are within the permissible range, stress is a concept of the greatest importance. Unfortunately stress is not easy to visualize. We can readily demonstrate forces by inserting a spring balance in a structural model (as in Figs. 1.10, 3.5, and A.4). We can also demonstrate strain by using a material such as rubber, which deforms visibly (see Fig. 6.5), or by means of a strain gauge, which can measure strains as small as 1×10^{-5} inches per inch (see Figs. 6.6 and 6.31). Stress cannot be measured directly.

We can, however, interpret stress in terms of strain. Hooke's Law tells us that a load is proportional to the elastic deformation it causes; consequently the load per unit area is proportional to the deformation per unit length. Hooke's Law may therefore be expressed as: *Stress is proportional to strain*, or

$$f = Ee \qquad (6.1)$$

where f is the stress, e is the strain, and E is a constant which is called the modulus of elasticity.

The modulus of elasticity is evidently the stress required to produce an elastic deformation equal to the original length. Materials which deform to this extent, even if they existed, would be

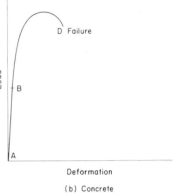

(b) Concrete

Fig. 6.1.
Typical load-deformation diagrams (not to scale). A = start of test; B = elastic limit; C = strain hardening begins; and D = test piece breaks.

(a) Structural steel is a ductile material with a definite yield point and an extended range of plastic deformation. This type of deformation is known as elasto-plastic. (b) Concrete is a brittle material.

useless in structures, and the modulus of elasticity is therefore always much higher than the maximum permissible stress—roughly a thousand times as much. The modulus of elasticity varies from material to material, but it is always the same for the same material.

We can thus think of stress as strain, multiplied by the constant *E*. The structure is deformed by the loads, which produce strains. These can be calculated from the geometry of the deformed structure or measured with strain gauges in the laboratory. The stresses are calculated from the strains, and the combined effect of the stresses produces the internal moments and forces which keep the loads in equilibrium.

There are two different types of stress and strain: *shear stress and strain, and direct stress and strain.*

Direct strain is deformation in tension and compression. Tensile strain increases the length (area or volume), while compressive strain reduces it. Thus tensile strain is usually considered to be positive, and compressive strain to be negative (Fig. 6.2). Like credits and debits in a bank balance, direct strains (provided they act in the same plane and in the same direction) can be added and subtracted. Thus a tensile strain of 100×10^{-6} added to a compressive strain of 60×10^{-6} produces a resultant tensile strain of 40×10^{-6}. (Since strain is a ratio of length/length, it is the same in British and metric units.)

Shear strains cause a distortion of the structural element. They can also be positive or negative, and the combined effect of a shear strain of $+100 \times 10^{-6}$ and -60×10^{-6} is a shear strain of $+40 \times 10^{-6}$. However, positive and negative shear strains are fundamentally of the same kind, whereas positive and negative direct strains manifest themselves as tension and compression.

Shear strains or stresses and direct strains or stresses cannot be added and subtracted by simple arithmetic; however, one can be transformed into the other (see Section 6.8). Thus a tensile force produces tensile stresses, say, in a piece of steel wire; these can be transformed into shear stresses, and the wire actually fails when the shear stresses become too high (see Example 6.7). Similarly a shear force produces shear stresses; these can be transformed into tensile and compressive stresses, and in the case of concrete, failure occurs when the tensile stresses become too high (see Example 6.8).

The direct strain (Fig. 6.2) is the change in length, divided by the original length

$$e = \frac{2x}{L}$$

and the direct stress is proportional to the direct strain

$$f = Ee$$

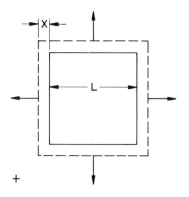

(a) Positive, or tensile, strain

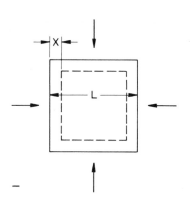

(b) Negative, or compressive, strain

Fig. 6.2.
Direct strain. Direct strain is defined as change of length/original length = $2x/L$. (a) Positive, or tensile, strain. (b) Negative, or compressive, strain.

Elasticity and Stress

where E is the direct modulus of elasticity. For steel, E is approximately 30,000,000 p.s.i., irrespective of the strength of the steel. For concrete, E varies from 2,000,000 p.s.i. (for low-strength concrete) to 4,000,000 p.s.i. (for high-strength concrete), after allowance has been made for creep (see Section 8.9).*

The shear strain (Fig. 6.3) is the change in angle divided by the original angle

$$\gamma = \frac{2\theta}{90°}$$

and the shear stress is proportional to the shear strain

$$v = G\gamma \tag{6.2}$$

where G is the shear modulus of elasticity. For steel, G is approximately 12,000,000 p.s.i., and for concrete, it varies from 1,000,000 to 2,000,000 p.s.i.** The relationship between the values of E and G is explained in most textbooks in the Strength of Materials (e.g., F. L. Singer, *Strength of Materials*, Harper, New York, 1962, p. 39).

The magnitude of the direct modulus of elasticity of steel wire can be determined by a simple experiment. We hang a piece of high-tensile (piano) wire about 6 ft (2 m) in length from a hook in the wall (Fig. 6.4). We fix a scale to the wire and a pointer to the wall as low as is convenient; to obtain greater accuracy we use a vernier scale and pointer, which enables us to read a change in length to a thousandth of an inch. We now hang weights at the end of the wire and take readings at increments of 10 lb (5 kg). The change in length read on the vernier gauge is divided by the length of the wire (the distance between the hook and the point where the vernier gauge is attached) to give the strain e. The load is divided by the cross-sectional area to give the stress f. The modulus of elasticity $E = f/e$.

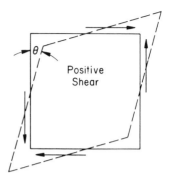

(a) Positive shear strain

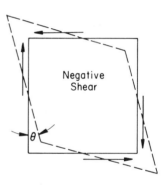

(b) Negative shear strain

Fig. 6.3.
Shear strain. Shear strain is defined as change of angle/original angle $= 2\theta/90°$.

6.3. Stress and Strain in Simple Tension and Compression

In simple tension the distribution of stress and strain is uniform, and the stress is therefore

$$f = \frac{P}{A} \tag{6.3}$$

where P is the tensile force and A is the cross-sectional area.

*In metric units, E for steel is taken as 200 GPa (200,000 MPa); E for concrete is between 15 and 30 GPa.
**G for steel is taken as 80 GPa, and G for concrete as 7 to 14 GPa.

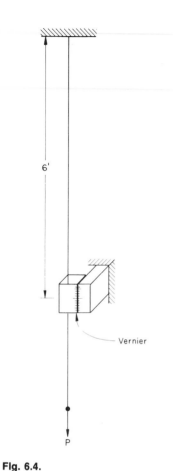

6'

Vernier

P

Fig. 6.4.
Experimental determination of modulus of elasticity. The stress is the weight hanging from the wire, divided by its cross-sectional area. The strain is the change in length (measured with a vernier scale), divided by the original length of the wire between the support and the scale. The modulus of elasticity is the linear change in stress divided by the linear change in strain.

The same applies to compression, provided that the compressed members (called struts) are short enough or have a cross-sectional geometry which avoids buckling (see Sections 2.6 and 7.3).

Evidently the problem of calculating the stresses in the members of a truss once the forces have been determined, as in Chapter 4, is a very simple one. We will examine the design of tension and compression members in Sections 7.3 to 7.5.

[†]Example 6.1. *Determine the cross-sectional area required for the tie in a steel truss, which is subjected to a force of 45,000 lb, if the maximum permissible stress is 22,000 p.s.i.*

The cross-sectional required is 45,000/22,000 = 2.045 sq in. An American Standard 3-in × 3-in × $\frac{3}{8}$-in angle, weighing 7.2 lb per ft, has a cross-sectional area of 2.11 sq in, which is sufficient.

6.4.   The Theory of Bending

The theory of bending is of the greatest importance, since most structural members are subjected to bending. However, since the distribution of stress and strain in beams is not uniform, it is far from simple, and it took several centuries to obtain the solution. The problem was first tackled by Leonardo da Vinci in the fifteenth century. Galileo Galilei gave the first, though incorrect, solution in 1638.

The error was corrected by M. H. Navier in 1826. He provided the key to the right answer by assuming that under a uniform bending moment, initially plane and parallel cross-sections remain plane during bending and converge on a common center of curvature. This can be visually demonstrated by drawing a square grid on a rubber beam (Fig. 6.5a) and then bending the beam (Fig. 6.5b). Navier's assumption has since been proved correct for all structural materials by strain measurements on test beams under load (Fig. 6.6); strain gauges are fixed at, say, each tenth of the depth of the beam, and the variation of strain throughout the section is plotted as the beam is loaded.

Navier's theorem follows from the assumption that plane sections remain plane. Consequently, under the action of a uniform bending moment M, two originally straight and parallel lines AC and EG (Fig. 6.7) remain straight, but converge on the common center of curvature, O. Evidently the beam is strained in compression at the top and in tension at the bottom, and somewhere between there is a line at which the strain is neither tensile nor compressive. This line of zero strain is called the *neutral axis*, BF.

Let us now draw a line through F parallel to AC. Since the planes started parallel, and are now converging on the point O, the

[†]This and following examples are worked in metric units in Appendix G.

Elasticity and Stress

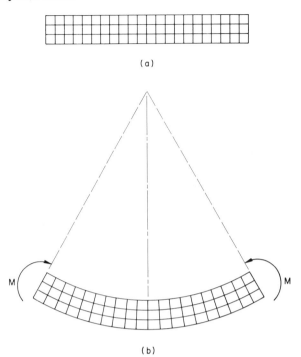

(a)

(b)

Fig. 6.5.
Navier's assumption. Originally plane and parallel sections (a) remain plane after bending (b), but converge onto a common center of curvature. This assumption can be illustrated with a rubber beam.

Fig. 6.6.
Proof of Navier's assumption with strain gauges. The strain is measured at several points on the same cross section, and the results for each load are plotted. So long as the beam is elastic, the neutral axis remains in the same position, and the strains vary proportionately with the distance from the neutral axis.

distance DE represents the tensile change of length at the bottom of the beam, and the distance GH represents the compressive change of length at the top of the beam. There is no change in length at the neutral axis, F.

Let us call the tensile change of length at the bottom of the beam DE = dx, and the original length AD = BF = x, then the tensile

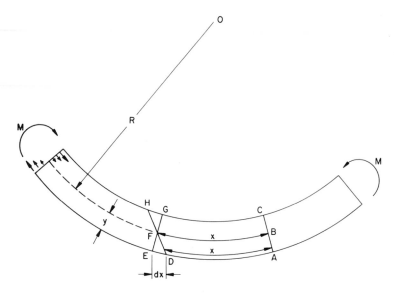

Fig. 6.7.
The theory of bending. The beam, under the action of a uniform bending moment M, bends into a circular arc, whose radius of curvature is $R = OF$. Originally plane and parallel sections ABC and EFG thus converge onto a center of curvature O. This causes compressive strains on top and tensile strains at the bottom. The neutral axis BF is the line of zero strain. The maximum tensile strain, at a distance y below the neutral axis, is $DE/AD = dx/x$. The corresponding stresses are shown on the left-hand side of the diagram.

strain at the bottom of the beam

$$e = \frac{dx}{x}$$

Let us call the radius of curvature, measured to the neutral axis, $FO = R$, and the distance of the bottom from the neutral axis $EF = y$. Since the triangles DEF and BFO have all their sides parallel, they are similar, and consequently

$$\frac{DE}{BF} = \frac{EF}{FO}$$

which gives the strain

$$e = \frac{dx}{x} = \frac{y}{R} \tag{6.4}$$

We now determine the corresponding stress by Hooke's Law.

$$f = Ee = \frac{Ey}{R} \tag{6.5}$$

The stress varies proportionately to the distance y from the neutral axis, and it is tensile below and compressive above for a positive bending moment (see Fig. 5.9).

The force acting on an infinitesimally small area dA at a distance

Elasticity and Stress

y from the neutral axis is

$$dP = f\, dA$$

and the moment of the force about the neutral axis

$$dM = y\, dP = y\, f\, dA = \frac{Ey^2}{R}\, dA \qquad (6.6)$$

The total resistance moment M of the section is the sum, or integral, of all the infinitesimally small elements dM.

$$M = \int \frac{Ey^2}{R}\, dA \qquad (6.7)$$

Since the modulus of elasticity is a constant, and the radius of curvature does not vary with the depth y, we can take them outside the integral (or summation) sign.

$$M = \frac{E}{R} \int y^2\, dA = \frac{E}{R}\, I \qquad (6.8)$$

where $I = \int y^2 dA$ is called the *second moment of area*, which is a purely geometric property of the section; it is independent of the type of material used. Since the same I also occurs in dynamic calculations, it is frequently called the *moment of inertia*.

The second moment of area is derived in most elementary books on geometry, the theory of structures, or the strength of materials (e.g., R. H. Trathen, *Statics and Strength of Materials*, McGraw-Hill, New York, 1954, p. 122). We will here confine ourselves to giving the answer to this purely geometric problem (Table 6.1).

From Eq. (6.5)

$$\frac{E}{R} = \frac{f}{y}$$

Substituting this into Eq. (6.8) we obtain

$$\frac{M}{I} = \frac{f}{y} = \frac{E}{R} \qquad (6.9)$$

which is *Navier's Theorem*.

In this equation, M is the bending moment at that particular section of the beam, R is the radius of curvature to which the beam is bent by the moment M, f is the stress at a distance y from the neutral axis, I is the second moment of area (moment of inertia) of the section, and E is the modulus of elasticity of the structural material.

In general, we are interested only in the maximum stress (which must be kept within the permissible range). This occurs at the greatest distance y from the neutral axis; in an unsymmetrical section there are two different values—one for the bottom (y_b) and

Table 6.1
Formulas for the Second Moment of Area
(Moment of Inertia)
and for the Section Modulus

Line No.	Dimensions	Second Moment of Area (Moment of Inertia) (I)	Section Modulus (S)
1	Rectangle	$bd^3/12$	$bd^2/6$
2	Circle	$\pi d^4/64$	$\pi d^3/32$
3	Triangle	$bd^3/36$	$bd^2/24$ (apex)
			$bd^2/12$ (base)
4	Standard steel sections	Listed in the *AISE Steel Construction Manual*	
5	Two steel plates fixed to the top and bottom flanges of a steel section	$Ad^2/4$	$Ad/2$
6	Rectangular box, I–, and channel sections	$(BD^3 - bd^3)/12$	$(BD^3 - bd^3)/6D$

one for the top (y_t), as shown in Fig. 6.8. It is therefore convenient to introduce a further geometric section property, the *section modulus*

$$S = \frac{I}{y} \tag{6.10}$$

In the case of an unsymmetrical section there are different section moduli for bottom and top:

$$S_b = \frac{I}{y_b} \quad \text{and} \quad S_t = \frac{I}{y_t} \tag{6.11}$$

Elasticity and Stress

Navier's theorem then becomes

$$M = f\,S \qquad (6.12)$$

The values of the section modulus for various shapes are also shown in Table 6.1. For other commonly used values of the second moment of area (moment of inertia) and the section modulus the reader is referred to the *AISC Steel Construction Manual*, published by the American Institute of Steel Construction, New York.

Example 6.2. *Determine the section modulus required for a steel beam subjected to a bending moment of 10,000 lb ft, if the maximum permissible steel stress is 24,000 p.s.i.*

The moment $M = 10,000$ lb ft $= 120,000$ lb in. The required section modulus $S = 120,000/24,000 = 5$ in^3.

From Section Tables, an American Standard 5-in beam (S 5 $\times$ 10) has a section modulus of 4.92 in^3, a width of 3.004 in, and weighs 10.0 lb per ft. A heavier section has a section modulus of 6.09 in^3, a width of 3.284 in, and weighs 14.75 lb per ft. The heavier S 5 $\times$ 14.75 section is required.

Example 6.3. *A compound steel beam consists of an American Standard 5 in $\times$ 3.284 in section (weighing 14.75 lb/ft), with a 6 in $\times$ $\frac{1}{2}$ in cover plate on each side. Determine its resistance moment, if the maximum permissible stress is 24,000 p.s.i.*

The second moment of area (moment of inertia) of the beam (from Section Tables) without cover plates is $I = 15.2$ in^4.

The clear distance between the cover plates is 5 in, and the center-to-center distance is 5.5 in. The additional second moment of area (from Table 6.1, line 5) is $2 \times 6 \times \frac{1}{2} \times 5.5^2/4 = 45.4$ in^4.

The total $I = 15.2 + 45.4 = 60.6$ in^4.

The overall depth of this symmetrical section is 6 in, and the section modulus

$$S = \frac{I}{y} = \frac{60.6}{\frac{1}{2} \times 6} = 20.2 \text{ in}^3.$$

Consequently the maximum permissible bending moment

$$M = f\,S = 24,000 \times 20.2 = 484,800 \text{ lb in}$$

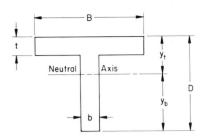

Fig. 6.8.
T-Section. Cross-sectional area:
$A = bD + (B - b)t$. Second moment of area (moment of inertia):
$I = \frac{1}{3}bD^3 + \frac{1}{3}(B - b)t^3 - Ay_t^2$.
In the case of an unsymmetrical section, there are different section moduli for the top: $S_t = I/y_t$, and for the bottom: $S_b = I/y_b$.

*6.5. Slope and Deflection of Beams

The slope and deflection of beams are important because they provide the key to the solution of statically indeterminate (hyperstatic) structures, and because excessive deformations are architecturally not acceptable. If the structure deflects too much, brittle finishes may crack, and door and windows may jam.

While the subject is most conveniently covered immediately following the theory of bending, it is a little harder to follow than the earlier sections. *It is therefore suggested that at a first reading*

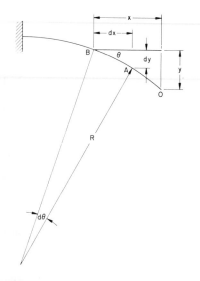

Fig. 6.9.
Slope and deflection of beams. The infinitely short length of beam, AB, subtends an angle $d\theta$ at the center of curvature, the radius of curvature being R. The horizontal and vertical components of AB are dx and dy, and the horizontal and vertical distances from the origin O are x and y.

the remainder of this chapter be omitted, and that readers proceed to Chapter 7.

Let us consider the shape of a deflected beam; Fig. 6.9 shows a cantilever, but the argument applies equally to a beam on two supports. The deflection is shown grossly exaggerated, because the small elastic deflection of a typical structural beam would not be sufficient to show in a diagram. Let us now take a short length of the beam AB, which subtends an angle $d\theta$ at the center of curvature, the radius of curvature being R. The beam between A and B forms a circular arc of length $R\ d\theta$. As a first approximation we may put AB equal to dx, since the deflections are assumed to be very small, so that the difference between AB and dx is negligible. We thus obtain

$$dx = R\ d\theta \tag{6.13}$$

Substituting for the radius of curvature from Navier's theorem, Eq. (6.9)

$$dx = \frac{EI}{M}\ d\theta$$

and the slope

$$\theta = \int \frac{M}{EI}\ dx \tag{6.14}$$

Thus the slope is the integral of the bending moment diagram, divided by EI, or, in geometric terms, the area of the bending moment diagram, divided by EI, from the origin O to x.

Since the differential coefficient is defined as the slope of a curve,

$$\theta = \frac{dy}{dx}$$

the deflection

$$y = \int \theta\ dx = \int \int \frac{M}{EI}\ dx\ dx \tag{6.15}$$

Thus the deflection is the integral, or area, of the slope diagram, and also the first moment of area of the bending moment diagram, divided by EI, from the origin O to x.

The solution of these equations for individual loading and support conditions is worked out in many books on the theory of structures or the strength of materials (e.g., F. L. Singer, *Strength of Materials*, Harper, New York, 1962, p. 198). In an architectural text it may suffice to give the answer without the detailed derivations (Table 6.2).

Example 6.4. *Determine the slope and deflection at the end of a cantilever, 10 ft long, carrying a concentrated load of 1,000 lb at the end (Example 5.1). The cantilever takes the form of an American Standard*

Elasticity and Stress

5-*in* × 3.284-*in steel beam* (*Example* 6.2)

$$W = 1,000 \text{ lb}, L = 10 \text{ ft} = 120 \text{ in},$$
$$M = -WL = -10,000 \text{ lb ft} = -120,000 \text{ lb in}.$$

For steel $E = 30,000,000$ p.s.i.
The second moment of area of the section (from steel section tables)
$I = 15.2 \text{ in}^4$.
From Table 6.2, the end slope

$$\theta = + \tfrac{1}{2} WL^2 / EI$$

$$= + \tfrac{1}{2} \times 1,000 \times 120^2 / 30,000,000 \times 15.2$$

$$= 0.0158 \text{ radians} = 0.0158 \times 180/\pi = 0.905°$$

$$= 54 \text{ min of arc}.$$

Furthermore, from Table 6.2, the end deflection

$$y = -\frac{1}{3}\frac{WL^3}{EI} = -\frac{1,000 \times 120^3}{3 \times 30,000,000 \times 15.2}$$

$$= 1.263 \text{ in} = \frac{L}{95}$$

A cantilever deflection of 1/95th of the span is too high if the steel section supports a floor or ceiling with a brittle finish (e.g., a plastered ceiling).

Example 6.5. *Determine the end slope and the mid-span deflection of a timber beam carrying a uniformly distributed load of* 1,000 *lb over a simply supported span of* 10 *ft* (*Example* 5.8).

The timber is Douglas Fir with a maximum permissible stress of 2,000 p.s.i., and a modulus of elasticity of 1,800,000 p.s.i.
The bending moment

$$M = \tfrac{1}{8} WL = \tfrac{1}{8} \times 1,000 \times 10 = 1,250 \text{ lb ft} = 15,000 \text{ lb in}$$

The section modulus required

$$S = \frac{M}{f} = \frac{15,000}{2,000} = 7.5 \text{ in}^3$$

We can supply the correct size by a narrow and very deep timber, such as a 1-in × 10-in rectangular section, or a shallow section such as a 4-in × 4-in. A narrow and deep timber needs cross-bracing for stability, and a shallow section uses more material, since S is proportional to b and to d^2. Let us compromise and select (from lumber section tables*) a 2-in × 6-in, which has a section modulus of 8.57 in.^{3} Its second moment of area $I = 24.10 \text{ in}^4$ (from lumber section tables*).

*According to American Lumber Standards, the dressed size of a 2-in × 6-in nominal section is $1\frac{1}{2}$ in by $5\frac{1}{2}$ in. However, most species of lumber are available dressed to $1\frac{5}{8}$ in × $5\frac{5}{8}$ in, and this size is listed in the *AITC Timber Construction Manual* (Wiley, New York, 1966, p. 7–14). The values of S and I used are for an actual section of $1\frac{5}{8}$ in × $5\frac{5}{8}$ in.

Table 6.2
Coefficients for the Maximum Values of the Shear Force, the Bending Moment, the Slope, and the Deflection of Statically Determinate Beams

Line No.	Loading Total Load = W Span = L	Maximum Shear Force	Maximum Bending Moment	Maximum Slope	Maximum Deflection
1		$-W$	$-WL$	$+\dfrac{1}{2}\dfrac{WL^2}{EI}$	$-\dfrac{1}{3}\dfrac{WL^3}{EI}$
2		$-W$	$-\dfrac{1}{2}WL$	$+\dfrac{1}{6}\dfrac{WL^2}{EI}$	$-\dfrac{1}{8}\dfrac{WL^3}{EI}$
3		$-W$	$-\dfrac{1}{3}WL$	$+\dfrac{1}{12}\dfrac{WL^2}{EI}$	$-\dfrac{1}{15}\dfrac{WL^3}{EI}$
4		$-W$	$-\dfrac{2}{3}WL$	$+\dfrac{1}{4}\dfrac{WL^2}{EI}$	$-\dfrac{11}{60}\dfrac{WL^3}{EI}$
5	CENTRAL LOAD	$\pm\dfrac{1}{2}W$	$+\dfrac{1}{4}WL$	$\pm\dfrac{1}{16}\dfrac{WL^2}{EI}$	$-\dfrac{1}{48}\dfrac{WL^3}{EI}$
6		$\pm\dfrac{1}{2}W$	$+\dfrac{1}{8}WL$	$\pm\dfrac{1}{24}\dfrac{WL^2}{EI}$	$-\dfrac{5}{384}\dfrac{WL^3}{EI}$
7		$l^a\ +\dfrac{1}{3}W$ $r\ -\dfrac{2}{3}W$	$+0.128WL$	$l^a\ -0.389\dfrac{WL^2}{EI}$ $r\ +0.444\dfrac{WL^2}{EI}$	$-0.01304\dfrac{WL^3}{EI}$
8	SYMMETRICAL	$\pm\dfrac{1}{2}W$	$+\dfrac{1}{6}WL$	$\pm\dfrac{5}{96}\dfrac{WL^2}{EI}$	$-\dfrac{1}{60}\dfrac{WL^3}{EI}$

Line No.	Loading Span = L End Moment = M	Maximum Shear Force		Maximum Bending Moment		Maximum Slope		Maximum Deflection
		1^a	r	1	r	1	r	
9	$-M$	$+\dfrac{3}{2}\dfrac{M}{L}$	$-\dfrac{3}{2}\dfrac{M}{L}$	$-M$	$-\dfrac{M}{2}$	$+\dfrac{ML}{4EI}$	0	$+\dfrac{ML^2}{27EI}$
10	$-M$	$+\dfrac{M}{L}$	$-\dfrac{M}{L}$	$-M$	0	$+\dfrac{ML}{3EI}$	$-\dfrac{ML}{6EI}$	$+0.0642\,\dfrac{ML^2}{EI}$
11	$-M$	0	0	$-M$	$-M$	$+\dfrac{ML}{2EI}$	$-\dfrac{ML}{2EI}$	$\dfrac{ML^2}{8EI}$
12	$-M_A$ $-M_B$	$+\dfrac{M_A - M_B}{L}$	$\dfrac{M_A - M_B}{L}$	$-M_A$	$-M_B$	$+\dfrac{2M_A + M_B}{6EI}L$	$-\dfrac{M_A + 2M_B}{6EI}L$	$+0.0642(M_A + M_B)\dfrac{L^2}{EI}$

^{a}l = left end; r = right end.

169

From Table 6.2 the end slope

$$\theta = \frac{WL^2}{24EI} = \frac{1,000 \times 120^2}{24 \times 1,800,000 \times 24.10}$$

$$= 0.0138 \text{ radian} = 48 \text{ min of arc}$$

and the central deflection

$$y = -\frac{5}{384}\frac{WL^3}{EI} = -\frac{5 \times 1,000 \times 120^3}{384 \times 1,800,000 \times 24.10}$$

$$= -0.519 \text{ in} = \frac{L}{231}$$

This is structurally acceptable, but too high if the timber beam supports a plastered ceiling. It would then be necessary to choose the next size, 2 in $\times$ 8 in (actual size $1\frac{5}{8}$ in $\times$ $7\frac{1}{2}$ in) which has an I of 57.13 in^4. This reduces the deflection to $L/547$.

*6.6. Shear Stress Distribution in Beams

In the great majority of architectural structures, beams are designed for bending moment. It is sometimes necessary to check for shear, but the shear stresses are rarely critical. The bending moment is the product of load *and* span, whereas the shear force is dependent only on the load. For the very short spans which are found, for example, in many engine parts, the shear force tends to be the controlling design factor. As the span increases (and most architectural beams have relatively long spans), the bending moment becomes the controlling factor.

The shear force and the bending moment are closely interrelated. Let us consider an element of a beam of length dx (Fig. 6.10). The bending moment to the left of the element is M, and to the right it is slightly larger, $M + dM$. Let us now cut the beam horizontally at a distance y above the neutral axis, and consider the equilibrium of the element of width b (at right angles to the sketch), of length dx, and of depth $\frac{1}{2}d - y$ (Fig. 6.11). Since the compressive stresses due

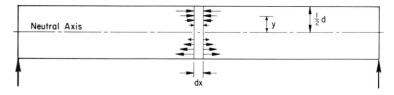

Fig. 6.10.
Direct stresses acting on an infinitesimally short length of beam, dx. The bending moment increases over the length dx from M to $M + dM$, and the direct stresses increase proportionately.

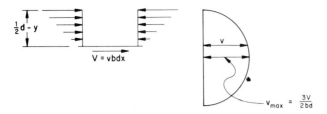

Fig. 6.11.
Horizontal shear force in a beam. Since the direct stresses increase over the length dx (see Fig. 6.10), a shear force V is required to balance the forces acting on the element of the beam, dx long and b wide. The shear stress acting on this area, v, increases parabolically from zero at the free surface to a maximum at the neutral axis.

to the bending moment $M + dM$ are greater than those due to the moment M, there is a resultant force tending to push the element to the right. This is resisted by the horizontal shear force V, acting along the cut which we made at a distance y above the neutral axis. There can be no shear at the top face of the beam, because at a free horizontal surface there is nothing to offer resistance to a horizontal shear force.

The shear force

$$V = v \, b \, dx$$

where v is the shear stress, b the width of the section, and dx the length of the element. The shear stress builds up with depth from the free surface, as the difference between the direct compressive forces increases. Since this shear stress is the sum (or integral) of the difference in compressive stress over the depth (which is a linear or first-order function), the shear stress varies parabolically with depth. The total shear force V equals the area of this diagram, and a parabola has an area two-thirds that of the enclosing rectangle. In a rectangular section of width b and depth d, the maximum shear stress (which occurs at the neutral axis) is therefore

$$v_{max} = \frac{V}{\frac{2}{3} b \, d} \tag{6.16}$$

A rigorous proof of this formula is given in many textbooks on the strength of materials (e.g., F. L. Singer, *Strength of Materials*, Harper, New York, 1962, p. 165).

In the case of an I-beam, there is a sharp change in shear stress at the junction of the flanges with the web, since v is inversely proportional to b. The shear stress distribution diagram is thus almost rectangular (Fig. 6.12b), and the maximum shear stress in a steel I-section is approximately

$$v = \frac{V}{b_w \, d}$$

where b_w is the width of the web.

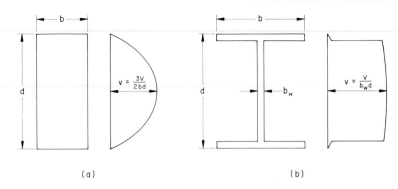

Fig. 6.12.
Shear stress distribution in rectangular and I-sections. The shear stress is inversely proportional to the width, b. Since the width changes abruptly in an I-section at the junction of the web and the flange, the shear stress changes in the proportion of b to b_w. Since the flange width, b, is so much greater than the thickness of the web, b_w, the shear force is almost entirely resisted by the web.

Evidently an I-section divides itself into the flanges, which resist practically the whole of the bending moment, and the web, which resists practically the whole of the shear.

Example 6.6. *Determine the maximum shear stress in the I-beam of Example 6.4.*

The shear force $V = 1,000$ lb. The section is 5 in. deep, and its web (from steel section tables) is 0.494 in. thick. Consequently, the maximum shear stress is approximately

$$v = \frac{V}{b_w\, d} = \frac{1000}{0.494 \times 5} = 405 \text{ p.s.i.}$$

This is well below the permissible shear stress for steel.

Example 6.7. *A rectangular timber section is to be built up from eight laminations glued together, as shown in Fig. 6.13. Determine the greatest magnitude of shear stress in the glued joints if the shear force at the section is 10,000 lb.*

The maximum shear stress, which occurs at half-depth, is

$$v = \frac{3V}{2bd} = \frac{3 \times 10,000}{2 \times 8 \times 13} = 144 \text{ p.s.i.}$$

Fig. 6.13.
Dimension of rectangular timber section built up from laminations (Example 6.7).

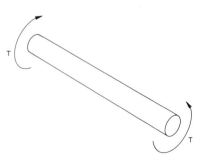

Fig. 6.14.
Torsion.

*6.7. Torsion

Torsion is a common design factor in engines; for example, in the shafts of rotary motors (Fig. 6.14). It causes one part of the shaft to shear relative to the other, and torsion is thus a form of shear (see Fig. 6.22d). If beams are subjected to torsion, then the torsional

Elasticity and Stress

shear stresses are additive to the shear stresses resulting from the shear force caused by bending (see Section 6.6).

The solution of torsion problems is simple for circular sections, but complicated for all others. Since circular sections are rarely used in buildings, torsion is discussed only in advanced structural texts (e.g., B. Bresler and T. Y. Lin, *Design of Steel Structures*, Wiley, New York, 1960, pp. 388–414; and H. J. Cowan and I. M. Lyalin, *Reinforced and Prestressed Concrete in Torsion*, St. Martin's Press, New York, 1965, 138 pp.).

Torsion invariably occurs when spandrel beams are loaded eccentrically (Fig. 6.15) in balcony girders (Fig. 6.16), in the spandrels of a building without corner columns (Fig. 6.17), and in spiral staircases. The reader should avoid these until he has gained substantial experience of structural design.

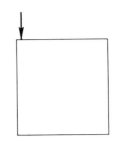

Fig. 6.15.
Torsion in spandrel beam caused by eccentric loading of outer wall.

*6.8. Analysis of Stress

As we mentioned in Section 6.2, stresses are directional, and stresses acting in different directions cannot be added by simple arithmetic (see Section 4.2). Furthermore, shear stresses are different from direct stresses.

Let us first consider the apparently simple case of a bar stressed concentrically in tension (Fig. 6.18a). If we make a cut at right angles to the tensile force, the stress

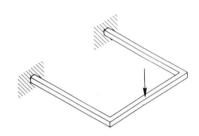

Fig. 6.16.
Balcony girder. A beam projecting beyond its supports—e.g., a balcony beam—is subject to combined bending and torsion.

$$f = \frac{P}{A} \tag{6.3}$$

as we stated in Section 6.3.

Let us now cut the bar at an angle α. The cross-sectional area is increased from A to $A/\cos\alpha$, and thus the stress is reduced to $f\cos\alpha$. The stress $f\cos\alpha$ has two components. One is at right angles to the cut, and therefore a direct stress; the other is parallel to the cut, and therefore a shear stress (Fig. 6.18b). Thus, in a ductile material (which has strong interatomic forces resisting the tension), the shear stress ultimately causes failure as one part of the crystal slides over the other (see Fig. 6.21a). The two components are

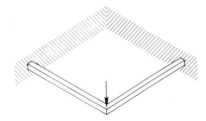

Fig. 6.17.
When two spandrels meet at the corner of a building without a supporting column, both are subjected to torsion.

$$f_1 = (f\cos\alpha)\cos\alpha = f\cos^2\alpha$$

and

$$v_1 = (f\cos\alpha)\sin\alpha = \tfrac{1}{2}f\sin 2\alpha$$

Let us now make a second cut, at right angles to the first (Fig. 6.18c). The stress across this second cut is $f\sin\alpha$, and its direct and

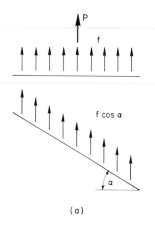

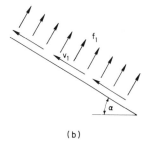

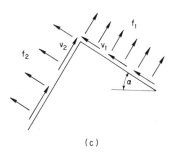

(a)

(b)

(c)

Fig. 6.18.
Direct and shear stresses in a tension member across a section at an angle α.

shear components are

$$f_2 = (f \sin \alpha)\sin \alpha = f \sin^2 \alpha$$

and

$$v_2 = (f \sin \alpha)\cos \alpha = \tfrac{1}{2} f \sin 2\alpha$$

Evidently the shear stresses along two mutually perpendicular cuts are the same, and we will call them v_{12}.

Let us now consider the stresses on a small unit element, at an angle α, in a bar subjected to a simple tensile stress f (Fig. 6.19). There are direct stresses f_1 and f_2 across the two perpendicular faces, and also shear stresses v_{12} along the same faces:

$$f_1 = f \cos^2 \alpha \tag{6.17a}$$

$$f_2 = f \sin^2 \alpha \tag{6.17b}$$

$$v_{12} = \tfrac{1}{2} f \sin 2\alpha \tag{6.17c}$$

Next we will consider a plate subjected to a tensile stress f_x in one direction, and another tensile stress f_y at right angles (Fig. 6.20).

The direct stress f_1 has a component $f_x \cos^2 \alpha$ from the x-direction, and additionally a component $f_y \sin^2 \alpha$ from the y-direction, so that

$$f_1 = f_x \cos^2 \alpha + f_y \sin^2 \alpha \tag{6.18a}$$

Similarly the direct stress

$$f_2 = f_y \cos^2 \alpha + f_x \sin^2 \alpha \tag{6.18b}$$

The shear component of f_x is $\tfrac{1}{2} f_x \sin 2\alpha$, and the shear component of f_y is $\tfrac{1}{2} f_y \sin 2\alpha$; but these act in opposite directions on all four faces. Consequently

$$v_{12} = \tfrac{1}{2}(f_x - f_y)\sin 2\alpha \tag{6.18c}$$

This may be either positive or negative, depending on whether f_x is larger or smaller than f_y. A negative shear is of the same kind as a positive shear; however, it acts in the opposite direction (see Section 6.2).

As we pointed out in Section 6.2, it is convenient to think of compressive stresses as negative tensile stresses. Thus Eqs. (6.18) apply to a plate subjected to compression in both the x and y directions, except that f_1 and f_2 are both negative, or compressive.

If the plate is subjected to a tensile stress f_x and a compressive

Elasticity and Stress

stress f_y, then the stress f_1 is tensile for small angles α, becomes zero when $f_x \cos^2 \alpha = f_y \sin^2 \alpha$, and thereafter is compressive.

The shear stress, however, is increased since the compressive stress f_y must now be added to the tensile stress f_x in Eq. (6.18c). Since ductile materials invariably fail in shear, and brittle materials often do (see Fig. 6.21), two direct stresses of opposite sign in Fig. 6.20 are more dangerous than two of the same sign.

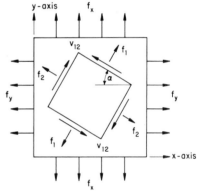

Fig. 6.19.
Direct and shear stresses on an element in a tension member.

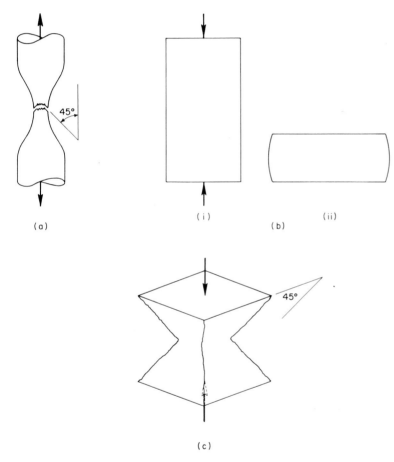

(a)

(b)

(c)

Fig. 6.20.
Principal stresses.

Fig. 6.21.
Shear failure of materials subjected to a direct force.

(a) Failure of a structural steel bar in tension commences by shear failure at 45°. When the yielding of the steel has reduced the cross-sectional area sufficiently, the material reaches its ultimate tensile strength, and it breaks at right angles to the direction of the tensile force.

(b) Failure of a structural steel bar in compression (i) also occurs by shear at 45°, but the material is squeezed out and does not fracture (ii).

(c) A concrete cube (the standard European test specimen) normally fractures due to shear at an angle of 45°. The 6-in × 12-in (150 × 300 mm) cylinder used in America and Australia shows the same type of failure, but because of the greater length, the angle of fracture often differs from 45°.

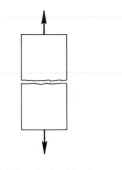

(a) Failure in direct tension

(b) Failure in tension due to bending

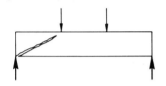

(c) Failure in diagonal tension due to shear

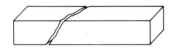

(d) Failure in diagonal tension due to torsion

Fig. 6.22.
Tension failure of concrete.

On the other hand, if $f_x = f_y$ and both are either tensile or compressive, then the shear stress is zero irrespective of the angle α. Thus, if we lower a piece of material to the bottom of the ocean, we subject it to high compressive stresses. However, since the pressure acts along all three axes, the shear stress is zero and the material is undamaged when it is pulled back to the surface.

From Eq. (6.18c) the *maximum shear stress* occurs when $\sin 2\,\alpha = 1$, i.e., when $\alpha = 45°$:

$$v_{max} = \tfrac{1}{2}\left(f_x - f_y\right) \tag{6.19}$$

Thus the *shear failure* of a tension or compression test specimen occurs at an angle of 45° (Fig. 6.21).

The shear stresses are zero when $\alpha = 0$ or 90°, since $\sin 2\,\alpha = 0$. At these angles, however, the direct stresses have their maximum values (Fig. 6.20):

$$f_{max} = f_x \quad \text{or} \quad f_y \text{ (whichever is the larger)}$$

These two stresses are called the *principal stresses*. One is the larger principal stress, and the other is the smaller principal stress. If one is tensile and one compressive, then these are the largest tensile and compressive stresses for any value of α (see Fig. 6.22).

Since the object of structural design is to ensure that the greatest tensile, compressive, and shear stresses in the structure are less than the maximum permissible tensile, compressive, and shear stresses, the *determination of the magnitude and direction of the principal stresses* is important for combined-stress problems. Let us therefore determine the principal stresses for any given stress combination of f_1, f_2, and v_{12}.

Adding Eqs. (6.18a) and (6.18b)

$$f_1 + f_2 = \left(f_x + f_y\right) \times \left(\cos^2 \alpha + \sin^2 \alpha\right)$$

Since $\cos^2 \alpha + \sin^2 \alpha = 1$

$$\tfrac{1}{2}\left(f_x + f_y\right) = \tfrac{1}{2}\left(f_1 + f_2\right) \tag{6.20}$$

Subtracting Eq. (6.18b) from (6.18a)

$$f_1 - f_2 = f_x\left(\cos^2 \alpha - \sin^2 \alpha\right) + f_y\left(\sin^2 \alpha - \cos^2 \alpha\right)$$

$$= \left(f_x - f_y\right)\cos 2\,\alpha \tag{6.21}$$

Substituting Eq. (6.21) into (6.18c)

$$v_{12} = \frac{\tfrac{1}{2}\left(f_1 - f_2\right)\sin 2\,\alpha}{\cos 2\,\alpha}$$

Elasticity and Stress

This gives the direction of the principal stresses:

$$\tan 2\,\alpha = \frac{v_{12}}{\frac{1}{2}(f_1 - f_2)} \qquad (6.22)$$

The tangent is shown graphically in Fig. 6.23, where $\tan 2\,\alpha = BC/AB$.

From the geometry of this right-angled triangle

$$AC = \sqrt{AB^2 + BC^2} = \sqrt{\tfrac{1}{4}(f_1 - f_2)^2 + v_{12}{}^2}$$

and

$$\sin 2\alpha = \frac{BC}{AC} = \frac{v_{12}}{\sqrt{\tfrac{1}{4}(f_1 - f_2)^2 + v_{12}{}^2}}$$

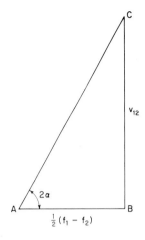

Fig. 6.23.
Solution of principal stress equation.

Substituting for $\sin 2\alpha$ into Eq. (6.18c)

$$\tfrac{1}{2}(f_x - f_y) = \sqrt{\tfrac{1}{4}(f_1 - f_2)^2 + v_{12}{}^2} \qquad (6.23)$$

Adding Eqs. (6.23) and (6.20)

$$f_x = \tfrac{1}{2}(f_1 + f_2) + \sqrt{\tfrac{1}{4}(f_1 - f_2)^2 + v_{12}{}^2} \qquad (6.24a)$$

Subtracting Eqs. (6.23) from (6.20)

$$f_y = \tfrac{1}{2}(f_1 + f_2) - \sqrt{\tfrac{1}{4}(f_1 - f_2)^2 + v_{12}{}^2} \qquad (6.24b)$$

Readers should note that there is a positive sign in the first term, and a negative sign in the second parentheses (under the square root).

Equation (6.24) gives the two principal stresses in terms of any combination of direct stresses f_1 and f_2 and the shear stress v_{12} in the same direction. Equation (6.22) gives the angle between the principal stress f_x and the direct stress f_1, and between the principal stress f_y and the direct stress f_2.

In shell structures there are generally two direct membrane forces and a membrane shear, and the principal tensile stress (which determines the amount and the direction of the reinforcement) is calculated from these.

In beams, there is at any point only one direct stress (due to the bending moment; see Section 6.4) and the shear stress (due to the shear force; see Section 6.6). The Eq. (6.24) becomes much simpler when $f_2 = 0$.

$$f_{x,\,y} = \tfrac{1}{2} f_1 \pm \sqrt{\tfrac{1}{4} f_1{}^2 + v_{12}{}^2} \qquad (6.25)$$

Example 6.8. *A steel bar is tested in tension and fails when the tensile stress is 50,000 p.s.i. Observation shows that the bar fails in shear, as shown in Fig. 6.21a. Determine the magnitude of the ultimate shear stress.*

From Eq. (6.19) the maximum shear stress

$$v_{max} = \tfrac{1}{2} \times 50{,}000 = 25{,}000 \text{ p.s.i.}$$

The bar fails at a shear stress of 25,000 p.s.i., because the tensile strength of the steel is more than 50,000 p.s.i.

Example 6.9. *Determine the magnitude of the diagonal tensile stress in a concrete beam due to a shear stress of 150 p.s.i.*

Let us consider a simply supported beam, as shown in Fig. 5.17. The shear force is a maximum just off the support, where the bending moment is zero. Thus, the principal stresses are those caused by the shear force alone.

From Eq. (6.25), the principal stresses are both

$$f_{x,y} = \pm \sqrt{150^2} = \pm 150 \text{ p.s.i.}$$

One of these stresses is compressive, and 150 p.s.i. is a very small compressive stress for concrete. However, a tensile stress of 150 p.s.i. is too high, and shear reinforcement is required (Figs. 6.22c and 6.24).

The direction of the principal tensile stress, from Eq. (6.22), is given by

$$\tan 2\alpha = \frac{150}{\tfrac{1}{2} \times 0} = \text{infinity}$$

so that $2\alpha = 90°$, and $\alpha = 45°$.

The shear reinforcement therefore consists of bars inclined in the direction of the diagonal tension, or of vertical stirrups and horizontal bars, which combine to produce a diagonal tensile force (Fig. 6.24b).

Note that the diagonal tensile stress is numerically equal to the shear stress. In Example 6.8 the diagonal shear stress was half the tensile stress.

Example 6.10. *The stresses in a shell roof due to the direct membrane forces in two mutually perpendicular directions are 200 p.s.i. tension and 300 p.s.i. compression, and the membrane shear stress in the same directions is 150 p.s.i. Determine the principal stresses.*

From Eq. (6.24) the principal stresses are

$$f_{x,y} = \tfrac{1}{2}(200 - 300) \pm \sqrt{\tfrac{1}{4}(200 + 300)^2 + 150^2}$$

$$= -50 \pm 100\sqrt{6.25 + 2.25} = -50 \pm 291$$

$f_y = 341$ p.s.i. compression, $f_x = 241$ p.s.i. tension.

The angle of inclination of the principal tensile stress is given by Eq. (6.22).

$$\tan 2\alpha = \frac{150}{250} = 0.600$$

$$\alpha = 15°29'$$

The angle of inclination of the principal compressive stress is 74°31'.

(a) Bars bent up at 45°

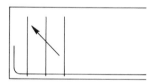

(b) Vertical stirrups and horizontal bars, which have a resultant diagonal tensile force

Fig. 6.24.
Shear reinforcement for concrete beams.

*6.9. Stress Trajectories, Stress Concentrations, and Experimental Stress Analysis

Stress trajectories, also called *isostatic lines*, show the direction of the principal stresses. There are two sets of lines, and they must always intersect at right angles, since at each point there are two principal stresses at right angles to one another (Fig. 6.25).

The stress trajectories give a clear visual pattern of structural behavior, and they are particularly helpful in the design of complex structures (see Chapter 9).

In reinforced concrete, steel is used to resist the tensile stresses in the structure, and the most efficient layout of the steel, from a purely mechanical point of view, is along the tensile stress trajectories. In practice, the labor cost of laying the steel along the complex curves of the stress trajectories is rarely warranted (Fig. 6.26), although it is sometimes economical to do so, for example, in cylindrical shells and folded plates (see Sections 9.4 and 9.6). P. L. Nervi has on several occasions made use of the stress trajectories, not so much because of their structural economy, but because of their interesting patterns (Fig. 6.27).

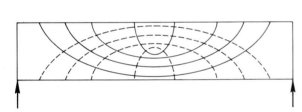

Fig. 6.25.
Stress trajectories in a beam carrying a uniformly distributed load.

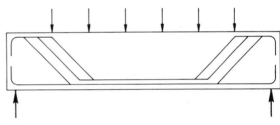

Fig. 6.26.
Bent-up bars serving as shear reinforcement. In practice it is uneconomical to have bars curved to follow the stress trajectories (see Example 6.9).

Since the trajectories are normally drawn at equal increments, a crowding of the lines indicates stress concentrations, just as the crowding of contours on maps of a mountainous region indicates a steep gradient.

Load concentrations, re-entrant corners, notches, and holes invariably produce stress concentrations (Fig. 6.28). These do not necessarily produce structural failure, even though the maximum stress may be, say, four times as high as the permissible stress. In structural steel a considerable strain is possible at a constant (or approximately constant) yield stress (see Fig. 6.1a). If the stress concentration is purely local, the excess stress can thus be distributed to adjacent portions of the material. Provided this *redistribution of stress* leaves the stresses in the greater part of the structural

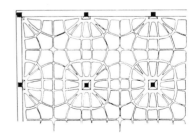

Fig. 6.27.
Concrete floor for a factory in Rome, with ribs following the lines of the stress trajectories (see also Fig. 7.20b).

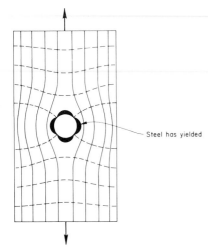

Fig. 6.28.
Stress trajectories in a steel tension plate with a hole, and stress redistribution in the region where the yield stress is exceeded.

member within the permissible range, it is perfectly safe, even though elastic analysis shows very high local stresses. This is an example of the use of ultimate strength considerations in elastic analysis (see also Section 7.9).

In concrete structures, stress concentrations are more serious, because excessive tensile stresses, even though localized, tend to produce cracks; cracks propagate over a period of time because of the high stress concentrations at the end of the cracks. The splitting of timber along the grain presents similar problems.

Since structural members in concrete and timber are much bulkier than those of metal, it is often possible to avoid stress concentrations by rounding sharp corners, or changing the section more gradually. Steel reinforcement should always be placed across the lines of potential cracks in concrete, and prestressing should be considered when cracking poses serious problems (see Section 7.13).

The drawing of stress trajectories from theoretical calculations is laborious, and experimental techniques are simpler for producing the patterns. If a structural member is coated with a *brittle lacquer* before being stressed, the cracks in the lacquer show up the tensile stress trajectories. Unfortunately, few lacquers are sufficiently sensitive, and the method has only limited application.

Photoelasticity is frequently used to produce stress patterns (Fig. 6.29). It is based on the property of many plastics (e.g., plexiglas) to break up light into two components polarized in the directions of the principal stresses. A simple demonstration-apparatus requires only a light source and two polarizing filters, oriented at right angles to one another (Fig. 6.30). If white light is used, the photoelastic fringes have all the colors of the spectrum, and they produce some fascinating patterns. However, the black-and-white fringes resulting from monochromatic light (e.g., from a sodium lamp) are more practical for design. Photoelasticity is particularly useful for studying stress concentrations.

If we make an exact scale model of the structure, using any suitable elastic material, such as plexiglas, we can measure the strains in the model with electric resistance strain gauges (Fig. 6.31)

Fig. 6.29.
Photoelastic patterns in a beam. Note the region of pure bending in the middle, the shear near the ends, and the stress concentrations under the four loads.

Elasticity and Stress

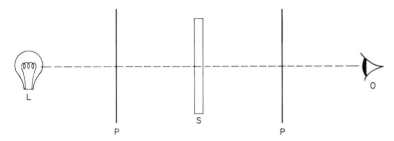

Fig. 6.30.
Photoelastic demonstration apparatus. L = lamp; P = polarizing filters, oriented at 90° to one another; S = structural model; O = observer.

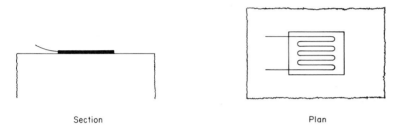

Section   Plan

Fig. 6.31.
Electric resistance strain gauge. The gauge consists of a length of very thin wire between two sheets of paper or plastic; this is laid zigzag to accommodate a greater length. Special alloys are used for the resistance wire, and special glue for fixing the gauge. As the wire extends with the model, it becomes thinner and passes less current. The reverse happens in compression. The change in electrical resistance can be determined with great accuracy, and the gauge is capable of measuring strains of 1×10^{-5}.

and deduce from these the stresses in the actual structure. This is particularly useful when the structure is too complex for an accurate theoretical solution (see Chapter 9).

The various techniques of experimental stress analysis which are useful for architectural design have been described in detail in a volume in the Architectural Science Series (H. J. Cowan *et al.*, *Models in Architecture*, Elsevier, London 1968, 228 pp.).

Suggestions for Further Reading

J. P. DEN HARTOG: *Strength of Materials*. Dover, New York, 1961. Chapters 2–5, and 10, pp. 14–108, and 199–226.

A. JENSEN: *Statics and Strength of Materials*. McGraw-Hill, New York, 1962. Book II, Chapters 1, 4, 7, 9, and 11, pp. 1–28, 80–95, 136–160, 187–219 and 243–252.

R. J. ROARK: *Formulas for Stress and Strain*. McGraw-Hill, New York, 1943, Chapters 1–9, pp. 1–185.

F. L. SINGER: *Strength of Materials*. Harper, New York, 1962. Chapters 1–6, and 9, pp. 1–244 and 312–363.

Section tables are generally available from the producers of the sections, i.e., steel and aluminum manufacturers, lumber millers, and precast concrete manufacturers. American readers may find the following helpful:

American Institute of Timber Construction: *Timber Construction Manual*. Wiley, New York, 1966. *Standard lumber and timber sizes*, p. 7, 3–21.

Hot Rolled Steel Shapes and Plates. United States Steel Corporation, Pittsburgh, Pennsylvania, 1966. 117 pp.

Problems

6.1. Describe and explain the various types of strain and stress which occur in architectural structures. Give some typical examples of each type. Why is it possible to measure strain, but not stress?

[†]**6.2.** Draw the bending moment and shear force diagrams of the beam shown in Fig. 6.32. If the beam is of timber, with a maximum permissible stress of 1,000 p.s.i., select a suitable size of cross-section.

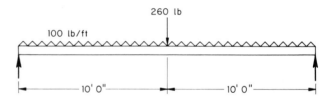

Fig. 6.32.
Problem 6.2.

6.3. Draw the shear force and bending moment diagrams for the beam shown in Fig. 6.33, neglecting its own weight.

If the width is 12 in, and the maximum permissible stress is 1,200 p.s.i., calculate the minimum depth of the beam.

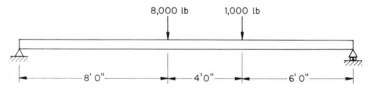

Fig. 6.33.
Problem 6.3.

6.4. A prestressed concrete box section has an overall depth of 48 in and an overall width of 36 in. The concrete walls are 6 in thick

[†]Similar problems in metric units are given in Appendix G.

on both the horizontal and the vertical parts of the box. Calculate the second moment of area (moment of inertia) and the section modulus.

6.5. A beam AB spans 12 ft. At 3 ft from the left-hand support an anticlockwise moment of 12,000 lb ft is applied, and at 3 ft from the right-hand support a 12,000-lb vertical force is applied to the bar. Draw the shear force diagram and the bending moment diagram.

If the beam is of timber, with a maximum permissible stress of 2,500 p.s.i. in bending, select a suitable cross-section for the beam to satisfy bending requirements.

6.6. If the beam of Problem 6.5 had to be strengthened to take an increase in load, and all that was available was a length of timber 12 ft long and wider than the width determined previously, where should this strengthening piece be placed to obtain maximum benefit from it? Indicate the position by means of a sketch.

6.7. Calculate the deflection of the box section of Problem 6.4 due to its own weight over a simply supported span of 100 ft, and the camber required to eliminate it. The weight of concrete may be taken as 144 lb per cubic foot, and its modulus of elasticity as 4,000,000 p.s.i.

6.8. The shear force at a section of a laminated timber arch is 25,000 lb. The arch at that section has a depth of 24 in and a width of 10 in. Calculate the maximum shear stress parallel to the laminations, and also the principal tensile and compressive stresses, if there is no bending moment at that section.

6.9. At a point in the web of a steel beam there is a horizontal tensile stress of 21,000 p.s.i. due to bending, and a shear stress of 14,000 p.s.i. Calculate the magnitude of the principal stresses.

6.10. At a certain position in a floor slab, which acts as a diaphragm, the following stresses occur: in the north-south direction, 200 p.s.i. compression; in the east-west direction, 100 p.s.i. compression; and in both directions, 200 p.s.i. shear.

Calculate the magnitude of the principal stresses, and draw a diagram illustrating the direction of the principal stresses in relation to the direct and shear stresses.

6.11. An element of a reinforced concrete shell roof, $3\frac{1}{2}$ in thick, is subject to a compressive force of 2,000 lb per foot width parallel to the span, to a tensile force of 3,000 lb per foot width perpendicular to the span, and shear forces of 4,000 lb per foot width parallel and perpendicular to the span.

Determine the maximum compressive stress in the concrete, the area of reinforcement required, and its inclination to the shell. It may be assumed that the whole of the tension is taken by the steel, and the maximum permissible steel stress is 20,000 p.s.i.

Effects of the Structural   **7**
Material on Design

Certain truths regarding the material [concrete]
are clear enough. First, it is a mass material;
second, an impressionable one as to surface; third,
it is a material which may be made continuous or
monolithic within very wide limits; fourth, it is a
material which can be chemicalized, colored, or
rendered impervious to water; fifth, it is a willing
material when fresh, fragile when still young, stubborn
when old, always lacking in tensile strength.

Frank Lloyd Wright

We derived the forces and moments in trusses and beams in
Chapters 4 and 5, and the stresses caused by these moments and
forces in Chapter 6. We will now examine how the character of the
structural materials affects the sizes of structural members and their
joints.

7.1. The Buckling Problem

As we have seen in Section 6.3, the maximum permissible load for a
piece of material stressed in tension is simply

$$P = f A \qquad (7.1)$$

where f is the maximum permissible tensile stress, and A the
cross-sectional area of the tension member.

The same equation also applies to compression members, provided they do not buckle. As we have seen in Section 2.6, concrete
compression members tend to be massive, and buckling is only a
minor design factor. However, it must usually be allowed for in
timber and steel. It assumes great importance in the design of
aluminum; this has a low modulus of elasticity, and we shall
presently see why this affects the problem.

Let us perform three simple experiments with long, slender steel
sections. In the first, we will use two pieces of identical length and

cross-sectional area, but of different shape: one piece is round, $\frac{1}{4}$ in (6.4 mm) diameter, and the other is thin, 1 in by 0.05 in (25.4 mm by 1.3 mm) (Fig. 7.1). The thin piece carries only 1/18th of the load of the round one, even though it has the same weight and cross-sectional area.

In the next experiment we will take two pieces of identical cross-section, but make one twice as long as the other (Fig. 7.2). The long piece carries only a quarter the load of the shorter piece.

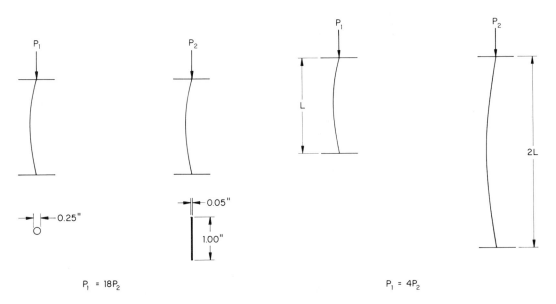

Fig. 7.1.
A thin strut has a lower buckling load. The cross-sectional area, the material, the length, and the support conditions are the same for both struts.

Fig. 7.2.
A long strut has a lower buckling load. The cross-sectional dimension, the material, and the support conditions are the same for both struts.

In the third experiment we will take three pieces of steel of identical length and cross-section. We will load one as a cantilever; i.e., one end is firmly fixed and the other is allowed to move freely. We will load one as a pin-ended column; i.e., the ends are allowed to rotate freely, but they are restrained to remain in line with the load. Another is built-in; i.e., its ends are not allowed to rotate, and it can move only in the direction of the load (Fig. 7.3). The built-in column carries four times as much as the pin-ended column, and sixteen times as much as the cantilever column.

Buckling evidently depends on the end restraint, the stiffness, and the length and the cross-sectional geometry (which determine the *slenderness*) of the column. It is an elastic phenomenon, and failure occurs whenever the energy required to recover the original shape is greater than the energy required to continue the buckling deformation.

Effect of the Structural Material on Design

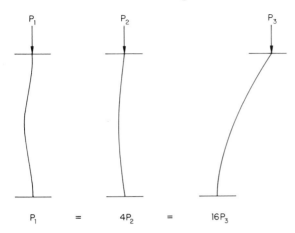

Fig. 7.3.
The more rigid the end restraints, the higher the buckling load. The first strut is built-in, the second is pin-ended, and the third is cantilevered. The cross-sectional dimensions, the material, and the length are the same for all three.

We will derive the ultimate buckling load of a slender column in the next section for those readers who are sufficiently acquainted with mathematics. However, let us look at the answer now.

$$P_e = \frac{\pi^2 EI}{l^2} \tag{7.2}$$

where P_e is the collapse load of a slender strut (or column), known as the Euler load, after the originator of the equation
   π is the circular constant 3.142
   E is the modulus of elasticity
   I is the least second moment of area (moment of inertia); evidently a rectangular section buckles at right angles to the thinner direction
and l is the effective length of the column, defined as the length of an equivalent strut with pinned ends.

Let us call the actual length of the column L. Then $l = L$ for a pin-ended strut, $l = 2L$ for a cantilevered strut, and $l = \frac{1}{2}L$ for a built-in strut (see Fig. 7.3).

The maximum permissible load for a slender column is obtained by dividing the buckling load by the factor of safety, S_f.

$$P = \frac{P_e}{S_f} = \frac{\pi^2 EI}{S_f l^2} \tag{7.3}$$

Since the buckling strength of the column depends on its slenderness, it is convenient to introduce the *slenderness ratio*. For this purpose we define the least radius of gyration

$$r = \sqrt{I/A}$$

This is simply a convenient short notation for $\sqrt{I/A}$, and it is a geometric property of the section. Thus for a rectangular section (see Table 6.1) the radius of gyration $r = d/\sqrt{12}$, and for a circular section $r = d/4$.

We can now write Eq. (7.3) in the form

$$P = \frac{\pi^2 E A r^2}{S_f l^2} = \frac{\pi^2 E A}{S_f} \left(\frac{r}{l} \right)^2 \qquad (7.4)$$

and the average stress

$$f = \frac{P}{A} = \frac{\pi^2 E}{S_f} \left(\frac{r}{l} \right)^2 \qquad (7.5)$$

The term l/r is called the *slenderness ratio*, and the strength of a slender strut is inversely proportional to its square.

In actual fact, columns which are slender enough to fail in accordance with the Euler theory, i.e., with Eqs. (7.4) and (7.5), are too flexible to be useful in buildings, although they do find application in some other structures. We shall consider the design of practical struts for buildings in Section 7.3, after the derivation of the Euler formula.

*7.2. Euler's Theory for Slender Columns

Leonard Euler derived the formula for the strength of a slender, pin-ended column in 1757. It is an important part of the theory of structures and is the oldest structural formula still in use. *However, the derivation requires an elementary knowledge of the theory of differential equations; readers unfamiliar with that subject are recommended to proceed straight to Section 7.3.*

Let us consider a pin-ended strut of length l, which buckles under the load P_e. The deflection from its original straight shape at the midpoint is a (Fig. 7.4). Let us now consider a point along the buckled strut, which is x to the right and y above the central point of the buckled strut. The load P_e is at a distance $(a - y)$ from that point, and it therefore causes a bending moment

$$M = P_e(a - y) \qquad (7.6)$$

in the strut.

As we saw in Section 6.5, the deflection caused by a bending

Effect of the Structural Material on Design

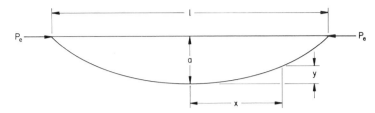

Fig. 7.4.
Euler's buckling theory.

moment M is

$$y = \int \int \frac{M}{EI} \, dx \, dx \qquad (6.15)$$

which gives

$$\frac{M}{EI} = \frac{d^2y}{dx^2} \qquad (7.7)$$

Combining Eqs. (7.6) and (7.7)

$$\frac{d^2y}{dx^2} = \frac{P_e}{EI}(a - y)$$

Since the term recurs throughout the next few equations, it is convenient to introduce

$$\frac{P_e}{EI} = k^2$$

which gives

$$\frac{d^2y}{dx^2} + k^2(y - a) = 0$$

This is a standard linear differential equation with constant coefficients, and the solution (which is given in most textbooks on differential equations, e.g., A. R. Forsyth, *Differential Equations*, Macmillan, London, 1943, p. 69) is

$$y = A \sin kx + B \cos kx + a \qquad (7.8)$$

where A and B are constants of integration.

We note that $x = 0$ for $y = 0$, which gives $B = -a$.

Differentiating, we obtain

$$\frac{dy}{dx} = k(A \cos kx - B \sin kx)$$

and we note that the slope $dy/dx = 0$ for $x = 0$, which gives $A = 0$.

Substituting into Eq. (7.8)

$$y = a(1 - \cos kx)$$

At the ends of the strut $y = a$ and $x = \pm \frac{1}{2} l$, so that $\cos kx = 0$.

This means that kx must be $\frac{1}{2}\pi$, or $\frac{3}{2}\pi$, or $\frac{5}{2}\pi$. Evidently the strut buckles at the lowest value, so that the higher values are never reached.

Consequently

$$\pm kx = \pm \frac{1}{2} l \left(\frac{P_e}{EI} \right)^{1/2} = \frac{1}{2}\pi$$

and the Euler load

$$P_e = \frac{\pi^2 EI}{l^2} \tag{7.2}$$

7.3. Practical Design of Compression Members

As we showed in Section 7.1, there are two formulas for the design of compression members. One applies to short columns

$$P = f A \tag{7.1}$$

which fail by overstressing of the material in compression. It depends on the strength of the material, but not on its modulus of elasticity or its slenderness ratio.

The other applies to slender columns

$$P = \frac{\pi^2 EA}{S_f} \left(\frac{r}{l} \right)^2 \tag{7.4}$$

which fail by buckling. It depends on the slenderness ratio and modulus of elasticity of the material but not on its strength. (The term π^2/S_f is a number, and the cross-sectional area A occurs in both equations.)

Many columns in buildings actually fail in accordance with Eq. (7.1); in particular, many concrete columns have low slenderness ratios and thus show no tendency to buckle. No column is so slender as to fail in accordance with Eq. (7.4), although some struts in aluminum trusses come very close to it. Aluminum has a low modulus of elasticity (10,000,000 p.s.i.) by comparison with steel (30,000,000 p.s.i.)*, but even steel struts usually show a tendency to buckle.

*The corresponding metric values are 70 GPa and 200 GPa.

Effect of the Structural Material on Design

In Fig. 7.5 we plot Eqs. (7.1) and (7.4). The first is a straight line, since P/A is independent of the slenderness ratio. The second is a hyperbola, since P/A is inversely proportional to $(l/r)^2$. Evidently the physical behavior of the strut is unlikely to imitate the sharp corner that occurs at the junction of Eqs. (7.1) and (7.4), and experiments show (as might be expected) a gradual transition, represented by the thick line in Fig. 7.5. Practical column formulas are based on this experimental curve, and they imitate it with more or less complex formulas, depending on the importance of the problem. These are described in most textbooks in steel construction (e.g., B. Bresler and T. Y. Lin, *Design of Steel Structures*, Wiley, New York 1960, p. 304). Steel codes and steel section books almost invariably tabulate the variation of the stress $f = P/A$ with the slenderness ratio l/r, corresponding to the prescribed formula, so that the designer need only look up the correct value in the table.

†Example 7.1. *Determine the permissible axial load for a 12 in $\times$ 12 in American standard wide-flange column weighing 85 lb per foot run; its effective length is 10 ft, and the material is structural steel with a minimum yield point of 36,000 p.s.i.*

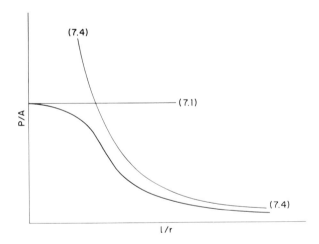

Fig. 7.5.
Practical design of compression members. Equation (7.1) represents the short-column formula, $P/A = f$, which is independent of the slenderness ratio. Equation (7.4) represents the Euler formula

$$\frac{P}{A} = \frac{\pi^2 E}{S_f} \left(\frac{r}{l} \right)^2$$

which varies hyperbolically with the slenderness ratio. Evidently experiments on compression members, represented by the heavy line, do not imitate the sharp junction between the two curves, and practical design formulas are based on this experimental curve.

†This and all following examples are worked in metric units in Appendix G.

From section tables, the least radius of gyration, $r = 3.07$ in, and the slenderness ratio is $l/r = 10 \times 12/3.07 = 39.1$.

From the tables contained in the specification of the American Institute of Steel Construction, the permissible stress conforming to the A.I.S.C. column formula for a slenderness ratio of 39.1 is 19,260 p.s.i.

The cross-sectional area of the section $A = 24.98$ sq in, and thus the permissible column load

$$P = 19{,}260 \times 24.98 = 481{,}000 \text{ lb}$$

Even this simple calculation is superfluous, since steel manufacturers usually supply tables showing the permissible column loads which their sections will carry for different effective lengths.

Example 7.2. Determine the maximum permissible axial load for the column of Example 7.1, if the 12 in $\times$ 12 in section is strengthened with two 16 in $\times$ 1 in plates of the same steel, one on each flange.

The coverage of compound sections by tables is more restricted, because there are so many possible combinations.

The second moments of area (moments of inertia) of the wide-flange section (from section tables) are 723.3 in^4 about the axis X–X (Fig. 7.6) and 235.5 in^4 about the axis Y–Y. To this we must add the second moment of area (moment of inertia) of the plates, which is (from Table 6.1) $A\,d^2/4$ $= 2 \times 16 \times 1 \times 13^2/4 = 1{,}352.0$ in^4 about the axis X–X, and $2b\,d^3/12$ $= 2 \times 1 \times 16^3/12 = 682.7$ in^4 about the axis Y–Y.

The smaller combined $I = 235.5 + 682.7 = 918.2$ in^4 about the axis Y–X.

From section tables the cross-sectional area of the wide-flange section is 24.98 sq in, so that $A = 24.98 + 2 \times 16 \times 1 = 56.98$ sq in.

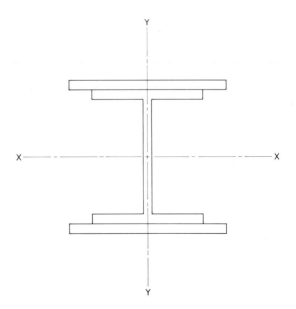

Fig. 7.6.
Compound column (Example 7.2).

Effect of the Structural Material on Design

The radius of gyration of the compound section $r = \sqrt{918.2/56.98}$ = 4.01 in.

The slenderness ratio $l/r = 120/4.01 = 29.9$, and the corresponding permissible stress is 19,950 p.s.i.

The permissible column load $P = 19,950 \times 56.98 = 1,367,000$ lb.

Example 7.3. *Determine the maximum permissible load for a 6 in × 6 in square tube column; its effective length is 10 ft, and the material is 36,000 p.s.i. steel.*

Tubes provide neat sections for lightly loaded columns, particularly in conjunction with timber beams or light-gauge steel floors.

For a 0.375 in thick section, the radius of gyration, $r = 2.2547$ in, and the slenderness ratio $l/r = 120/2.2547 = 53.2$. The permissible steel stress is 18,060 p.s.i.

The cross-sectional area is 7.954 sq in, and the permissible column load is thus 143,700 lb.

Example 7.4. *Determine the maximum permissible load for a double-angle strut in a roof truss, consisting of two 4 in × 3 in × $\frac{1}{2}$ in angles, separated by a $\frac{3}{8}$ in space. The effective length is 6 ft, and the material is 36,000 p.s.i. steel.*

Double angles are commonly used in the rafters of roof trusses, because the flange of a T-section does not offer enough space for connecting the other members (see Section 7.4). A gusset plate is placed between the angles, to which all the members meeting at the joint are connected (see Figs. 4.7 and 7.7). Double angles are therefore a standard form of truss construction, and their properties (acting as a pair) are available in tabular form. The compound section has two radii of gyration: one about the Y–Y axis, and one about the X–X axis. The least value for this compound section (from tables) is $r_x = 1.25$ in.

The slenderness ratio $l/r = 72/1.25 = 57.6$, and the corresponding permissible stress is 17,660 p.s.i.

The cross-sectional area is 6.50 sq in, and the permissible column load is thus 114,800 lb.

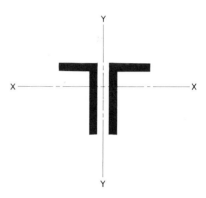

Fig. 7.7.
Double-angle strut for truss (Example 7.4).

Example 7.5. *Determine the maximum permissible load for an 8 in × 10 in Douglas fir column, 15 ft long.*

The specifications of the American Institute of Timber Construction provide for two basically different curves, as in Fig. 7.5. One holds good for short columns; the other, for slender struts, is a Euler curve, based on the modulus of elasticity, which is 1,760,000 p.s.i. for Douglas fir.

Since timber sections are generally rectangular, and since the radius of gyration of a rectangular section is $d/\sqrt{12}$, it is convenient to use as the slenderness ratio l/d (where d is the smaller side of the rectangle) rather than l/r.

The dressed size of an 8 in × 10 in section is $7\frac{1}{2}$ in × $9\frac{1}{2}$ in, and the ratio $l/d = 15 \times 12/7.5 = 24$. From timber design charts, the permissible stress is the smaller of 1,300 p.s.i., the permissible stress for short Douglas fir columns, and 900 p.s.i., the stress corresponding to $E = 1,760,000$ p.s.i. and $l/d = 24$.

Consequently the permissible column load is $900 \times 7.5 \times 9.5 = 64,125$ lb.

Although the formulas used in the British, Canadian, and Australian timber codes are more complicated (and differ from one another), the procedure is the same, since it is only necessary to ascertain the permissible stress corresponding to the appropriate l/d ratio from a table or a chart.

7.4. The Jointing Problem

As we noted in Section 2.3, the joints were frequently the weakest part of traditional structures. It is normal practice in engineered construction to make the joints as strong as the structural members they join, since this is the most economical way of designing a structure.

This is easily achieved in concrete structures cast on the site in one piece, or *monolithically*, and it is one of the great advantages of concrete.

Timber, steel, and aluminum are produced as linear pieces, which must be joined to form a structure. This adds to the cost, and it requires space. In designing the structure it is necessary to allow enough room for the joints and to consider their appearance. Some methods of jointing are invisible (e.g., "secret" nailing of timber), and others are unobtrusive (e.g., welds in light steel structures); some add interest to the look of the structure, and others are disturbing and should be hidden.

A joint of two (tension or compression) members under direct stress (as in a truss, Chapter 4) is almost invariably a piece of metal in shear. This applies to timber structures which are usually joined by nails, screws, or metal connectors, and to steel and aluminum structures which are joined by riveting, welding, or bolting.

Rivets are the oldest form of connecting steel structures. They are driven red-hot and, on cooling, the shanks of the rivets contract and thus press the two members together. The resulting friction greatly contributes to the strength of the joint, which otherwise depends on the shear strength of the rivets. As Fig. 7.8a, b, and c shows, the rivets are in shear, whether both members are in tension, or both in compression, and this applies equally with and without cover plates.

Since the handling of red-hot rivets is unpleasant, this form of jointing has declined with improving labor conditions. It survives in countries with an old-established steel fabricating industry, but is rarely used elsewhere.

Welding also relies on shear (Fig. 7.8d), but this can be distributed unobtrusively over a long weld. Since the weld metal is deposited liquid—i.e., at a temperature above the melting point of steel—it affects the metallurgical structure of the steel with which it is in contact. It is thus necessary to preheat thick sections to reduce

Effect of the Structural Material on Design

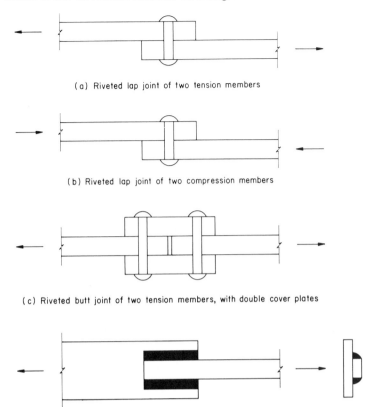

(a) Riveted lap joint of two tension members

(b) Riveted lap joint of two compression members

(c) Riveted butt joint of two tension members, with double cover plates

(d) Fillet-welded lap joint of two tension members

Fig. 7.8.
Joints in structural steel members.

temperature stresses and to avoid the use of steels which because of their crystal structure are "not weldable." Some high-tensile steels are produced by cold-working, which raises the strength by distorting the crystal structure; the high-tensile properties can thus be lost by welding, and the weldability of high-tensile steels must always be investigated before they are specified.

It is difficult and expensive to perform welding and riveting on the site, and site joints are usually made with bolts. Until recently the choice was between black bolts which are cheap—but because of their rough fit allow some slipping of the pieces joined—and fitted bolts—which because of the machining required are expensive. The development of high-tensile bolts thus marked an important advance. These press the pieces together, and thus add frictional forces to the shear strength of the bolt. Evidently there is a danger that the high tension applied to the bolt might overstress the material in the bolt itself. The bolts are therefore tightened with a torsion wrench, which applies a specified force to the bolt, but slips when this is exceeded. The wrench requires calibration at regular intervals.

Joining the members of aluminum structures is accomplished in the same way. However, riveting has much wider use, because small-size rivets can be driven cold, which is cheap and convenient. Moreover, inaccessible parts of the structure can be joined with explosive rivets; these contain a small charge which explodes when the accessible head of the rivet is hit, and causes the other (inaccessible) part to blow out and fix the rivet.

Some aluminum alloys cannot be welded because the heat damages the crystal structure. Thus the form of jointing should be considered when selecting the alloy to be used. Since the range of available aluminum alloys is much larger than for steel, this should present no difficulty.

Bolted joints are comparatively rare in aluminum structures. Steel bolts may be used if they are plated with cadmium, or at least galvanized (i.e., zinc-plated), to avoid electrolytic corrosion.

Example 7.6.   *Two steel plates are to be joined by two $\frac{1}{4}$-in fillet welds, 6 in long, as shown in Fig. 7.9a. Determine the strength of the weld if the maximum permissible shear stress in the weld metal is 15,800 p.s.i.*

In determining the effective size of the weld, the penetration of the weld metal into the structural material and the "reinforcement" of the weld are ignored. The cross-sectional dimensions of the weld are thus its length and its throat (Fig. 7.9a and b).

Consequently the strength of the weld is $15,800 \times 2 \times 6 \times 0.177 = 33,560$ lb.

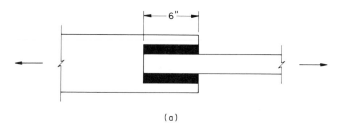

(a)

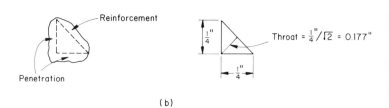

(b)

Fig. 7.9.
Design of fillet weld (Example 7.6). The dimensions of the weld are its length and its throat, neglecting the reinforcement and the penetration of the weld metal.

Effect of the Structural Material on Design

This calculation is very simple; but the design of welds is a matter of experience, rather than mechanics. Badly designed welds can distort a thin metal structure, or cause metallurgical damage to the thick parts of a metal structure. Readers should visit a steel fabricating plant, and talk to people experienced in the practice of welding.

Example 7.7.   *Determine the strength of a joint between two $\frac{1}{4}$-in plates (a) lap-jointed with two $\frac{3}{4}$-in diameter high-strength bolts, and (b) butt-jointed with two cover plates, each connected by two $\frac{3}{4}$-in diameter high-strength bolts.*

The joint is shown in Fig. 7.10.

Taking the maximum permissible shear stress for high strength bolts as 15,000 p.s.i., the value of one $\frac{3}{4}$-in diameter bolt sheared through once only (or in *single shear*) is $15,000 \times 0.441 = 6,615$ lb.

In the lap joint the bolts are in single shear, and thus the strength of the joint is $2 \times 6,615 = 13,230$ lb.

In the butt joint with two cover plates, each bolt would have to shear through *twice* before the joint could fail, and the bolts are thus in *double shear*. Consequently the strength of the joint is $4 \times 6,615 = 26,460$ lb.

The $\frac{3}{4}$-in bolts require holes $\frac{1}{16}$ in larger in the plate, so that the area of $\frac{13}{16} \times \frac{1}{4} = 0.203$ sq in must be deducted from the plate to give its effective cross-sectional area as a tensile member.

Example 7.8.   *Determine the strength of the same joint if $\frac{3}{4}$-in diameter hot-driven rivets are used.*

A high-strength bolt holds the plates together by friction, so that bearing of the bolt against the hole in the plate need not be considered; a rivet fills the hole, and failure may occur if the bearing pressure between the plate and the rivet is too high.

Taking the maximum permissible shear stress of rivets as 15,000 p.s.i., and the maximum permissible bearing stress as 48,500 p.s.i., the strength of a $\frac{3}{4}$-in rivet in single shear is, as before, 6,615 lb. Its bearing on a $\frac{1}{4}$-in plate is $48,500 \times \frac{3}{4} \times \frac{1}{4} = 9,094$ lb.

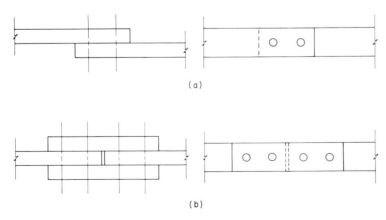

(a)

(b)

Fig. 7.10.
Design of riveted and bolted joints (Examples 7.7 to 7.9)

Evidently bearing is not critical for the joint in Fig. 7.10a, which is governed by single shear. Its strength is 13,230 lb.

Bearing is critical for the joint in Fig. 7.10b, since the bearing strength is less than the strength in double shear. Consequently the strength of the joint is $2 \times 9,094 = 18,188$ lb.

A $\frac{3}{4}$-in rivet requires a hole of $\frac{7}{8}$-in diameter, and this must be deducted from the plate to give its effective cross-sectional area as a tension number.

Example 7.9. *Determine the strength of the same joint if $\frac{3}{4}$-in diameter common bolts are used.*

Common bolts do not provide the same friction grip as high-strength bolts, so that bearing must be considered. Furthermore the maximum permissible shear stress is lower than for high-strength bolts and for rivets. Taking the maximum permissible shear stress as 10,000 p.s.i., and the maximum permissible bearing stress as 48,500 p.s.i., the bolt has a strength in single shear of 4,410 lb, and in bearing on a $\frac{1}{4}$-in thick plate of 9,904 lb.

Shear is thus critical for both types of joint. The lap joint has a strength of 8,820 lb, and the butt joint of 17,640 lb.

7.5. The Design of Simple Structures in Timber, Steel, and Aluminum

We have now covered all the basic elements in the design of these structures. The principal problem is that of determining the forces in trusses or the bending moments in beams. The translation of the forces and moments into material sizes is relatively simple, since tables are available for the purpose; they allow for buckling of compression members and for the reduction in the size of tension members by rivet or bolt holes in the tension members.

Beam design tables usually allow for shear and deflection by means of a limiting line to mark off those spans that produce excessive deflections or shear stresses, even though the stresses caused by the bending moment are within the permissible range. Allowance is also made for reduced stresses in the compression flange where the slenderness ratio is high enough for failure by buckling.

Since the calculations are always of the same kind, and only the dimensions of the particular structure vary, the design of simple structures will be done increasingly by computer. At the time of writing this is already so for a large proportion of structures in steel, which is the most standardized material.

The performance of the detailed calculations is therefore not nearly as important as the understanding of the principles involved. The basic structural decisions must still be made by the designer, since the computer can work only within the limits of the instruction given to it.

The computer can, however, assist in making a choice between alternative forms of construction. For example, it would be intolerably tedious to run through the calculations for the same structure in steel, timber, and concrete (see Sections 7.9 to 7.13) using a slide rule or desk calculator. Moreover, it would be necessary to repeat most of the calculations. The computer, however, can re-use the same program, so that a considerable part of the work need not be repeated. Since the complete solution of a simple structure occupies only a few minutes of computer time, repetition is worthwhile. The designer can then compare the various solutions, and choose the best. The computer can even be programmed to optimize the solution for cost; however, the decision on intangible factors, such as appearance, still rests with the designer.

7.6. Open-web Steel Joists

These are light welded trusses (Fig. 7.11) of the Warren type (see section 1.5). The web consists of a light member, usually a circular section, which alternates between the top and bottom flange at 45°, and the strength of this web is often critical for design, because of its lightness as compared with the solid plate. Otherwise, the joist behaves like an ordinary steel beam. Because of their lightness and standard dimensions, they are particularly useful for single-story buildings of modular design.

Fig. 7.11.
Open-web steel joist.

Open-web steel joists are mass-produced and the manufacturers supply tables of permissible loads, which usually indicate whether a particular figure is governed by the bending moment, the shear force, or the admissible deflection (usually limited to $1/360$ of the span). The designer need only look up the appropriate joist in the table.

7.7. Light-gauge Steel and Aluminum as Structural Materials

Light-gauge steel structures are cold-formed from steel sheet (Fig. 7.12). These may consist of shapes imitating hot-rolled angles, T-sections, channels or I-sections, or completely new shapes. Light-gauge steel, because of its thinness, has a tendency to buckle (see Section 7.1), e.g., the outstanding legs of the I-section of Fig. 7.12d) form cantilevers, and the compression flange may show a local buckling failure before the bending strength of the section is reached. The strength is increased by strengthening the cantilevered ends of the flange (Fig. 7.12e). A "top hat" section (Fig. 7.12f) is particularly useful in buildings because when it is fixed to a bottom surface it provides a duct for wiring, etc.

As the length of the top flange increases, its slenderness ratio can be reduced by rolling in a corrugation (Fig. 7.12g). It can be turned into a closed box before installation, to hold some of the services (Fig. 7.12h).

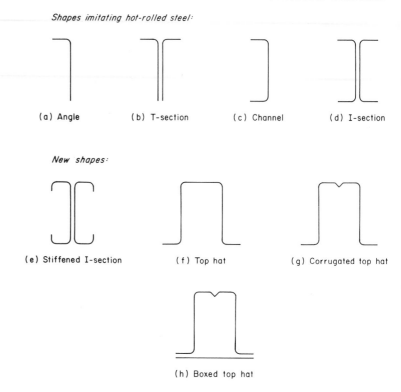

Shapes imitating hot-rolled steel:

(a) Angle   (b) T-section   (c) Channel   (d) I-section

New shapes:

(e) Stiffened I-section   (f) Top hat   (g) Corrugated top hat

(h) Boxed top hat

Fig. 7.12.
Light-gauge steel sections.

The design of light-gauge steel structures is exactly the same as for those assembled from hot-rolled sections, except that buckling becomes much more important as a design factor due to the high slenderness ratios employed; this can be reduced in floors by using a concrete topping, properly bonded through corrugations in the sheet. The cost of fabrication may be higher, but there is considerable saving in weight, which is particularly valuable when handling is a problem. Service ducts can be incorporated into a light-gauge steel floor, and fittings can be easily attached to the sheet.

Aluminum can also be cold-formed from sheet; but a wide range of sections can be more economically produced by extrusion—i.e., by squeezing an aluminum ingot through a die. This is done at a much lower temperature than that required for the rolling of steel, so that the aluminum surface has a smooth finish which does not require cleaning or painting. Since aluminum has a modulus of elasticity only one-third that of steel, buckling enters more prominently into its design.

In locations where its corrosion resistance is important, such as window frames and curtain walls, aluminum has largely displaced steel because stainless steel is more expensive, and ordinary steel has a high maintenance cost. For structural purposes, however, it is only marginally economical in buildings. It has particular value where lightness is important, e.g., in church steeples lifted into position by helicopter, or in prefabricated buildings moved to inaccessible places by air transport.

7.8. Stress-graded and Laminated Timber

Timber has traditionally suffered from the uncertainty of its natural defects. A large, loose knot in a 2-in by 3-in section can reduce the strength so seriously that a normal factor of safety would not be sufficient. It has thus been normal practice to grade timber by visual examination, but some defects such as gum pockets or sloping grain, are not always visible on the surface. The maximum permissible stresses of ordinary timber are much below the strength of defect-free timber, to allow for undetected flaws.

Stress grading is performed by subjecting every single piece, after milling, to a bending test. Experiments show a good correlation between deflection and strength, and the machine marks each piece according to its deflection. The timber is then graded into strength groups by its marking. This procedure greatly increases the certainty of the strength estimate, and it is thus possible to use a lower factor of safety, i.e., a higher permissible stress (see Section 2.2).

Even assuming that the strength of timber could be assessed with perfect precision, it would still have the weakness across the grain which results from its fibrous structure. This can, however, be removed by laminating timber in alternate layers at right angles. Thus plywood has essentially the same properties as other elastic materials. The structural use of plywood is dependent on the strength and permanence of the glue, and the development of structural plywood is thus as much an achievement of the plastics industry as of the timber industry.

Lamination of timber is useful even when the pieces are glued together with the grain in the same direction. Very large pieces of timber are difficult to obtain, because trees grow only to a certain size, and it may take centuries to reach the maximum. Even when large timbers are obtainable, they are difficult to transport. Large sizes can, however, be built up from planks by gluing (see Fig. 6.13). Moreover, thin planks can be bent easily; thus curved sections can be produced in laminated construction which would be impossible to produce without glue, because thick sections cannot be bent, and curved shapes cannot be cut from straight beams (Fig. 7.13)

Laminating successfully overcomes most of the traditional weaknesses of timber, since it is unlikely that several laminations would have natural defects at the same section. If there are sufficient laminations, then a local weakness in one is of little importance.

Fig. 7.13.
Three-pinned arch formed from laminated timber.

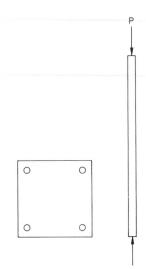

P

Fig. 7.14.
Reinforced concrete column.

*7.9.   Reinforced Concrete Columns

Since concrete design is a more complex subject, we suggest omission of the remainder of this chapter at a first reading.

Concrete is an inexpensive and bulky material, and most concrete columns have therefore a low slenderness ratio. Buckling is not a major design factor. However, the interaction of the concrete and the steel reinforcement poses a new kind of problem.

Let us consider a column symmetrically loaded by a force P (Fig. 7.14). The concrete has a cross-sectional area A_c, and it is reinforced with four bars, area A_s. Consequently

$$P = f_c A_c + f_s A_s \qquad (7.9)$$

where f_c and f_s are the stresses in the concrete and the steel.

Since the two materials are elastically and uniformly compressed by the same force, the strain in the steel and the concrete is the same.

$$e_c = e_s \qquad (7.10)$$

Substituting stresses by using Hooke's law

$$\frac{f_c}{E_c} = \frac{f_s}{E_s} \qquad (7.11)$$

The ratio of the moduli of elasticity of steel and concrete is generally lower than the ratio of the maximum permissible stresses for steel and concrete; elastic design consequently does not take full advantage of the strength of the steel. Reinforced concrete columns are therefore designed by the ultimate strength method.

Let us consider the behavior of the column beyond the elastic limit (Fig. 7.15). Since concrete has a lower modulus of elasticity, its stress-strain diagram is less steep, but both concrete and steel reach their elastic limit, B, at roughly the same strain, viz. 1×10^{-3}, approximately. The concrete then deforms beyond the elastic limit and disintegrates at D, at a strain of roughly 3×10^{-3}.

The steel yields at B, at a strain of about 1×10^{-3}, and then continues to deform at a constant yield stress, f_y, until C. It then strain-hardens, and eventually breaks at D. However, this occurs at a strain of about 200×10^{-3}, i.e., more than 60 times the ultimate concrete strain, and well outside Fig. 7.15 as drawn.

Evidently the greatest load that the concrete can sustain occurs at the strain e_u, and it is $0.85 f'_c A_c$. The load carried by the steel at the same strain is $f_y A_s$. Although the stress in the steel eventually increases beyond f_y, this does not influence the strength of the column, since the concrete has started to disintegrate prior to that. The steel in a reinforced concrete column is not acutally broken

Effect of the Structural Material on Design

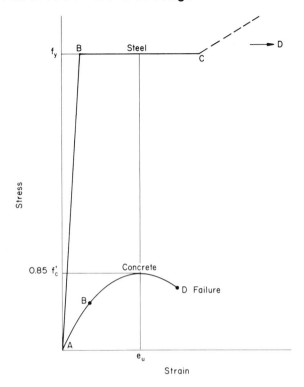

Fig. 7.15.
Reinforced concrete design is based on ultimate strength.

The stress-strain diagrams of steel and concrete are plotted to the same scale. The steel and concrete are elastic from A to B. As it happens, the elastic limit of both is at a strain of approximately 1×10^{-3}. The steel then yields, and its ultimate strain is very large, about 200×10^{-3}. The concrete fails at a strain of about 3×10^{-3}.

The maximum load of a column occurs at the strain e_u.

when the column collapses, and its maximum useful stress is the yield stress.

The maximum column load is therefore

$$P' = 0.85 \, f'_c A_c + f_y A_s \tag{7.12}$$

In this equation f'_c is the strength of the concrete, as obtained from cylinder tests. The strength of concrete in columns and beams is about 15% lower, because the strength of brittle materials is reduced as the size of the test pieces is increased. This is because failure occurs at an incipient crack, and the larger the test piece the greater is the probability of a flaw that might start a crack at a lower load.

The factors to be used to convert the load at which the column fails into the maximum permissible load are specified in the *Building Code Requirements for Reinforced Concrete (ACI 318-71)*, published by the American Concrete Institute in 1971. This is commonly called the *ACI Code*.

We multiply the load P' by a *capacity reduction factor* $\phi = 0.70$, which allows for simplifications in the structural analysis, variations in the quality of the materials, and inaccuracies in the formwork or the placement of the reinforcement, which produce minor variations in the cross-sectional dimensions.

We multiply the actual dead and live loads carried by the structure by *load factors*, which are 1.4 for the dead load, P_D, (which is known with considerable accuracy) and 1.7 for the live load, P_L.

$$P_u = 1.4P_D + 1.7P_L$$

$$= \phi P' = 0.70P' = 0.70\left(0.85\,f_c'A_c + f_yA_s\right) \qquad (7.13)$$

Example 7.10. *Calculate the live load permissible for a column 12 in square, reinforced with four No. 8 bars. The specified (ultimate) compressive strength of the concrete is 4,000 p.s.i., the specified yield stress of the reinforcement is 60,000 p.s.i., and the actual dead load is 120,000 lb.*

We have: $P_D = 120,000$ lb; $f_c' = 4,000$ p.s.i.; $f_y = 60,000$ p.s.i.; and A_s (from Table C1, Appendix C) = 3.141 sq in.

The cross-sectional area of the concrete is the gross cross-sectional area of the column minus the cross-sectional area of the steel

$$A_c = 12^2 - 3.141 = 140.86 \text{ sq in}$$

From Eq. (7.13)

$$1.4 \times 120,000 + 1.7P_L = 0.70(0.85 \times 4,000 \times 140.86 + 60,000 \times 3.141)$$

This gives the permissible (or service) live load for the column

$$P_L = 176,000 \text{ lb}$$

Even when the entire column is too short to buckle, the individual reinforcing bars have very high slenderness ratios. They must thus be restrained by ties at regular intervals (Fig. 7.16).

The size and spacing of these ties is specified in Clause 7.12.3 of the ACI Code. For this column we require No. 3 bars at 12 in centers.

In actual fact, the design of reinforced concrete columns is much more complicated, because the ACI Code requires that each column shall be designed for a minimum load eccentricity of $0.10\,h$ (where h is the length of side of a square column, or of the shorter side of a rectangular column). This is because it is as difficult to achieve perfectly concentric column loading as it is to balance a needle on its point.

The design of eccentrically loaded reinforced concrete columns for ultimate strength is very complicated, and the theory is beyond the scope of this book. It may be found in any book specifically dealing with reinforced concrete design (e.g., P. M. Ferguson,

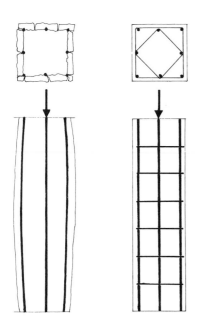

Fig. 7.16.
Ties are needed at frequent intervals in reinforced concrete columns to prevent buckling of the longitudinal bars and consequent bursting of the cover. Multiple ties are needed if there are more than four longitudinal bars, to ensure that every bar is adequately restrained.

Effect of the Structural Material on Design

Reinforced Concrete Fundamentals, Wiley, New York, 1973, p. 497 ff.). Because of the complexity of the equations, reinforced concrete columns are rarely designed from first principles. Instead, column design charts or tables are used and a large number of tables or charts is needed to cover all likely combinations of the variables. Column Tables are included in the *Design Handbook* published by the American Concrete Institute (Publication SP 17-73, Detroit 1973, Volume I, pp. 267–323).

The ACI Code uses ultimate strength design only for the reinforced concrete sections, but not for the entire structure. The direct and shear forces and the bending moments are calculated by the elastic method, not by ultimate strength (see Section 2.2 and Chapter 8). This mixture of elastic and ultimate strength design is not logical, and it may be rather confusing to students; however, it produces concrete columns which are perfectly safe and use far less material than those designed by the elastic Eq. (7.9) for the column cross-section.

We do not use ultimate strength for the entire reinforced concrete frame, because of the brittle nature of concrete (see Section 6.8 and Figs. 6.21 and 6.22). Small steel frames can be designed entirely by ultimate strength (this is described in specialized textbooks, e.g., L. S. Beedle, *Plastic Design of Steel Frames*, Wiley, New York, 1958, pp. 1–104). This "plastic design" method utilizes the experimental fact that plastic "hinges" form at the most-highly stressed sections as the steel reaches the yield point. These hinges allow rotation like the hinges considered in Chapters 4 and 5 without actual rupture of the steel. When enough hinges have formed to turn the structure into a mechanism, it collapses.

In a reinforced concrete frame, the concrete is liable to disintegrate before enough hinges have formed to turn the structure into a mechanism. Hence the collapse load calculated from the plastic theory may not be reached. The ACI Code therefore does not permit the use of ultimate strength design for determining the moments and the forces in a reinforced concrete frame, but it does allow the design of the cross sections of columns, beams, and slabs by ultimate strength, and this has become normal procedure.

*7.10. Design of Rectangular Reinforced Concrete Beams and Slabs

Let us consider a rectangular beam of width b and overall depth D, reinforced with an area of steel A_s, at a depth d (Fig. 7.17). Since the overall depth, D, does not enter into the strength calculations, whereas the depth, d, to the center of the reinforcement does, d is commonly called the *effective depth*.

As in Section 6.4, we assume that plane sections remain plane, so

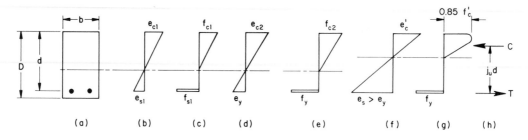

Fig. 7.17.
Elastic and ultimate strength theory for reinforced concrete beams and slabs.

(a) Cross-sectional dimensions of the section.

(b) and (c) Variation of strain and of stress in the elastic range. The concrete stress f_{c1} and steel stress f_{s1} are both elastic.

(d) and (e) Variation of strain and stress when the steel begins to yield at the strain e_y. The steel has just commenced plastic yield at a constant yield stress f_y, but the concrete stress is still elastic at a stress f_{c2}.

(f) and (g) Variation of strain and stress at the ultimate load, when the concrete strain reaches its ultimate value e'_c. The maximum concrete stress is $0.85\,f'_c$. The steel strain is above e_y, but the steel stress is still equal to f_y.

(h) Resultant horizontal forces at the ultimate load. The force C acts at the center of gravity of the compressive stress diagram in (g), and the tensile force T acts at the center of the reinforcement.

that strain varies proportionately with the distance from the neutral axis (see Fig. 6.7). Because of the asymmetry of the section, the neutral axis is not halfway down, but at some unknown distance.

At the bending moment corresponding to the maximum permissible load (or service load), the beam behaves elastically (Fig. 7.17b and c). The strain varies from e_{c1} at the compression face of the concrete to e_{s1} at the center of the reinforcing steel. The stress in the concrete is proportional to the strain, with a maximum compressive stress of f_{c1}. As we have seen in Section 6.1, the strength of the concrete in tension is only about one-tenth of that in compression. We overcome this weakness by introducing steel reinforcement, and we assume that the steel takes the whole of the tension; the error in neglecting the tensile strength of the concrete is negligibly small. Thus the variation of stress is represented by Fig. 7.17c.

If we were designing the beam elastically, this is the diagram we would use. The 1971 ACI Code still permits elastic design, but design aids are now normally in terms of ultimate strength design which has become normal procedure since 1971. We will therefore consider the behavior of the beam when the bending moment is increased beyond the elastic range.

The failure of reinforced concrete due to bending may be initiated by the crushing of the concrete (primary compression failure) or by the yielding of the steel (primary tension failure). In the first case the failure occurs suddenly when the concrete crushes at the point D (Fig. 7.15). This is an undesirable type of failure because it does not give warning of impending collapse.

In the second case, the steel starts to yield at the point B (Fig. 7.15). The steel then continues to yield along the line BCD on the steel diagram, while the strain in the concrete increases until it reaches the point D on the concrete diagram, when the concrete is crushed and the beam fails. The steel is still in one piece, because its breaking strain is very high (at least 100×10^{-3}, or 30 times as much as the crushing strain of the concrete), so that it is never reached in reinforced concrete. Failure is therefore slow, while the neutral axis gradually rises until the concrete is crushed. In the case of a primary tension failure, it is therefore likely that there will be sufficient warning to evacuate the building and perhaps repair the structure before actual collapse takes place.

A primary tension failure occurs when there is only a little reinforcing steel to resist the tensile force T in Fig. 7.17h, so that the steel is highly stressed. As the amount of steel is increased, the tensile force is distributed over a greater cross-sectional area, and the steel stress is reduced relative to the concrete stress. At a certain ratio of the area of tension reinforcement to that of concrete, the concrete reaches its crushing strain, e'_c, at the same load that initiates yielding of the steel at the strain e_y. This is called a balanced reinforcement ratio, ρ_b, and it defines the transition from a desirable primary tension failure to an undesirable primary compression failure. The ACI Code requires that the actual reinforcement ratio, ρ, in a beam or slab may not exceed $0.75 \rho_b$. In practice, this presents no problems in the design of slabs and few problems in the design of beams since ρ_b is uneconomically high for most purposes. Design charts and tables for reinforced concrete slabs and beams normally have a cutoff line at $\rho = 0.75 \rho_b$.

We will next consider the primary tension failure of the beam shown in Fig. 7.17. When the strain in the steel reaches the yield strain, e_y, (corresponding to point B on the steel diagram in Fig. 7.15), the beam ceases to behave elastically. The strain and stress distribution is now as shown in Fig. 7.17d and e. When the bending moment is further increased, the steel strain increases beyond e_y, but the steel stress remains equal to the yield stress, f_y, as long as we are on the portion BD of the steel stress-strain curve in Fig. 7.15.

The tensile force, T, in the beam is the stress in the reinforcement, f_y, acting on the area of tension steel, A_s.

$$T = f_y A_s \tag{7.14}$$

The compressive force, C, in the beam acts at the center of gravity of the compressive stress diagram in Fig. 7.17d. The distance between C and T is called the lever arm. For horizontal equilibrium the two forces C and T must be equal.

The resistance moment of the section is the product of the (equal) forces C or T, and the lever arm. But we noted that the steel stress remains constant and equal to f_y, and the tensile force, T, and the compressive force, C, therefore remain constant. The resistance moment can thus increase only if the lever arm gets longer. This means that the neutral axis rises.

The maximum resistance moment is attained when the concrete reaches the crushing strain, e'_c (D on the concrete diagram in Fig. 7.15). We then have the strain and stress distribution shown in Fig. 7.16f and g. The maximum resistance moment

$$M' = T\,j_u d = f_y A_s\,j_u d \tag{7.15}$$

where $j_u d$ is the length of the lever arm.

We must now, as we did for columns in Section 7.10, multiply by a capacity reduction factor which, for bending, is $\phi = 0.90$, so that the (factored) ultimate resistance moment

$$M_u = \phi M' = 0.90\,f_y A_s\,j_u d \tag{7.16}$$

The derivation of the formula for j_u is beyond the scope of this book, but it may be found in any book on reinforced concrete design (for example, Ferguson, *op. cit.* p. 41). The formula is

$$j_u = 1 - \frac{0.59\,f_y A_s}{bd f'_c} \tag{7.17}$$

For normal reinforced concrete slabs $j_u = 0.90$ is sufficiently accurate.

Example 7.11. *Determine the amount of reinforcement required for a concrete slab, $5\frac{1}{2}$ in thick, which is subjected to a bending moment due to the dead load of 50,000 lb in per ft width, and to a bending moment due to the live load of 30,000 lb in per ft width. The specified (ultimate) compressive strength of the concrete is 4,000 p.s.i., and the specified yield strength of the reinforcement is 60,000 p.s.i.*

To determine the effective depth, which is measured to the *center* of the reinforcement, we must assume the bar diameter. However, the error caused by a slightly erroneous assumption is not serious. Let us assume that the bars are No. 5 (nominal diameter $\frac{5}{8}$ in). The reinforcement requires concrete cover to protect it from rusting, and to ensure proper bond between the concrete and the steel. The minimum cover for a slab is $\frac{3}{4}$ in. The effective depth is thus (Fig. 7.18)

$$d = 5.50 - \tfrac{3}{4} - \tfrac{1}{2} \times \tfrac{5}{8} = 4.44 \text{ in}$$

The load factors are the same for all structural members. As we noted in Section 7.9, the load factor for the dead load is 1.4 and the load factor for

Effect of the Structural Material on Design

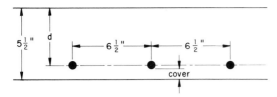

Fig. 7.18.
Cover in reinforced concrete slabs. (Example 7.11). The overall depth = effective depth + bar radius + cover.

the live load is 1.7. Therefore the ultimate resistance moment

$$M_u = 1.4 \times 50,000 + 1.7 \times 30,000 = 121,000 \text{ lb in}$$

Let us assume an ultimate lever arm ratio $j_u = 0.90$. From Eq. (7.16)

$$121,000 = 0.90 \times 60,000 \times A_s \times 0.90 \times 4.44$$

which gives $A_s = 0.561$ sq in per ft.

From Table C2 (Appendix C) we chose No. 5 bars at $6\frac{1}{2}$ in centers (0.565 sq in per ft). No. 4 bars would be at 4 in centers, and No. 6 bars at 9 in centers; the first is too close for placing the concrete satisfactorily, and the second might leave a little too much unreinforced concrete between the bars.

We will check our assumption for j_u. Since the reinforcement is stated as an area per foot width, $b = 1$ ft = 12 in. From Eq. (7.17)

$$j_u = 1 - \frac{0.59 \times 60,000 \times 0.565}{12 \times 4.44 \times 4,000} = 0.906$$

Since concrete slabs are very common structural elements, charts and tables for determining their depth and reinforcement are readily available in textbooks, in handbooks, and in brochures available from material suppliers.

It has become common practice to support concrete slabs directly on the columns (Fig. 1.59). This *flat plate* structure is neat in appearance and very economical to construct. It requires only a flat surface for formwork, and partitions can be placed in any position under the slab (provided the weight of movable partitions has been allowed for in the design). However, the computation of the bending moments (see Section 8.7) is more complicated, and special shear reinforcement may be required around the columns, to prevent them from punching through the slab (Figs. 1.60 and 7.19).

The shear stresses around the columns can be reduced by using enlarged column capitals or large column heads as illustrated in Fig. 1.59. This *flat slab* structure is suitable for parking garages and warehouses, where the heavy loading produces particularly high shears. The column capitals can only be used in open interiors,

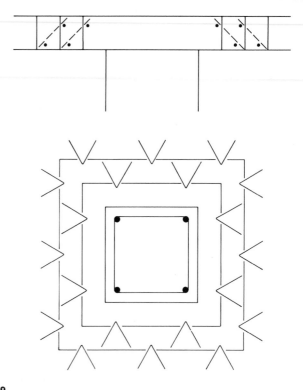

Fig. 7.19.
Flat plates sometimes require shear reinforcement around the column circumference. This can be avoided by using thicker slabs, larger columns, or enlarged column capitals.

where they look appropriate and do not interfere with partitions. The flat slab can be thickened around the column by using a *dropped panel*, which surrounds the column head. The dropped panel increases the floor thickness where the shear force is highest, and thus reduces the shear stresses.

The traditional type of concrete construction, still widely used, supports the slabs on beams, and the slabs may span in one direction (Fig. 1.48) or in both directions (Fig. 1.49). The beams may be supported directly on the columns, as in Fig. 1.59, or on larger primary beams (Fig. 1.49).

Supporting the slab on beams allows an increase in the column spacing or (for the same spacing) a reduction in the thickness of the slab, because the bending moments and shear forces are reduced. Thus a flat plate is usually 6 to 10 in (150 to 250 mm) thick, while a beam-supported slab tends to be 4 to 6 in (100 to 150 mm) thick. Reduction in thickness is desirable for very tall buildings, because the weight of reinforced concrete equals a force of approximately 12 lb per sq ft for each inch (or 23 N per sq m for each mm) of thickness of the slab, and the column sizes would become excessive

Effect of the Structural Material on Design

for thick slabs. On the other hand, a thick concrete slab provides excellent sound insulation, and in low-rise residential buildings a thick structural slab is the cheapest solution to the noise problem; if the insulation is required in any case, then it is economical to utilize the structural slab for the purpose (see Section 8.7).

Since the concrete below the neutral axis is assumed to be cracked in tension, and not to contribute to the strength of the slab, it can be partially removed to reduce the weight of the slab. We thus obtain the ribbed slab for spanning in one direction and the waffle slab for spanning in both directions (Fig. 7.20). Instead of

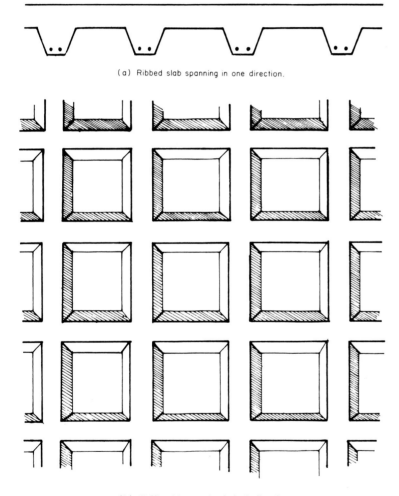

(a) Ribbed slab spanning in one direction.

(b) Waffle slab spanning in both directions.

Fig. 7.20.
Reinforced concrete slabs can be made lighter by removing some of the concrete from the tension face (where it is presumed to have cracked). (See also Figs. 1.48, 1.49, and 6.27.)

having relatively closely spaced reinforcement, two bars of larger size are used in each rib. The ribs can be shaped by reusable plastic forms, and the waffle pattern provides a good ceiling for heavy structures, such as parking garages.

Another way of reducing the weight of the concrete is to use lightweight aggregate. This is more expensive than gravel or crushed rock, but the extra space gained from the reduced column sizes may compensate for this if the building is tall.

*7.11. Reinforced Concrete Beam-and-slab Floors

Equations (7.14) to (7.17) can be used for rectangular beams, as well as slabs; but in practice beams of this type are rare. It is one of the advantages of reinforced concrete that it is cast in one piece, so that the slab and the supporting beams interact, and the slab on top of the beam also contributes to the bending resistance. Although the slab spans between the beams at right angles, the resulting stresses have no components in the direction of the beam, because the cosine of 90° is zero (see Sections 3.2 and 6.8). Thus we can use the same concrete twice over.

The effective width of the beam,—i.e., the extent to which the slab contributes to its bending resistance,—has been determined experimentally, and it is defined in the ACI Code (ACI 318-71, Clause 8.6). Since we have a very large width of beam, we have far more concrete on the compression face than we need, and its stress is not critical for the design. We may thus make two simplifications:

(i) we neglect the concrete in the web (i.e., the beam proper)—even that in compression,—since the width is small and the stresses near the neutral axis are low;

(ii) we assume that the resultant compressive force acts halfway down the slab, although actually the center of gravity of the stress diagram is a little nearer to the top of the slab: this approximation is on the safe side. The resistance arm is thus $d - \frac{1}{2}t$ (Fig. 7.21). Beams of this type are termed T-beams.

In many T-beams the neutral axis falls inside the flange, so that the beam is a rectangular beam of width b (i.e., the full width of the flange). This increases the length of the lever arm, and reduces the area of reinforcement. However, the use of $d - \frac{1}{2}t$ for the lever arm gives a good approximation which errs on the safe side.

The maximum resistance moment

$$M' = T\left(d - \tfrac{1}{2}t\right) = f_y A_s\left(d - \tfrac{1}{2}t\right) \tag{7.18}$$

and the (factored) ultimate resistance moment

$$M_u = \phi M' = 0.90\, f_y A_s\left(d - \tfrac{1}{2}t\right) \tag{7.19}$$

Effect of the Structural Material on Design

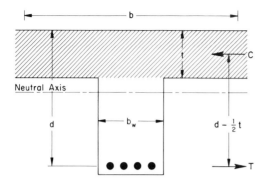

Fig. 7.21.
Strength of reinforced concrete beam-and-slab floor at mid-span. The resistance moment is provided by the concrete in the compression flange, shown shaded, and the tension reinforcement. The effective width is the width of the flange, b, which is provided by the floor slab. The much smaller width of the beam itself, b_w, does not enter into the design for bending. Casting the entire concrete floor in one piece thus greatly adds to its strength.

Example 7.12. *Determine the reinforcement required for the T-beams of a reinforced concrete floor structure, if the ultimate bending moment at mid-span is 1,350,000 lb in. The beams have an overall depth of 12 in, and an effective width of 48 in; the slab is 4 in thick. The specified yield stress of the reinforcement is 60,000 p.s.i.*

The cover required in beams is $1\frac{1}{2}$ in, and the reinforcing bars are likely to be approximately 1-in diameter. The effective depth is thus

$$d = 12 - 1\tfrac{1}{2} - \tfrac{1}{2} \times 1 = 10 \text{ in}$$

and the length of the lever arm is

$$d - \tfrac{1}{2}t = 10 - \tfrac{1}{2} \times 4 = 8 \text{ in}$$

From Eq (7.19)

$$1,350,000 = 0.90 \times 60,000 \times A_s \times 8$$

which gives $A_s = 3.125$ sq in. This is satisfied by four No. 8 (1-in nominal diameter) bars in each beam (Table C1, $A_s = 3.141$ sq in). The cross section of the beam is shown in Fig. 7.22.

In actual fact, this beam, if analyzed precisely, is a rectangular beam whose neutral axis falls within the flange. The area of reinforcement required is therefore slightly less.

The beams develop negative bending moments near the supports (see Section 8.4), which may require compression reinforcement. Shear reinforcement is generally needed near the supports (see Example 6.8 and Figs. 6.24 and 6.26). The deflection of the beams under the service loads, and the anchorage of the reinforcement

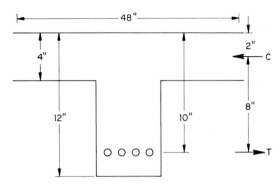

Fig. 7.22.
Design of mid-span reinforcement for beam (Example 7.12).

must also be checked. There is thus a great deal more to the design of reinforced concrete than we can cover in this introductory text, and the proper study of this subject requires a book in itself.

However, this chapter is sufficient for the design of simple reinforced concrete slabs, which do not require compression or shear reinforcement.

*7.12. Eccentrically Loaded Columns

We have so far considered only structural members which are *either* in compression *or* in bending. When a column is loaded eccentrically, or forms part of a rigid frame or arch (see Sections 5.7 to 5.9), it is subjected to combined compression and bending (Fig. 7.23), since the eccentricity causes a bending moment $M = Pe$.

Stress is directly proportional to force and to moment, so that we

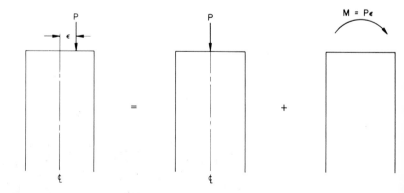

Fig. 7.23.
Eccentrically loaded column. The eccentric load is equivalent to the combined effect of a concentric load and a bending moment.

Effect of the Structural Material on Design

can consider the effect of each separately, and then add the stresses (see Sections 6.3 and 6.4).

The stress due to the compressive force is

$$f_c = \frac{P}{A} \qquad (6.3)$$

and the stress due to the bending moment varies from a maximum compressive stress

$$f_c = \frac{M}{S_c} \qquad (6.12)$$

to a maximum tensile stress

$$f_t = \frac{M}{S_t}$$

where S_c = section modulus relative to the compression face

and S_t = section modulus relative to the tension face

Thus the maximum and minimum stresses in the eccentrically loaded column are

$$f_c = \frac{P}{A} + \frac{M}{S_c} \qquad (7.20)$$

and

$$f_t = \frac{P}{A} - \frac{M}{S_t} \qquad (7.21)$$

(In a symmetrical section $S_c = S_t = S$.)

If both values of f are positive, there is no tension in the section. This happens when the eccentricity is relatively small, so that the section behaves essentially as a column. As the eccentricity increases, the section behaves more like a beam, and tension develops (which is indicated by a negative f).

Example 7.13. *A 12-in square column carries a load of 100,000 lb (a) at its center, (b) 1 in left off center, (c) 2 in left off center, and (d) 6 in left off center, along one center-line of the column. Determine the stresses at the left-hand and right-hand side of the section in each case.*

The cross-sectional area A = 144 sq in, and the section modulus $S = S_t = S_c = 12 \times 12^2/6 = 288$ in.^{3}

The column load P = 100,000 lb. The bending moment varies from 0 to 600,000 lb in.

Thus the top and bottom stresses from Eq. (7.21) are (in p.s.i):

Eccentricity	Left-hand Stress	Right-hand Stress
0	694 C^a	694 C
1 in	1,042 C	347 C
2 in	1,389 C	0
6 in	2,778 C	1389 T

^{a}C = compression; T = tension.

The column is in compression when the load is in the middle third (4 in), i.e., when the eccentricity is ± 2 in (see Section 3.7). Tension develops when $P/A = M/S$, i.e., when $\epsilon = S/A$.

Relatively small eccentricities evidently cause a substantial redistribution of stress. The designer must therefore allow for eccentricities which may result from constructional inaccuracies or rearrangement of the load (see Section 7.9).

*7.13. Prestressed Concrete

We noted in Section 7.10 that concrete is too weak in tension to be used satisfactorily in beams by itself, and we introduced steel reinforcement to take the tensile component of the bending moment. However, *concrete* cracks at a *strain* of approximately 3×10^{-4}; the modulus of elasticity of steel is 30,000,000 p.s.i., and the concrete thus cracks at a *steel stress* of $3 \times 10^{-4} \times 30,000,000 = 9,000$ p.s.i. Since we ordinarily permit a steel stress equal to about half the yield stress (i.e., 20,000 p.s.i.* or more) under the actual service loads, cracking of the concrete is unavoidable in reinforced concrete (Fig. 7.24 top). We do, however, ensure that the cracks do not become too large by limiting the yield stress of steel to be used in reinforced concrete to 75,000 p.s.i.* Thus we cannot use high-tensile steel in reinforced concrete.

Reinforced concrete is by far the most important and widely used material for architectural structures, because it is cheap, versatile, durable, and rigid. However, we can avoid the cracks in the concrete, and we can, if necessary, utilize high-tensile steel in conjunction with concrete by the use of prestressing (Fig. 7.24 b).

We may illustrate the effect of prestressing with a model consisting of separate woodblocks, with thin pieces of rubber between. If we compress the blocks with a screw, we enable them to carry their own weight, as well as a considerable superimposed load. Eventually the tension due to bending overcomes the compressive prestressing force, the joints open up at the bottom, and the prestressed beam collapses into separate blocks.

*The corresponding metric values are 140 MPa and 525 MPa.

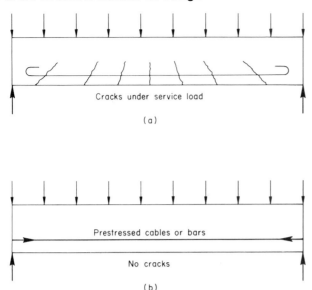

Cracks under service load

(a)

Prestressed cables or bars

No cracks

(b)

Fig. 7.24.
Reinforced concrete cracks under service loads; prestressed concrete cracks only under an overload.

The efficiency of the beam is greatly increased by lowering the point of application of the prestressing force, i.e., by applying an eccentric load to the beam (Fig. 7.25); but we can break the beam by overprestressing. As we increase the pressure of the screw, the joints begin to open up on top; and if we continue to turn the screw the beam collapses.

Eccentric prestressing greatly increases the load-bearing capacity of the beam for the same prestressing force; we can put a much greater load on the beam, before the joints open up at the bottom. It is economical to prestress with as much eccentricity as we can safely employ, since the eccentricity does not cost anything. Since buildings do not have the firm anchorages of the model against which to press, we supply the prestressing force by tensioned steel. Doubling the prestressing force doubles the amount of steel (see Eq. 7.26).

The cost of the prestressing operation has to be added to the cost of the steel and the concrete, and for most architectural applications, prestressed concrete is, at the present time, more expensive than reinforced concrete; but it is advantageous for certain purposes.

There are two distinctly different methods of prestressing. We can stress the steel tendons with a hydraulic jack against anchorages in a stressing bed, and cast the concrete around it. When the concrete has gained sufficient strength, the tendons are released from the outside anchorages, and the force is transferred to the concrete. The tensioned steel thus prestresses the concrete. (Both

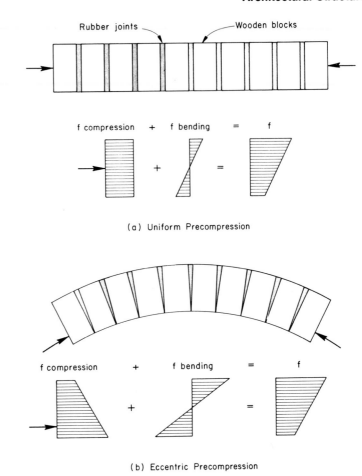

Fig. 7.25.
Model demonstration of prestressing.

the steel and the concrete shorten elastically, and there is consequently some loss of prestress.) This method is called *pretensioning*. Because of the need for external anchorages, it is generally limited to the precasting yard.

The need for outside anchorages can be avoided by using a *post-tensioning* system. A duct with unstressed steel tendons (wires or bars) is placed in the formwork, and the concrete is cast around this duct. Bearing plates are usually embedded in the ends of the molds around the ducts. After the concrete has gained sufficient strength, the tendons are tensioned using a hydraulic jack pressing on the bearing plates, and concrete is therefore compressed at the same time as the steel is tensioned. Wedge-like anchors are used to hold the tendons in a stressed condition after the jack is removed. The tendons may be left *unbonded* in the ducts; however, *bonded tendons*, produced by injecting cement grout into the ducts through apertures in the anchorages, increase the ultimate strength of the beams. Post-tensioning can be done on the building site on cast-in-place concrete, and the tendons can be "draped" to conform to the curve of the bending moment diagram, which greatly increases the load-bearing capacity (see Eq. 7.27).

Since there are no cracks, high-tensile steel with a permissible stress of 150,000 p.s.i. and more can be used. We thus require less than 20% of the steel needed for reinforced concrete. Consequently prestressed concrete made fast progress immediately after World War II, when the steel shortage made economical use of steel more important than cost.

Because high-tensile steel is used, we can also economically employ higher-strength concrete, so that the structure is lighter than one of normal reinforced concrete (but heavier than a steel structure). This is helpful when we deal with long spans, where the weight of the structure is the biggest single load, and it is helpful in precast concrete construction, where it reduces the weights to be handled.

Let us now consider the design of a prestressed beam. For simplicity we will take a symmetrical I-section.

The prestressing force produces a uniform compressive stress P/A. Since it acts eccentrically, it also produces a bending moment $P\epsilon$, where ϵ is the distance of the prestressing force from the center of gravity of the section. The moment $P\epsilon$ causes tension on top and compression on the bottom.

To this we may add the bending stresses produced by the weight of the beam, M_G/S (where M_G is the mid-span bending moment due to the beam's self-weight), since it helps to reduce the stresses during prestressing. We must, of course, ensure that the beam carries its own weight, and does not rest on the ground; but since the bottom of the beam contracts during prestressing, and the beam curves upward, this normally happens automatically.

The stress on top must remain compressive (i.e., positive, or larger than zero), and the stress on the bottom must not exceed the maximum permissible concrete stress, f_c. Thus we obtain (Fig. 7.26a)

Top:
$$\frac{P}{A} - \frac{P\epsilon}{S} + \frac{M_G}{S} \geqslant 0 \qquad (7.22)$$

Bottom:
$$\frac{P}{A} + \frac{P\epsilon}{S} - \frac{M_G}{S} \leqslant f_c \qquad (7.23)$$

When the full load is acting, there is an additional mid-span bending moment M_S, due to the superimposed load. This produces additional compression on top and tension on the bottom. Under full load the top stress must be within the permissible compressive

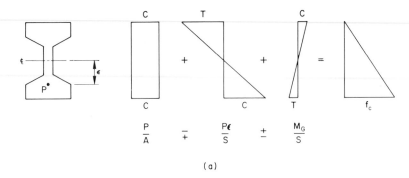

(a)

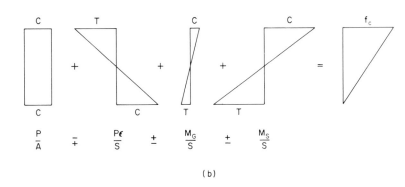

(b)

Fig. 7.26.
Stress distribution in eccentrically prestressed beam.
(a) during pre-stressing;
(b) under full load.
C = compression; T = tension.
The aim is to produce zero stress on top and the full permissible compressive stress at the bottom of the section during prestressing, and to reverse this condition under full load.

range, and the bottom stress must not be tensile. Thus we obtain (Fig. 7.26b):

Top:
$$\frac{P}{A} - \frac{P\epsilon}{S} + \frac{M_\mathrm{G}}{S} + \frac{M_\mathrm{S}}{S} \leqslant f_\mathrm{c} \qquad (7.24)$$

Bottom:
$$\frac{P}{A} + \frac{P\epsilon}{S} - \frac{M_\mathrm{G}}{S} - \frac{M_\mathrm{S}}{S} \geqslant 0 \qquad (7.25)$$

To supply the necessary prestress, we require an area of steel

$$A_s = \frac{P}{f_s} \qquad (7.26)$$

where f_s is the maximum permissible stress for prestressing steel.

The economy of prestressing can be greatly increased by employing the principle of *load-balancing*. We can balance a bending

Effect of the Structural Material on Design

moment M with a prestressing force P and an eccentricity ϵ by making

$$M = P\epsilon \qquad (7.27)$$

If the beam is simply supported, and the load is uniformly distributed, the bending moment varies parabolically from M at mid-span to zero at the supports. If we now drape the cable so that its eccentricity varies parabolically from zero at the supports to a maximum eccentricity ϵ at mid-span, we have a bending moment due to the prestress flexing the beam upward which is exactly equal and opposite to the bending moment due to the load flexing the beam downward. Under the combined action of the load and the prestressing the beam is therefore purely in compression and the net bending moment is zero along the entire length of the beam.

If load-balancing is used in prestressed concrete design, the load to be balanced is usually the dead load plus half the live load. The beam may thus be quite highly stressed during the prestressing operation, but these stresses are greatly reduced as soon as the full dead load is carried by the beam. The beam is flexed upward when there is no live load, and it is flexed downward when the full live load is carried. At half the live load there is no bending moment, and therefore no vertical deflection. This is particularly valuable for structural members, such as flat plates (see Section 7.10), which are liable to large deflections, and thus cause creep problems (see Section 8.9).

Example 7.14. A concrete I-section, with the dimensions shown in Fig. 7.27, is prestressed with 5 sq in of steel, tensioned to 100,000 p.s.i. Determine the stresses during prestressing and under full load.

The section has a cross-sectional area $A = 360$ sq in, and a section modulus $S = 3{,}120$ in.^{3}

The prestressing force, from Eq. (7.26) is

$$P = 5 \times 100{,}000 = 500{,}000 \text{ lb}$$

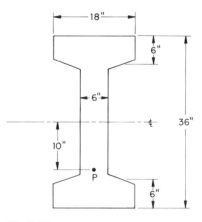

Fig. 7.27.
Dimension of section, and eccentricity of prestressing force (Example 7.14).

Let us assume that the bending moment due to the weight of the beam has been calculated as $M_\text{G} = 1{,}000{,}000$ lb in, and the bending moment due to the superimposed load as $M_\text{S} = 8{,}000{,}000$ lb in.

At the end of the prestressing operation the concrete stresses are at the top (Eq. 7.22)

$$\frac{500{,}000}{360} - \frac{500{,}000 \times 10}{3{,}120} + \frac{1{,}000{,}000}{3{,}120}$$

$$= 1{,}389 - 1{,}603 + 321 = 107 \text{ p.s.i. (compression)}$$

and at the bottom (Eq. 7.23)

$$1{,}389 + 1{,}603 - 321 = 2{,}671 \text{ p.s.i. (compression)}$$

221

Under full load, at the top (Eq. 7.24)

$$107 + \frac{8,000,000}{3,120} = 107 + 2,564 = 2,671 \text{ p.s.i. (compression)}$$

at the bottom (Eq. 7.25)

$$2,671 - 2,564 = 107 \text{ p.s.i. (compression).}$$

In practical prestressed concrete design allowance must be made for the loss of prestress due to shrinkage and creep of the concrete (see Section 8.9), elastic shortening of the concrete (in pretensioned beams only), relaxation of the steel, slip at the anchorages, and friction between the tendons and the duct lining (in post-tensioned beams only).

Furthermore it is necessary to check the strength of prestressed concrete beams both for the elastic (service load) condition *and* for ultimate strength. The behavior of prestressed beams at the ultimate load is similar to that of normal reinforced concrete beams, since the concrete is then cracked. The mechanics of prestressed concrete at the ultimate load is thus quite different from that of elastic, un-cracked beams at the service load.

The proper study of prestressed concrete design thus requires a book in itself, or at least several chapters in a textbook on concrete design; however, this simple example illustrates the basic principle.

Suggestions for Further Reading

This chapter deals only with the *principles* of design in timber, metal, and concrete. The reader who wishes to go into the design details should refer to a specialized book on the subject. There is a large choice, and the following may be suggested:

American Institute of Timber Construction: *Timber Construction Manual.* Wiley, New York, 1966. (This includes the design of laminated timber.) ca. 600 pp.

B. BRESLER and T. Y. LIN: *Design of Steel Structures.* Wiley, New York, 1960. (This includes the design of light-gauge steel.) 710 pp.

G. WINTER et al.: *Design of Concrete Structures.* McGraw-Hill, New York, 1964. (This includes prestressed concrete design.) 660 pp.

A brief elementary treatment for architects, covering all three, is:

H. PARKER: *Simplified Engineering for Architects and Builders.* Wiley, New York, 1967. 361 pp.

Section tables for timber and hot-rolled steel have been listed at the end of the suggested reading list for Chapter 6.

Section tables for light-gauge steel are given in:

Light Gauge Cold-Formed Steel Design Manual, American Iron and Steel Institute, New York, 1956. 91 pp.

Effect of the Structural Material on Design

Section tables for aluminum are given in:

Aluminum Standards and Data, The Aluminum Association, New York, 1968. 173 pp.

Reinforced concrete design is greatly simplified by the charts and tables contained in:

Design Handbook, American Concrete Institute, Detroit, Michigan, 1973. 403 pp.

Problems

†7.1. A sound reflector measuring 20 ft by 30 ft is to be suspended above a concert platform by four symmetrically placed steel rods. Determine the cross-sectional area of each rod, if the combined live and dead load is 75 lb per sq ft, and the maximum permissible steel stress is 25,000 p.s.i.

7.2. Discuss the factors which affect the buckling of a long slender strut. How should a steel strut be fabricated to give it the highest possible buckling strength without unduly increasing its weight?

7.3. Describe the basis of practical formulas for the design of compression members. How does the design of compression members in aluminum, light-gauge steel, hot-rolled steel, laminated timber, and reinforced concrete differ from one material to another?

7.4. Design a joint between a 3 in $\times$ 3 in $\times \frac{3}{8}$ in angle carrying a tensile force of 30,000 lb and a $\frac{1}{2}$-in thick gusset plate (a) using welding, (b) using rivets, (c) using black bolts, and (d) using high-strength bolts. Compare the space occupied by the different joints.

7.5. The open-web joist shown in Fig. 7.28 has a span of 20 ft, and a depth of 10 in between the centers of chords. It is subjected to a uniform loading equivalent to 200 lb per foot run. Calculate the forces in the most highly stressed portion of the bottom chord, and the most highly stressed tension and compression diagonals.

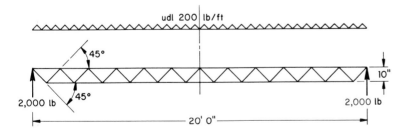

Fig. 7.28.
Problem 7.5.

†Similar problems in metric units are given in Appendix G.

7.6. Discuss the comparative advantages and disadvantages of reinforced concrete flat plate floors and of light-gauge steel floors with a concrete topping.

7.7. Describe the extent to which stress-graded and laminated timber overcome the traditional weaknesses of structural timber. What are the potential future applications of timber in architectural structures?

7.8. Ignoring in this example the eccentricity of the column load, determine the amount of longitudinal reinforcement required for the 24-in square columns on the ground floor of a 40-story flat-plate structure if each column supports a floor area 15 ft square. The dead load is 125 lb per sq ft, and the live load is 50 lb per sq ft. Assume that the column need only be designed to carry 85% of the full live load. The specified compressive strength of the concrete is 5,000 p.s.i. and the specified yield strength of the steel is 60,000 p.s.i.

7.9. A 4-in thick concrete slab cantilevers 6 ft beyond the facade of a building. The greatest ultimate bending moment is 100,000 lb in per ft width. Determine the amount of the main reinforcement required if the specified compressive strength of the concrete is 4,000 p.s.i., the specified yield strength of the steel is 60,000 p.s.i., and $j_u = 0.90$. On what face of the concrete is the reinforcement placed?

7.10. Discuss the relative advantages of (a) one-way slabs supported on beams; (b) two-way slabs supported on beams; (c) flat slabs with enlarged column heads; and (d) flat plates without enlarged column heads. Give a typical contemporary example of the use of each type of concrete slab, indicating its useful range of loading and span. Comment on the reasons for the present popularity of flat plates.

7.11. The ultimate negative bending moment at some point in a 5-in thick flat plate is 120,000 lb in per ft width. Determine the area of reinforcement at that section, using the same constants as in Problem 7.9.

7.12. Figure 7.29 shows the plan of a brick pier. The brickwork

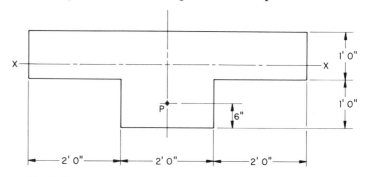

Fig. 7.29.
Problem 7.12.

weighs 120 lb per cu ft, and the pier is 20 ft high. It also carries a point load of 6,000 lb in the position *P*. Calculate the properties of the section about $X - X$ (which passes through the center of gravity), and calculate the stress distribution at the base of the pier.

7.13. The three-hinged frame shown in Fig. 7.30 is subject to a load of 200 lb per foot run, uniformly distributed in plan. Calculate the reactions and the maximum bending moment.

If the timber column is 15 in × 6 in (after dressing), determine the stress distribution in the column immediately below the point B.

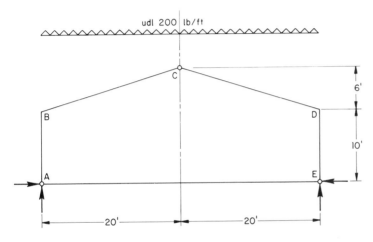

Fig. 7.30.
Problem 7.13.

7.14. A rectangular prestressed concrete beam, 12 in wide by 6 in deep, is prestressed with an eccentricity of 2 in. The initial prestressing force is 72,000 lb. Ignoring loss of prestress due to creep, etc., calculate the variation of concrete stress at the end of the prestressing operation and under full load. The mid-span bending moments are 72,000 lb in due to the weight of the beam, and 144,000 lb in due to the superimposed load.

Design of Building Frames 8

After having spent years in striving to be accurate,
we must spend many more in discovering when to
be inaccurate.

                                        Samuel Butler

In Chapters 4 to 7 we considered the design of the individual members which compose a building frame. We will now discuss the design of the entire frame, using three alternative concepts. In the first we assume that the members are pin-jointed, so that the frame is statically determinate; this is the traditional method for small steel frames. In the second we assume that the floors are restrained, but are designed separately from the columns; this is the traditional method for small reinforced concrete frames. In the third we assume that the frame is rigidly jointed, and the floors and columns are interacting; this is essential for high-rise buildings.

8.1. The Pinned-Frame Concept

Small steel frames are usually designed with flexible connections, unless they are built in an earthquake zone (see Section 8.7). Angle connectors are fitted to the web of the beam in the fabricating shop and to the column on the site (see Section 7.4). This provides a joint that resists the end shear force, but allows rotation of the beam relative to the column and is thus a pin joint (Fig. 8.1).

We may think of the frame as a series of columns cantilevering from the ground which carry simply supported beams. The frame is thus broken up by pin joints into a series of statically determinate units (Fig. 8.2).

Although it is difficult to make columns longer than 30 ft (10 m or 3 stories) in the shop, the vertical joints in the column, made on the building site, are sufficiently rigid to enable the columns to behave as cantilevers.

We have already considered all the elements of pin-jointed frames under vertical loading in Chapters 4 to 7, and the concept of Fig. 8.2 is sufficient for the design of the frame under vertical loading. It is not, however, stable under horizontal loading, and all frames are subject to some horizontal loads due to wind (see Section 2.7). In single-story industrial buildings with only light cladding, the wind

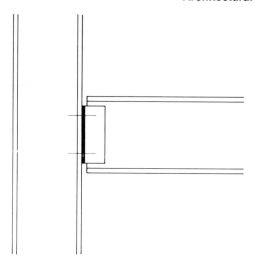

Fig. 8.1.
Pin-jointed beam-column connection in a steel frame. The angles are fitted to the web of the beam in the fabricating shop, and to the column on the site; only the site connection is shown. Sometimes a beam seat is added for ease of construction. This is an angle rigidly fitted to the column, but connected to the beam with only one line of bolts, so that it does not provide a rigid restraint. A small clip angle on top of the beam is also admissible in a pin joint.

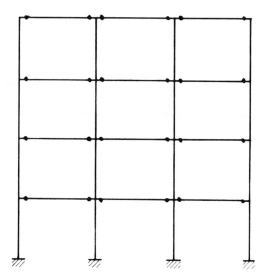

Fig. 8.2.
The pinned-frame concept for vertical loading. The columns are regarded as cantilevers, simply supporting the beams. If the steel frame has flexible connections between the columns and the beams, this is close to reality.

loads are normally resisted by diagonal braces placed inside the cladding or under the roof covering (Fig. 1.38). In small residential and office buildings only a few stories in height, the wind loads can often be resisted by block or brick partitions. When these are not available, the pin-jointed connections must be replaced by semi-rigid or rigid connections (see Section 8.5).

8.2. The Restrained-Floor Concept

The pinned-frame concept can be used for small precast concrete frames; but it is both uneconomical and unsafe for reinforced concrete frames cast on the site.

It is normally cheaper to cast concrete on the site, and thus produce a rigid frame; it is also better, since the frame has lower bending moments under vertical loading and better resistance to horizontal forces.

If the concrete slab is supported on substantial beams, the floor structure is much more rigid than the columns, and the floor can then be regarded as continuous over the columns, and simply supported on them (see Section 7.11).

Figure 8.3 shows the difference between the behavior of a beam continuous over four spans, and four beams simply supported over the same spans. Whereas the curvature of the simply supported beams is concave throughout, and the beams are thus subject to positive bending moments only (see Fig. 5.10), the continuous beam has both convex and concave curvature, and the bending moment thus changes from positive to negative in each span. The tensile strains in the beam are thus on top over the supports and at the bottom near mid-span. Thus is readily illustrated with a continuous beam of timber or plastic, into which slots have been cut to demonstrate the strains (Fig. 8.4).

In a reinforced concrete beam, the reinforcement is required

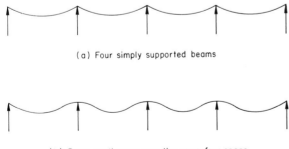

(a) Four simply supported beams

(b) Beam continuous over the same four spans

Fig. 8.3.
Simply supported and continuous beam.

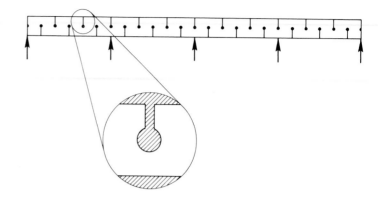

Fig. 8.4.
Demonstration of tensile and compressive strain distribution in a continuous model beam. The strains are shown up by slots cut into the plywood.

where the tension is (see Section 7.10) and the pinned-frame concept of Fig. 8.2 is unsafe because it implies that reinforcement is needed only on the bottom face of the beams.

*8.3. The Theorem of Three Moments

We will now derive the theory of continuous beams, and *suggest that readers omit this section at a first reading, and proceed to the next section, 8.4.*

Let us consider any two adjacent spans in a multispan continuous beam (Fig. 8.5). Because of the continuity, the slope at the support B, just inside the span 1, θ_{B1}, and the slope at B just inside the span 2, θ_{B2}, is the same except that from one side of B the slope points up and from the other down. Thus

$$+ \theta_{B1} = - \theta_{B2} \tag{8.1}$$

The continuity of the curvature is achieved by the restraint which each span exercises at the supports on the two adjacent spans. Thus we have at each support a restraining moment whose magnitude we must determine for the solution of the continuous-beam problem.

Each span is thus acted on by two load systems. One is the ordinary uniformly distributed load which causes a mid-span bending moment of $+ WL/8$, reducing to zero at the supports. From Table 6.2, line 6, the end slope in span 1 is

$$\theta_{w1} = \frac{W_1 L_1{}^2}{24EI}$$

and the end slope in span 2 is

$$\theta_{w2} = \frac{W_2 L_2{}^2}{24EI}$$

Design of Building Frames

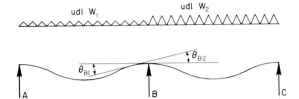

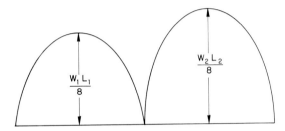

(a) Dimensions of two interior spans, L_1 and L_2

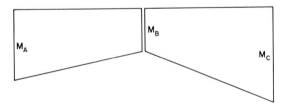

(b) Statically determinate bending moments

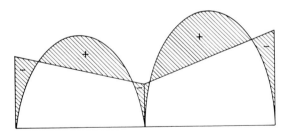

(c) Statically indeterminate bending moments

(d) Combined bending moment diagram

Fig. 8.5.
The Theorem of Three Moments. The slope of the beam at B is θ_B, which is downward to the left of B and upward to the right of B (or vice versa, depending on the relative spans and loads). The slope at B in the span L_1 we will call θ_{B1}, and the slope at B in the span L_2 we will call θ_{B2}.

The second system consists of the moments M_A, M_B, and M_C, which are negative and therefore produce slopes of opposite sign (hogging). In span 1 the moment varies uniformly from M_A to M_B, and at the point B, from Table 6.2, line 12, the slope is

$$\theta_{m1} = \frac{M_A + 2M_B}{6EI} L_1$$

In span 2 the moment varies uniformly from M_B to M_C, and at the point B the slope is

$$\theta_{m2} = \frac{2M_B + M_C}{6EI} L_2$$

The first load system is statically determinate, but the second is statically indeterminate, since we do not know the magnitude of M_A and M_B. However, at each support, where there is an unknown reaction (see Fig. 5.1), there is also a geometric statement like Eq. (8.1), so that there are as many equations as there are unknowns to be determined.

Let us now solve the problem for the (typical) support B. The slope on each side of B

$$\theta_B = \theta_m + \theta_w$$

so that Eq. (8.1) becomes

$$\theta_{m1} + \theta_{w1} + \theta_{m2} + \theta_{w2} = 0$$

which gives

$$\frac{M_A + 2M_B}{6EI} L_1 + \frac{W_1 L_1^2}{24EI} + \frac{2M_B + M_C}{6EI} L_2 + \frac{W_2 L_2^2}{24EI} = 0$$

Rearranging the terms, and multiplying through by $6\,EI$, we obtain the Theorem of Three Moments

$$M_A L_1 + 2M_B(L_1 + L_2) + M_C L_2 = -\tfrac{1}{4}\left(W_1 L_1^2 + W_2 L_2^2\right) \quad (8.2)$$

This was derived by B. P. E. Clapeyron in 1857, and became the basis of normal reinforced concrete design in the early twentieth century.

Unless there is a good reason to the contrary, the column spacing is symmetrical, so that $L_1 = L_2 = L_3$, etc. The theorem then simplifies to

$$M_A + 4M_B + M_C = -\tfrac{1}{4}(W_1 + W_2)L \quad (8.3)$$

Equations (8.2) or (8.3) give the negative (statically indeterminate)

Design of Building Frames

restraining moments of Fig. 8.5c. To these we must add the positive (statically determinate) moments of Fig. 8.5b. The net bending moment is obtained by superimposing one on the other (Fig. 8.5d).

[†]Example 8.1. *Determine the maximum positive and negative bending moments for the beam shown in Fig. 8.6.*

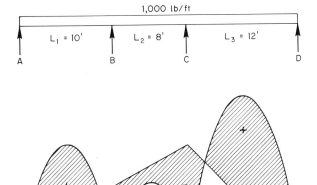

Fig. 8.6.
Bending moment diagram for continuous beam (Example 8.1).

Let us first consider spans 1 and 2. We have $M_A = 0$, because there is no bending moment at the end of a freely supported beam. From Eq. (8.2)

$$0 + 2M_B(10 + 8) + M_C \times 8 = -\tfrac{1}{4}(10{,}000 \times 10^2 + 8{,}000 \times 8^2)$$

$$36M_B + 8M_C = -378{,}000 \tag{i}$$

Let us next consider spans 2 and 3. Since $M_D = 0$,

$$M_B \times 8 + 2M_C(8 + 12) + 0 = -\tfrac{1}{4}(8{,}000 \times 8^2 + 12{,}000 \times 12^2)$$

$$8M_B + 40M_C = -560{,}000 \tag{ii}$$

Multiplying Eq. (i) by 5

$$180M_B + 40M_C = -1{,}890{,}000 \tag{iii}$$

Subtracting Eq. (ii) from Eq. (iii)

$$172M_B = -1{,}330{,}000$$

and

$$M_B = -7{,}730 \text{ lb ft}$$

[†]This and all following examples are worked in metric units in Appendix G.

233

From Eq. (i)

$$M_C = \frac{-378,000 + 36 \times 7,730}{8} = -12,500 \text{ lb ft}$$

The statically determinate positive moments for the spans 1, 2, and 3 are:

$$M_1 = +\tfrac{1}{8} \times 1,000 \times 10^2 = +12,500 \text{ lb ft}$$

$$M_2 = +\tfrac{1}{8} \times 1,000 \times 8^2 = +8,000 \text{ lb ft}$$

$$M_3 = +\tfrac{1}{8} \times 1,000 \times 12^2 = +18,000 \text{ lb ft}$$

These moments are plotted in Fig. 8.6. Evidently M_B and M_C are the maximum negative moments. The maximum positive moments can be obtained by scaling from an accurately drawn diagram. Alternatively, the value of x at which the bending moment is a maximum can be determined from $dM/dx = 0$; the bending moment for this value is then calculated.

This continuous beam is only slightly unsymmetrical; we have merely displaced support C by 2 ft to the left from a 10-ft regular column spacing. However, this is sufficient to eliminate the positive moment in the middle span entirely, and turn it into a negative moment, so that mid-span reinforcement in a concrete floor is required on top, and not as normally on the bottom. Evidently, small irregularities in column spacing can have serious consequences in continuous structures, whereas they have only minor effects in pinned frames.

8.4. Design of Concrete Structures as Restrained Floors

We have seen in the previous section that for equal spans and uniformly distributed loads the relation between the statically inde-terminate support moments in any two adjacent spans of *a continuous beam* is given by the Theorem of Three Moments (Fig. 8.5).

$$M_A + 4M_B + M_C = -\tfrac{1}{4}(W_1 + W_2)L \tag{8.3}$$

Similar equations can be derived for concentrated loads. These may be worked out for each structure; however, they are easily tabulated, since few buildings have more than five spans, and there is little variation as the number of spans increases beyond five. Table 8.1 gives the maximum positive and negative bending moments (calculated as in Example 8.1) for equal spans. Apart from uniformly distributed loading, consideration is given to one or two symmetrically placed concentrated loads, which allow for moving loads, for partitions, and for the reactions transmitted to the primary beams by one or two secondary beams (Fig. 1.49).

As we pointed out in Section 2.7, the dead loads must always be acting, because if we remove them we remove a part of the building. However, the live loads may or may not act on any part of the floor. In practice it is sufficient to consider two cases: (i) adjacent spans loaded, and (ii) alternate spans loaded.

From Eq. (8.3) it is evident that the biggest support moment at B occurs when W_1 and W_2 have their maximum value, i.e., when the two spans adjacent to it are loaded; this gives us the condition for the maximum negative moments in the continuous beam. If we load alternate spans, i.e., if we put $W_2 = 0$ in Eq. (8.3), we reduce the negative moments; but this gives us the maximum positive moment, since the net positive moment (see Fig. 8.5d) is largest when that span is fully loaded, and the negative moment is as small as possible. Table 8.2 gives maximum positive and negative moments for live loads.

Example 8.2. *From Tables 8.1 and 8.2, determine the maximum positive and negative bending moments for a beam continuous over three equal spans of 10 ft, carrying a dead load of 1,000 lb per foot run, and a live load of 500 lb per foot run.*

The maximum negative moment occurs at the inner supports

$$M_- = -0.100 \times 10{,}000 \times 10 - 0.117 \times 5{,}000 \times 10 = -15{,}850 \text{ lb ft}$$

The maximum positive moment occurs in the outer spans

$$M_+ = 0.080 \times 10{,}000 \times 10 + 0.101 \times 5{,}000 \times 10 = +13{,}050 \text{ lb ft}$$

When the restraints on the floor are substantial, the moments are nearer to those for a *built-in beam* (Fig. 8.7).

If the beam is fully restrained at the ends, the slope at the restraints A is zero, because the restraint prevents all rotation.

This means that the slope due to the uniformly distributed, statically determinate, load is equal and opposite to the slope due to the restraining moments, and they cancel one another out.

$$\theta_A = \theta_w + \theta_m = 0 \tag{8.4}$$

The slope at the ends of a simply supported beam, from Table 6.2, line 6, is

$$\theta_w = \frac{WL^2}{24EI}$$

and the slope due to two equal restraining moments M_A, from line 11, is

$$\theta_m = \frac{M_A L}{2EI}$$

From Eq. (8.4)

$$\frac{M_A L}{2EI} = -\frac{WL^2}{24EI}$$

and the end-restraining moments required for a firmly built-in beam

235

Table 8.1
Continuous beams. Bending moment coefficients for dead loads (assuming all spans are loaded). $M = \text{coefficient} \times WL$

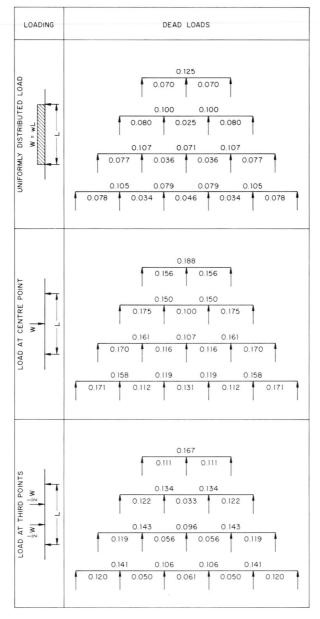

Table 8.2
Bending moment coefficients for live loads (assuming either alternate or adjacent spans are loaded, whichever gives the higher result).
$M = \text{coefficient} \times WL$

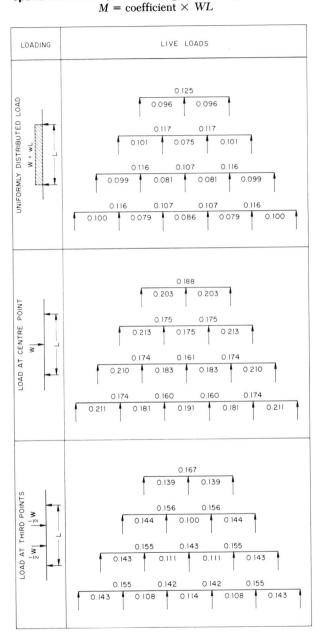

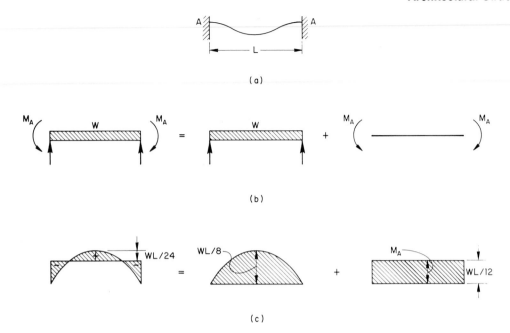

Fig. 8.7.
Built-in beams.
   (a) Deformations.
   (b) The combined load system is composed of the sum of the statically determinate and the statically indeterminate loads.
   (c) The total bending moment diagram is the sum of the statically determinate and the statically indeterminate bending moment diagrams.

carrying a uniformly distributed load

$$M_A = -\frac{WL}{12} \tag{8.5}$$

The maximum positive moment at mid-span (Fig. 8.7c) is

$$M_+ = \frac{WL}{8} - \frac{WL}{12} = +\frac{WL}{24}$$

We can derive the restraining moments (which are also the maximum negative moments) for built-in beams with different loads by the same method. The result is listed in Table 8.3.

Example 8.3. *Determine the maximum positive and negative bending moments for a primary beam, 27 ft long and built in at the ends, which carries two secondary beams at third-points. The reaction of each beam is 10,000 lb.*

From Table 8.3, line 2, the restraining moment is

$$M_A = -\frac{20,000 \times 27}{9} = -60,000 \text{ lb ft}$$

The end reactions are 10,000 lb each, and thus the maximum positive

Table 8.3
End Moments in Built-up Beams

Both Ends Fixed	
Type of loading	Maximum bending moment

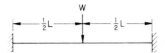

$$-\tfrac{1}{8}\,WL$$

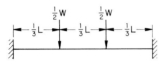

$$-\tfrac{1}{9}\,WL$$

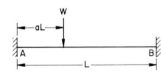

$$-\tfrac{5}{48}\,WL$$

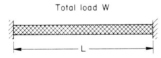

$$M_A = -a(1-a)^2\,WL$$
$$M_B = -a^2(1-a)\,WL$$

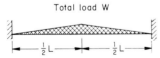

$$-\tfrac{1}{12}\,WL$$

$$-\tfrac{5}{48}\,WL$$

$$M_A = -\tfrac{1}{10}\,WL$$
$$M_B = -\tfrac{1}{15}\,WL$$

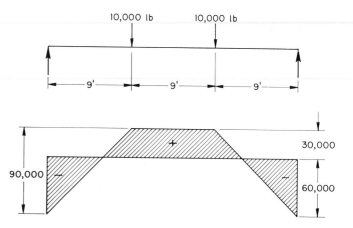

Fig. 8.8.
Bending moment diagram for built-in beam (Example 8.3).

moment (Fig. 8.8) in a *simply supported* beam carrying the same load over the same span is $10,000 \times 9 = +90,000$ ft. The maximum positive moment in the restrained beam is therefore

$$M_+ = 90,000 - 60,000 = +30,000 \text{ lb ft}$$

We can base the design of reinforced concrete slabs and small beams on bending moment coefficients, which are empirically adjusted from the theory of built-in beams and the theory of continuous beams. The most widely used table is that given in Clause 8.4 of the ACI Code (*Building Code Requirements for Reinforced Concrete, ACI 318-71*, American Concrete Institute, Detroit, 1971) to which reference should be made. The maximum negative moments range from $WL/9$, for the center support of a two-span continuous beam, to $WL/24$, for the support of a beam or slab on a spandrel; the positive moments vary from $WL/11$ to $WL/16$.

Similar coefficients are given in the British, Australian, Canadian, and Indian concrete codes.

8.5. The Rigid-Frame Concept—Approximate Solution

Obtaining bending moments from a general formula $M = WL/n$ where n is a number from a table, is evidently very simple, and it is quite satisfactory for small structures under precisely defined conditions (e.g., equal spans, uniformly distributed loads). However, the moment coefficients are conservative and are therefore not economical. For a major building a more accurate form of analysis should be used.

An important factor is the restraint exercised by the supports. Standard solutions exist for continuous beams on pinned supports and for fully restrained beams. Neither is accurate, however, when

Design of Building Frames

the columns exercise a *partial* restraint on the floor structure. Evidently the best approach is to design a rigid frame as such.

We will first consider an approximate method widely used in the earlier years of this century because it is the only one which is easy to understand. We will then proceed to the more complex methods in current use.

A rectangular frame deflects roughly in the way shown in Fig. 8.9 when subjected to a lateral load. Evidently there is a change of curvature near the mid-span of the beams and the columns, and at this *point of contraflexure* the slope θ has a maximum value. The bending moment is zero every point of contraflexure.

We can prove this statement to those readers who are familiar with differential calculus and who have read Section 6.5. When a function has a maximum value, its first derivative is zero (Fig. 8.10).

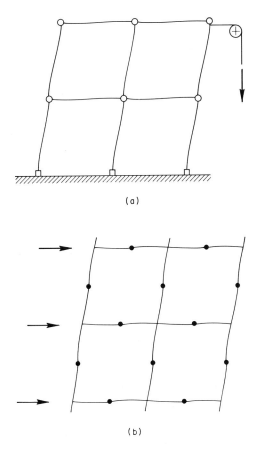

(a)

(b)

Fig. 8.9.
Deformation of rectangular frame under horizontal loading.
(a) The deformation can be illustrated with a Plexiglas frame.
(b) Deformation of two bays and two stories of the frame. Points of contraflexure are formed approximately at the mid-span of each beam and each column.

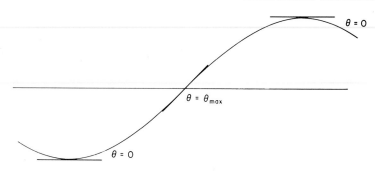

Fig. 8.10.
The slope passes through a maximum at a point of contraflexure.

From Eq. (6.14)

$$\frac{d\theta}{dx} = \frac{M}{EI}$$

Since E and I are constants, which can never reach infinity, $d\theta / dx = 0$ when $M = 0$; i.e., when the slope has a maximum value (or point of contraflexure), the bending moment is zero.

We defined a pin joint in Section 4.1 as a joint which does not transmit bending moments, and noted that in the twentieth century actual pins are rarely employed to produce pin joints. A point of zero bending moment is therefore equivalent to a pin joint. We do not insert actual pin joints at the points of contraflexure, because they correspond only to the wind load, and the structure would be unstable if a flexible joint was inserted at every point where the bending moment is zero for one particular loading condition. However, we can think of points of contraflexure as hypothetical pin joints, and treat them in the analysis like any other pin joints (Fig. 8.11), which transmit the full shear force, but no bending moment.

Example 8.4. Figure 8.12 shows an interior bay of a steel frame or bent, 50 ft high, 15 ft wide, and 180 ft long. The frame is subject to wind pressure above the ground floor and basement, which are sheltered. The frames are spaced 15 ft apart; i.e., the beams which bear on the frame span 15 ft in both directions The horizontal wind pressure is 20 lb per sq ft.

The wind load is distributed as positive pressure acting on the windward face of the building (approximately 60%) and as negative pressure (or suction) on the leeward face (approximately 40%). However, for the purpose of this analysis we may assume that the whole force acts on the windward side, so that

$$W = 20 \times 15 \times 10 = 3,000 \text{ lb}$$

The shear force due to this wind load on the top-floor columns (Fig. 8.12b) is thus $\frac{1}{2} W = 1,500$ lb, and it increases W at each of the two floors below,
and finally $\frac{1}{2} W$. The maximum shear force equals the total horizontal force of 9,000 lb.

Design of Building Frames

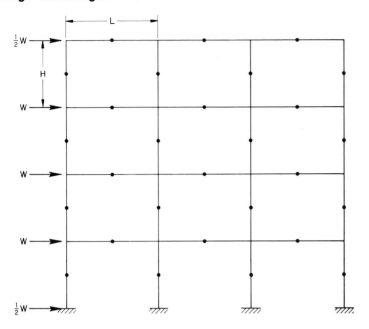

Fig. 8.11.
The pinned-frame concept for horizontal loading. Since the bending moment is zero at a point of contraflexure, we can insert a hypothetical pin joint at each.

Let us now consider the corresponding column bending moments in the top story. At the hypothetical pin joints (Fig. 8.11), $M = 0$, and under the influence of a constant shear force it increases linearly to a maximum value of

$$M = 1,500 \times 5 = 7,500 \text{ lb ft}$$

The rectangular beam-column junction transmits this moment to the roof beam. It is then reduced to 0 at the pin joint and increases again to a maximum value of 7,500 lb ft at the far end of the roof beam, where it is transmitted to the leeward column.

In the columns of the story next to the top, the maximum column moment

$$M = 4,500 \times 5 = 22,500 \text{ lb ft.}$$

Both this moment and the moment in the column above transmit the curvature to the adjoining floor beams of the top story, which thus has a maximum bending moment

$$M = 7,500 + 22,500 = 30,000 \text{ lb ft}$$

The complete bending moment diagram, shown in Fig. 8.12c, is obtained by the successive taking of moments in the columns, descending from top to bottom story, using the increasing shear force from Fig. 8.12b. The end moments of both adjacent columns are transmitted into the beams.

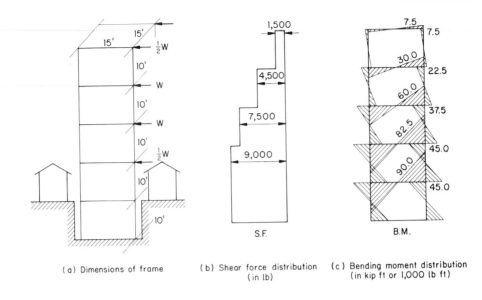

(a) Dimensions of frame

(b) Shear force distribution (in lb)

(c) Bending moment distribution (in kip ft or 1,000 lb ft)

Fig. 8.12.

Distribution of shear force and bending moment due to wind loads in a steel structure designed by the pinned-frame concept (Example 8.4).

The two principal methods in current use for the design of rigid frames are the moment distribution method, which requires only simple arithmetic and is discussed in Section 8.6, and analysis by digital computer, which is beyond the scope of this elementary book.

*8.6 The Moment Distribution Method

It is suggested that readers omit this section on first reading, and proceed to Section 8.7.

We examined in Section 5.7 how rigid frames can be made statically determinate by inserting three pins. We will now analyze the same type of frame without any pin joints.

As we pointed out, the simply supported beam corresponds to the (in practice unrealistic) portal with one end on rollers; however, another statically determinate solution is possible by inserting three pins (Fig. 8.13a and b). A portal pinned only at the base is statically indeterminate with one redundancy (the identical horizontal reactions). A fully restrained portal is statically indeterminate with three redundancies (Fig. 8.13c and d). These are the horizontal reactions (which must always be equal for horizontal equilibrium when the loads are purely vertical), and the two moment reactions (which are unequal when the load is unsymmetrical).

It is simple to put a steel or concrete frame into a hole, fill this up with concrete, and thus produce a rigid joint. A pinned joint requires more labor and subsequent maintenance.

Design of Building Frames

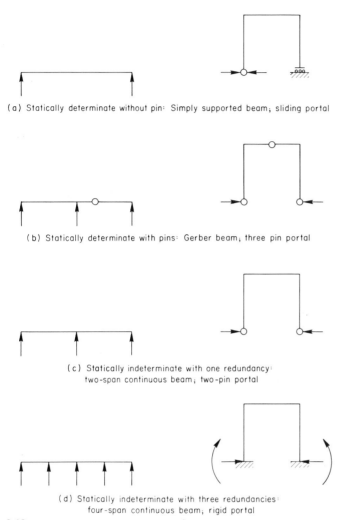

(a) Statically determinate without pin: Simply supported beam; sliding portal

(b) Statically determinate with pins: Gerber beam; three pin portal

(c) Statically indeterminate with one redundancy:
two-span continuous beam; two-pin portal

(d) Statically indeterminate with three redundancies:
four-span continuous beam; rigid portal

Fig. 8.13.
Comparison of beam and portal frame.

Once we get outside the range of statically determinate structures (which accommodate themselves to uneven settlement, temperature movement, etc.) we might as well employ rigid frames, which need less material because the maximum bending moments are lower.

Most methods for solving statically indeterminate structures require one or more additional equations for each redundancy, and they rapidly become more complicated as the number of redundancies produces more simultaneous equations. Moment distribution has the advantage that it is based on fully rigid joints, which are released in turn to distribute the moments; it is thus as simple for rigid as for pin-jointed frames.

It has two further advantages: it involves less mathematical theory than any other method, and solution is by a process of successive approximation, so that we need only as as accurate as required; e.g., we can obtain an approximate solution more quickly for a preliminary design.

In the moment distribution method, all the joints are initially assumed to be clamped. The beams and the columns of the frame are therefore assumed to be built-in at the ends, and the moments are then determined accordingly. There are, consequently, unbalanced moments at the joints; in a rectangular frame carrying vertical loads only, for example, there are substantial moments in the beams and no moments in the columns. Each of the joints is then released in turn, and the unbalanced moments are distributed.

The process of releasing the joints in turn and distributing the moments can be continued indefinitely, and the unbalanced moments become smaller and smaller. They never vanish completely, but we can discontinue calculations when the difference between the moments in beams and columns connected at each joint becomes negligible. In practice, a sufficient degree of accuracy is usually obtained after three cycles of distribution. The procedure need not, therefore, be laborious, and it is possible to adjust the amount of work to the importance of the problem.

The process of distribution consists of two parts. First, the unbalanced moment at each joint is distributed to the other members framing into the joint. If they are all of equal stiffness, the moment is divided equally; if, however, some of the members are stiffer, they take a correspondingly larger proportion of the moment. Stiffness is in this context defined by Eq. (8.6), and its use is explained after Example 8.5.

Second, the rotation of the joints induces moments at the far ends of the members (see Fig. 8.14c). These *carry-over* moments upset the moment equalization previously achieved, and the process of distribution must be continued until the residue is small enough to be neglected.

The moment distribution method is best illustrated by a couple of simple examples. Let us, in the first place, consider a frame of uniform stiffness.

Example 8.5. *Determine the maximum positive and negative bending moments in the beam and columns of the frame shown in Fig. 8.14a.*

Let us assume that the second moment of area (moment of inertia) of the beam and the columns is the same; then the stiffness of the beam and the columns is the same, since they have the same length.

We will now firmly clamp the joints A, B, C, and D. The beam is then a built-in beam. It carries a uniformly distributed load of 12,000 lb, so that the bending moment at B (from Table 8.3, line 5) is

$$M_\text{B} = -\frac{WL}{12} = -\frac{12,000 \times 16}{12} = -16,000 \text{ lb ft}$$

Design of Building Frames

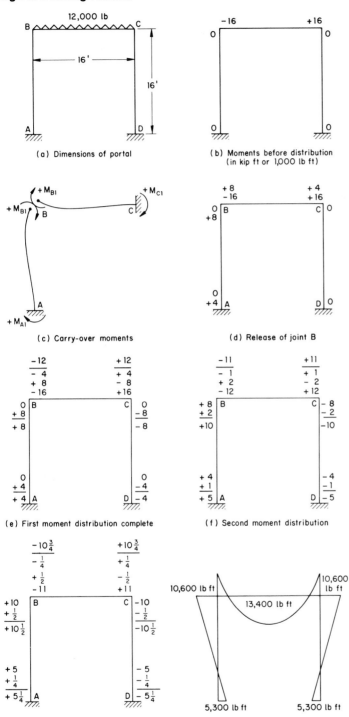

(a) Dimensions of portal

(b) Moments before distribution
(in kip ft or 1,000 lb ft)

(c) Carry-over moments

(d) Release of joint B

(e) First moment distribution complete

(f) Second moment distribution

(g) Third moment distribution

(h) Bending moment diagram

Fig. 8.14.
The moment distribution method (Example 8.5).

With all joints clamped, we thus have "hogging" bending moments of 16,000 lb ft at both ends of the beam, and zero at both ends of each column, where there is no bending moment (Fig. 8.14b).

In order to carry the distribution through without error, we propose to adopt the sign convention which is normal for this type of computation. We will call a moment positive when it is applied in a clockwise direction, and negative when it is counterclockwise. The value of this convention lies in the interaction of the beam and the columns; if the end of the beam rotates clockwise, it also causes the adjoining end of the column to rotate clockwise. We thus obtain moment of $-16,000$ lb ft on the left-hand end of the beam, and $+16,000$ lb ft at the right-hand end of the beam. Please note that this convention differs in character from that used in previous chapters (see Fig. 5.9), where we classified moments by their effect on the deformation of the member.

Let us now release joint B and thus equalize the beam and column moments at that joint. Since the stiffness of the beam and the column is the same, we distribute the moment equally between both; i.e., the beam and the column each receive a clockwise moment $M_{B1} = +8,000$ lb. ft. This is added to $M_B = -16,000$ lb ft in the beam, and to $M_B = 0$ in the column.

In this process we carry over moments to A and C (Fig. 8.14c). The moment M_{B1} acting on the column BA at B produces a moment M_{A1} at A, which is called the *carry-over moment*, because it is produced as a carry-over from the change in moment at B. Since the joint A is still firmly clamped, the slope at A is zero. From Table 6.2, line 12 (using our new sign convention)

$$\theta_A = 0 = \frac{2M_{A1} - M_{B1}}{6EI}$$

Since E and I are constants which cannot be infinity

$$M_{A1} = +\tfrac{1}{2}M_{B1} = +4,000 \text{ lb ft}$$

We thus have a *carry-over factor of* $+\tfrac{1}{2}$; i.e., a change in moment at B produces a change of half that moment at A, rotating in the same direction. Similarly $M_{C1} = +\tfrac{1}{2}M_{B1}$.

The initial moments and those produced by the release of joint B are shown in Fig. 8.14d. We now clamp joint B, and release joint C. The result is similar, since the structure is symmetrical, and the moments now stand as shown in Fig. 8.14e. We cannot release joints A and D, since the structure is built-in at those points, and this is therefore the end of the first moment distribution. At joints B and C, we started with beam moments of 16,000 and column moments of 0, an error of infinity. We have improved these figures to 12,000 and 8,000, an error of 50%.

We will proceed to the second moment distribution (Fig. 8.14f), using the same procedure. The moments at the beam-column junction are now 11,000 and 10,000 lb ft, an error of 10%.

After the third distribution (Fig. 8.14g), the moments at the beam-column junction are 10,750 lb ft and 10,500 lb ft, an error of 3%, which is sufficient. However, if for some reason high precision is needed, we can make another distribution. For preliminary design, two distributions, with an error of 10%, would be sufficient (see Appendix A.6). We now average the remaining differences, and make the bending moments 10,600 and 5,300 lb ft respectively (Fig. 8.14h). Readers may have noticed that the distribution proce-

Design of Building Frames

dure provides an immediate check on the accuracy of the result, which must converge at the joint.

We can now draw the bending moment diagram for the portal frame. The statically determinate bending moment varies parabolically from 0 to

$$\frac{WL}{8} = \frac{12,000 \times 16}{8} = 24,000 \text{ lb ft}$$

Thus the maximum positive bending moment at the mid-span of the beam is $24,000 - 10,600 = 13,400$ lb ft. The bending moment in the columns varies uniformly (Fig. 8.14h).

Let us now consider the effect of stiffness, which is defined as

$$K = \frac{EI}{L} \tag{8.6}$$

where E is the modulus of elasticity, I is the second moment of area (moment of inertia) and L is the effective length (defined in the same way as for buckling in Section 7.1). The greater the resistance to elastic deformation (E), the stronger the section in bending (I), and the shorter it is ($1/L$), the greater is the stiffness. The stiffer the member, the more resistance does it offer to bending. Consequently the moment distributes itself in the ratio of the stiffness of the member to the stiffness of all the members meeting at the joint:

$$\frac{K}{\Sigma K}$$

In this elementary text we will confine ourselves to moment distribution in single-bay portals of one story, so that only two members meet at any joint. However, in plane rectangular building frames, four members meet at most joints, and in nonrectangular frames the number can be higher. In rectangular space frames, six members meet at most joints, but two of these are in torsion (see Section 6.7), and we require the torsional stiffness for these members.

Example 8.6. Determine the maximum positive and negative bending moments in the beam and columns of the frame shown in Fig. 8.15.

In this frame the beam is subjected to a far larger moment, and thus requires a deeper section. It is one of the disadvantages of statically indeterminate design that assumptions must be made about the members *before* they have been designed. This can usually be done from past experience (drawing on reference books and the data of colleagues); but if the initial assumption is seriously in error, we must repeat the calculations with more accurate assumptions.

In this example we will assume that the second moment of area (I) of the beam is three times as large as that of the column. The structural material is the same for both, and one span is 1.5 times the other. Consequently

$$\frac{K_{beam}}{K_{column}} = \frac{3}{1.5} = 2$$

$$\frac{K_{beam}}{\Sigma K} = \frac{2}{3}$$

$$\frac{K_{column}}{\Sigma K} = \frac{1}{3}$$

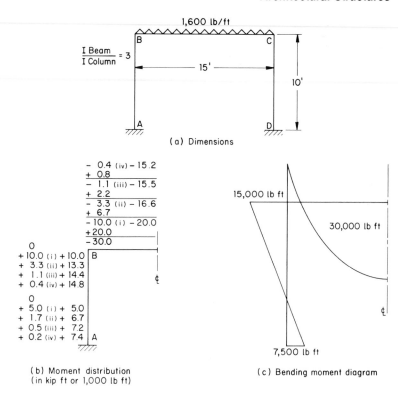

(a) Dimensions

(b) Moment distribution
(in kip ft or 1,000 lb ft)

(c) Bending moment diagram

Fig. 8.15.
Moment distribution (Example 8.6).

The total load on the beam $W = 1,600 \times 15 = 24,000$ lb, and the fixing moment for a built-in beam $WL/12 = 24,000 \times 15/12 = 30,000$ lb ft.

The moment distribution is shown in Fig. 8.15b, and four distributions are needed because of the greater stiffness of the beam.

The statically determinate bending moment in the beam is $WL/8 = 45,000$ lb ft, and the complete bending moment diagram for half the portal frame is shown in Fig. 8.15c.

The carry-over factor of $+\frac{1}{2}$ does not apply to all moment distribution problems. If a frame sways sideways through a distance δ, either because of an asymmetry of loading or because it is subjected to a horizontal load (e.g., due to an earthquake or due to wind), the carry-over moment is equal and opposite to the moment induced by the sidesway. Thus the carry-over factor is -1 (Fig. 8.16).

The same argument also applies to the Vierendeel truss (see Section 1.6), in which the diagonals are replaced by rigid joints. This frame is particularly useful when it is desired to eliminate the ground-floor columns in a multistory building and carry the weight of the upper floor on a girder which has the depth of a full floor. The diagonals of a triangulated, truss would then be in the way of windows and corridors.

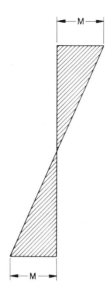

Fig. 8.16.
Carry-over factor for sidesway.

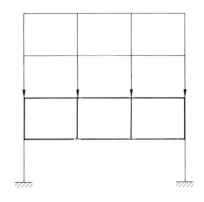

(a) Truss supporting the two center columns
on the upper floors

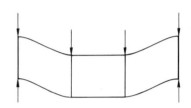

(b) Deformation of truss under load

Fig. 8.17.
Vierendeel truss.

Let us consider a multistory building with four lines of columns, and eliminate the two center lines of columns at the ground floor. Under symmetrical loading the girder deforms as shown in 8.17b. The deformations are like those in sidesway (Fig. 8.16), and the carry-over factor is -1. Consequently the convergence of the moment distribution is slower, and more cycles of distribution are required.

8.7. Design of Concrete and Steel Structures as Rigid Frames

Evidently the design of a complete rigid frame by moment distribution is a major undertaking, and the amount of work is greatly reduced by considering each floor separately (Fig. 8.18); the columns above and below the floor are included in the frame and considered to be rigidly restrained at their far ends. While this is not quite as accurate as the analysis of the complete frame, it is sufficiently close for the majority of low-rise buildings.

Flat-plate structures in particular cannot be designed by the restrained-floor concept of Section 8.2. The floor is less stiff because there are no supporting beams or enlarged column heads (see Sections 1.9 and 7.10), and the columns are larger because of the increased bending stresses and the shear stresses around the slab-column junction. Since K_{column}/K_{floor} is higher, the floor cannot be designed satisfactorily without considering the interaction of the columns and the floor.

The same also applies to buildings in earthquake zones, where the column-floor junctions have to be made rigid to resist the horizontal forces due to earthquakes (see Section 2.7).

Substantial floor thicknesses, which give rise to appreciable floor-column interaction, can also be caused by nonstructural considerations. One of the best ways of improving the sound insulation between floors (e.g., in residential or office buildings) is to use a concrete slab of substantial thickness. Even when the thickness is made greater than the structure requires, this may be the most economical method of satisfying the functional requirements. A thick monolithic concrete floor, whether due to structural or non-structural reasons, increases the stiffness of the floor in relation to the columns. The same argument applies when a thick concrete slab is required to satisfy fireproofing requirements.

The method shown in Fig. 8.18 is not sufficiently accurate for the analysis of high-rise buildings, which because of their height are subject to large horizontal wind-forces. Such frames behave essentially as cantilevers (Fig. 8.19). The outer columns act as the flanges of the cantilever, and the floor structure forms the web.

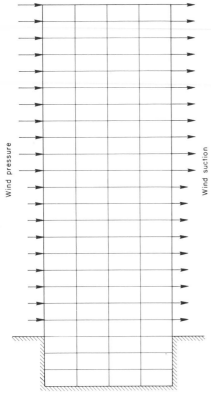

Fig. 8.19.
Under wind load the tall frame behaves essentially as a cantilever.

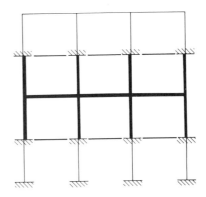

Fig. 8.18.
The rigid-frame concept. When computations are carried out without a computer, it may be sufficient to consider each floor separately.

In high-rise buildings the wind load is a major design factor. For frames of this size, computer solutions are not only more accurate but are actually much cheaper.

The traditional methods for the solution of rigid frames are the strain energy method (see Section 9.2), the moment distribution method (described in Section 8.6), and the slope deflection method (an extension to rigid frames of the method used in the solution of continuous beams in Section 8.3). All can be adapted to computer programming, but in recent years rigid frame analyses have been written especially for evaluation by computer. These are either framed in terms of the redundant actions (*flexibility methods*) or in terms of the unknown joint displacements (*stiffness methods*). Readers with a knowledge of matrix algebra are referred to specialized texts on the matrix analysis of frames (e.g., A. S. Hall and R. W. Woodhead, *Frame Analysis*, Wiley, New York, 1967).

8.8. Some Variations on the Rectangular Frame

The wind load can be resisted by the frame itself, by shear walls, or by diagonals. The last is normal practice in *single-story buildings with trussed roofs* (see Chapter 4 and Section 1.5). The roof trusses have sufficient stiffness to take wind loads in their own plane and are designed to do so, but at right angles to them it is necessary to provide walls or partitions to act as shear panels, or else provide diagonal bracing (Fig. 8.20).

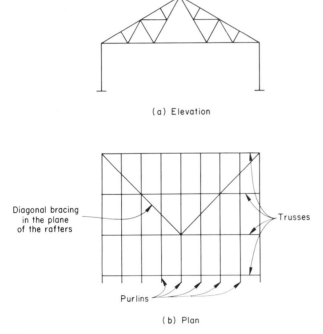

(a) Elevation

(b) Plan

Fig. 8.20.
Wind bracing for single-story industrial buildings. The roof truss offers wind-resistance in its own plane, but the purlins are designed only for vertical loadings. Diagonal bracing or a solid wall is required for wind bracing at right angles to the truss lines (see also Fig. 1.38).

Since the wind load is resisted by the trusses in one direction and by shear walls or diagonal bracing in the other, the purlins need only be designed to carry the roof sheet. They can be treated as simply supported beams, as continuous beams (see Section 8.3), or as Gerber beams (see Section 5.6). Simply supported beams give the highest bending moments ($WL/8$). The bending moments in continuous beams and Gerber beams are the same (Fig. 8.21).

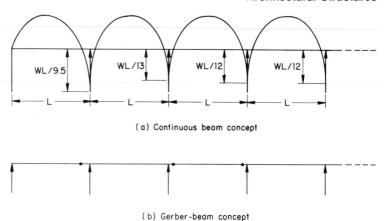

(a) Continuous beam concept

(b) Gerber-beam concept

Fig. 8.21.
Design of purlins for single-story industrial building.

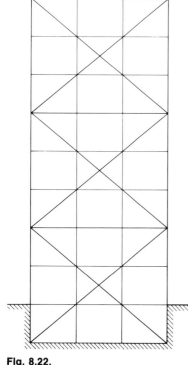

Fig. 8.22.
Diagonal bracing in high-rise buildings. The columns are designed for vertical loading; the diagonal bracing resists the wind loads.

The *diagonal-bracing concept* may be applied to high-rise frames (Fig. 8.22). The columns then take only the vertical loading, and the horizontal forces are resisted by the bracing. As the height of the building increases, the diagonals require large steel members, and special care is needed in the detailing of the walls if they are to be obscured, which is normal practice, while retaining freedom for elastic deformation. The alternative is to turn the diagonals into a dominating visual feature, as in the John Hancock Center in Chicago (Fig. 1.74).

Shear walls are widely used because most buildings have in any case permanent partitions or external walls that are available to resist the wind loads as shear panels (Fig. 8.23). Although window and door openings materially weaken shear walls, the amount of material required in solid concrete walls for architectural reasons is often far greater than that needed to resist wind-forces; window and door openings are acceptable is the concrete and the reinforcement between openings is sufficient to resist the vertical shear forces (Fig. 8.23b).

Another solution is to design the *services core* of the building to provide the resistance to horizontal forces. The elevators require a substantial structure, partly because the guides must be kept in perfect alignment for proper operation, and partly because in most building types the elevators must be enclosed in concrete for fire protection. There are constructional advantages in building the core before the rest of the building, because the elevators can then be used for transporting the materials. Whether the services core is designed to take the full horizontal forces, or only a part thereof, the outer columns are reduced in size, and the design of the facade is simplified.

254

Design of Building Frames

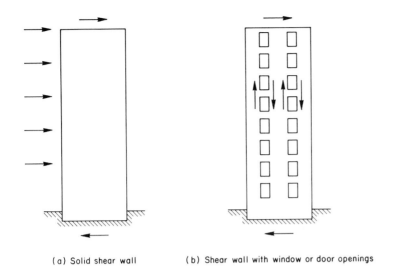

(a) Solid shear wall   (b) Shear wall with window or door openings

Fig. 8.23.
Shear wall. In a shear wall with window or door openings, there must be sufficient
material to resist the vertical shear.

When there are two service cores in a building, they may be
designed to form the flanges of a vertical girder, with the floors
serving as webs (Fig. 8.24); this greatly reduces the structural
thickness required, because the distance between the two cores
provides a resistance arm, a.

Another possible source of structural economy is the use of the
scaffolding employed for the construction of the service core as
reinforcement for the concrete columns of the structure surrounding

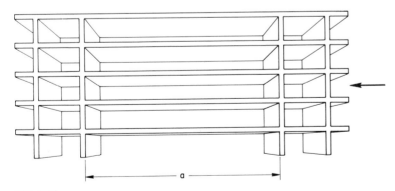

Fig. 8.24.
Wind bracing with twin service cores. Each service core provides one flange of the
"beam," and the floors provide the web; a = resistance arm.

the core. This requires a compromise between the need for fairly substanial scaffolding and well-distributed reinforcement for the concrete. A welded lattice of steel angles can meet both requirements.

The top of a multistory building is generally used for the plant of the elevators, water supply, air conditioning and electrical services. Windows are not required, and it is thus possible to stiffen the top story with a reinforced concrete wall which acts as a "top hat" to the core shear walls, and substantially increases the resistance to the horizontal loads (Fig. 8.25).

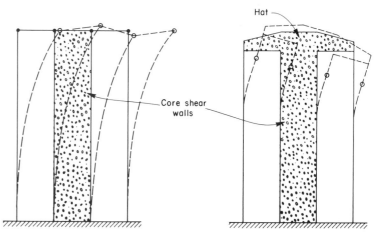

Fig. 8.25.
A "top hat"—consisting of an extension of the service core shear wall into the topmost service story—greatly increases the stiffness of a tall building.

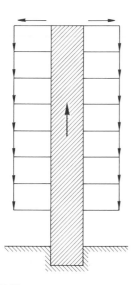

Fig. 8.26.
High-rise building with floors supported on hangers from roof cantilevers.

Finally, we should draw attention to the possibility of using the service core for the complete vertical support of the building, and hanging the floors from brackets suspended from the top of the core (Fig. 8.26). This cannot be justified in structural terms, since we are first transferring the weight of the floors to the top of the building with tensioned hangers and then transferring it down again in the service core. However, it has been used both for entire buildings and for glass-enclosed staircases, because it gives an impression of lightness.

The new generation of very tall buildings has utilized the entire facade of the building (Fig. 1.73), which acts like a perforated tube resisting both the vertical and the horizontal loads. This tube is formed by closely spaced external columns and by the spandrel beams. Generally the service core is also utilized as a load-bearing element, and this forms an inner tube; the result is a "tube-in-tube" structure.

Design of Building Frames

In the Sears Tower in Chicago which, at the time of writing, is the world's tallest completed building (1,450 ft or 442 m), the elevators are distributed over the building, and this makes it possible to design the structure as a bundle of nine tubes which can be terminated at different heights (Fig. 8.27).

8.9. Effect of Temperature, Shrinkage, Creep, and Settlement

Rigid frames are, by definition, incapable of accommodating movement of the structure without producing elastic stresses. If a statically determinate frame (see Figs. 4.1 and 5.1) has the length of one of its members reduced by shrinkage, the frame changes shape very slightly, but no stresses are induced. The change in shape is insignificant in most cases, since *shrinkage movement* is very small. For *concrete* it has an average value of 3×10^{-4}, which means that a member 100 ft long shrinks a barely noticeable 0.36 in. If, however, the member is firmly restrained at both ends, then a tensile stress is produced

$$3 \times 10^{-4} \times 3,000,000 = 900 \text{ p.s.i.}^*$$

(assuming $E = 3,000,000$ p.s.i. for concrete), and this is far in excess of the tensile strength even of a high-strength concrete. In practice the restraint is never complete, so that the actual shrinkage stresses are lower. However, they are normally sufficient to produce cracks in concrete (which has an ultimate tensile strength of 300 to 500 p.s.i.).** It is therefore important to insert reinforcement across all lines of potential shrinkage cracks. One typical example is the ordinary concrete slab spanning in one direction (Example 7.12) which requires reinforcement in one direction only to satisfy the requirements of structural mechanics. The ACI Code specifies reinforcement at least equal to 0.2% of the cross-sectional area to resist shrinkage and temperature stresses. This is a small amount of steel; but the tensile force due to shrinkage and temperature movement are not large. The problem arises only because the tensile strength of the concrete is very low.

Shrinkage is caused by the drying out of concrete. Concrete swells again when it is wetted, but only part of the shrinkage is recoverable in this way. Thus shrinkage consists of two parts: the initial irreversible drying out of the liquid concrete and the cyclic moisture movement which changes with moisture content.

The same occurs in *timber*. Tensile stresses due to skrinkage are relatively harmless along the grain, but they may cause splitting across the grain. For example, we normally lay timber floor boards with tongue-and-groove joints to permit movement of the timber

*Assuming $E = 20$ GPa, the tensile stress is $3 \times 10^{-4} \times 20,000 = 6$ MPa.
**The corresponding metric stresses are 2 to 3.5 MPa.

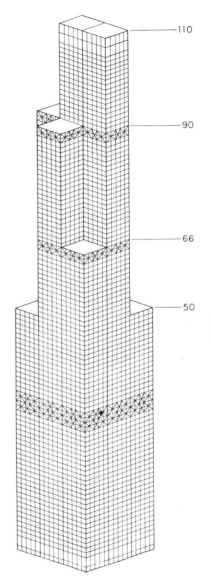

Fig. 8.27.
The "bundled-tube" concept of the Sears Tower in Chicago, designed by Dr. F. R. Khan, a partner of Skidmore, Owings and Merrill. Two of the nine perforated tubes are terminated at the 50th story, two at the 66th story, three at the 90th story, and two go to the full height of 110 stories. The diagonals are extra stiffeners surrounding service stories.

257

with temperature and moisture changes. Timber shrinkage also consists of an irreversible part, constituting the moisture lost from the green timber during the seasoning process, and a cyclic part, which is recovered when the timber becomes damp. Since we cannot reinforce timber like concrete, it is important that it should be properly seasoned, or dried out, lest the contractions due to shrinkage build up high tensile stresses which may cause failure, particularly of the connections. This is not the invariable practice. For example, in some parts of Australia, where hardwoods are commonly utilized for domestic roofs, the timber is used green, because it is too hard to nail when dry.

The effect of *temperature stresses* is very similar to that of shrinkage stresses (see Section 2.4). Thus the most adverse conditions in concrete and timber structures are dry cold or humid heat. *Metal structures* have a larger temperature movement but no moisture movement.

The phenomenon of *creep* occurs in both concrete and timber. It is caused by the squeezing of water from the pores by a sustained load. Thus creep is caused mainly by dead loads; live loads, because of their shorter application, cause far less creep.

Because of creep, the stress-strain diagram is slightly curved, instead of being a straight elastic line. The extent of the curvature depends on the time of application of the load (Fig. 8.28). While the true elastic modulus is f/e_e, it is convenient to use an effective

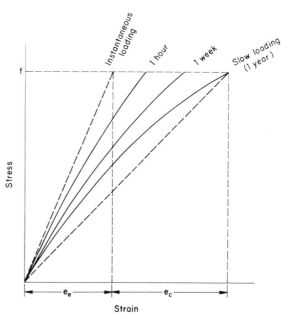

Fig. 8.28.
Effect of creep on the effective modulus of elasticity of concrete; e_e = elastic strain, e_c = creep.

Design of Building Frames

modulus $f/(e_e + e_c)$, which represents the secant to the curve; e_e is the elastic strain and e_c is the creep strain.

Creep received little consideration in the design of buildings until *flat plate structures* (see Sections 1.9, 7.10, and 8.7) became popular. Because of their simplicity they now represent a large proportion of the total volume of reinforced concrete construction. Flat plates have a low stiffness and consequently high deflections. The effective modulus of elasticity to be inserted into Eq. 6.15 therefore becomes of great importance.

Unless adequate allowance is made for creep deflection, it may cause cracking of brittle finishes, cracking of partitions (because of the deflection of the ceiling above), and jamming of doors and windows. Since creep deflection takes time, trouble starts only several months after the completion of the structure, and repairs may be costly. It is now widely accepted that creep strains may be more than twice as great as elastic strains.

One answer to the creep deflection of flat plates is to reduce the spans, since deflection is proportional to $L.^4$ Another is to prestress the plates and adjust the prestress so that it produces an upward deflection exactly equal and opposite to the deflection due to the dead load (see Section 7.13). The net deflection is then zero, and creep cannot alter a zero deflection.

Creep and shrinkage are also important in *prestressed concrete*. Since the prestress is sustained throughout the whole life of the structure, it causes large contractions along the axis of the beam (see Section 7.13). When the concrete beam shortens, the prestressing cables inside it shorten by the same amount. Creep has an average value of 5×10^{-4} in prestressed concrete. To this we must add the shrinkage of 3×10^{-4}. For a steel modulus of elasticity of 30,000,000 p.s.i., the loss of prestress due to shrinkage and creep is thus of the order of

$$(5 + 3) \times 10^{-4} \times 30,000,000 = 24,000 \text{ p.s.i.*}$$

We mentioned in Section 7.13 that it is possible to use high-tensile steel in prestressed concrete because of the absence of cracks. Evidently this is also essential. We must have a steel with a permissible stress much higher than 24,000 p.s.i., if a useful portion of the prestress is to remain after the inevitable losses have occurred. Fortunately steels with a strength of about 270,000 p.s.i.** can be produced by cold-working. These steels are so hard that they cannot be machined or cut, except with flame cutters, and thus they can be utilized only for cables in prestressed concrete, suspension structures, etc. (see Section 1.3).

---

*Assuming $E = 200$ GPa, the loss of prestress is $(5 + 3) \times 10^{-4} \times 200,000 = 160$ MPa.

**The corresponding metric stresses are 160 MPa and 1,860 MPa.

Finally we should draw attention to the effect of the *settlement of foundations* on statically indeterminate structures. Foundations may settle because they rest on compressible clays and were not properly designed. They may also settle in areas where mining operations have been carried on. Even settlement of the entire foundation is harmless; but differential settlement can give rise to an appreciable redistribution of stress.

Let us consider a concrete beam carrying a load of 1,000 lb per foot run, which is continuous over two equal spans of 10 ft. From Table 8.1, the maximum bending moment at the center support is

$$- \frac{WL}{8} = - \frac{10,000 \times 10}{8} = -12,500 \text{ lb ft}$$

Let us assume that the center support sinks sufficiently to turn the beam into a simply supported beam spanning 20 ft. The maximum bending moment is now, with $L = 20$ ft, and $W = 20,000$ lb

$$+ \frac{WL}{8} = + \frac{20,000 \times 20}{8} = +50,000 \text{ lb ft}$$

Evidently, if the beam had been properly designed for a moment of 12,500 lb ft, it would fail under a moment of 50,000 lb ft. Moreover, the reinforcement would be on top, whereas it is now needed on the bottom.

Let us assume that the beam is of a concrete, with a modulus of elasticity of 3,000,000 p.s.i., and that the second moment of area (moment of inertia) of the beam is 1,000 in^4 (which is roughly what is required for a moment of 12,500 lb ft). Then the deflection of the beam shown in Fig. 8.29b is, from Table 6.2

$$\frac{5}{384} \frac{WL^3}{EI} = \frac{5 \times 20,000 \times 20^3 \times 12^3}{384 \times 3,000,000 \times 1,000} = 1.2 \text{ in*}$$

Thus a settlement of 1.2 in would turn the continuous beam of Fig. 8.29a into case (b), and cause certain failure. A much smaller settlement would do serious damage. This can be avoided by using two separate, simply supported beams, as in Fig. 8.29c, which can settle differentially without altering the stresses. Since the maximum bending moment for the two simply supported beams is $+ WL/8$, exactly the same as for the continuous beam, this does not, in this instance, require more material.

*For a load of 90 kN, a span of 6 m ($= 6 \times 10^3$ mm), a second moment of area (moment of inertia) of 400×10^6 mm^4, and a modulus of elasticity of 20 GPa ($= 20$ kN/mm^2), the corresponding deflection in metric units is

$$\frac{5}{384} \frac{WL^3}{EI} = \frac{5 \times 9 \times 6^3 \times 10^9}{384 \times 20 \times 400 \times 10^6} = 32 \text{ mm}$$

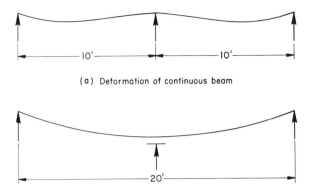

(a) Deformation of continuous beam

(b) Deformation of continuous beam after large settlement of center support

(c) Deformation of two simply supported beams

Fig. 8.29.
Effect of large differential settlement on a two-span continuous beam.

The arguments are therefore not all in favor of rigid frames. Statically indeterminate structures are more rigid and stronger than statically determinate structures, particularly in reaction to wind and earthquake forces. They require less structural material because the maximum moments are generally lower. However, they cannot easily accommodate movement due to temperature, shrinkage, creep, and settlement. Where these are likely to cause serious stress redistribution, consideration should be given to the use of a statically determinate structure.

Suggestions for Further Reading

F. B. BULL and G. SVED: *Moment Distribution in Theory and Practice*. Macmillan, New York, 1964. Chapters 1–4, pp. 1–94.

A. J. S. PIPPARD and J. F. BAKER: *The Analysis of Engineering Structures*. Arnold, London, 1968. Chapters 8–10, pp. 154–235.

W. G. RAPP: *Construction of Structural Steel Building Frames*. Wiley, New York, 1968. 340 pp.

J. J. TUMA: *Theory and Problems of Structural Analysis*. McGraw-Hill, New York, 1969. 292 pp.

Problems

8.1. Explain how the Theorem of Three Moments is derived.

8.2. Discuss the structural significance of the monolithic character of reinforced concrete and the value of the symmetry of column spacing and of loads for the solution of continuous beams.

†8.3. Determine the maximum positive and negative bending moments for a beam continuous over two equal spans of 20 ft, carrying a uniformly distributed dead load of 1,000 lb per foot run. Sketch the complete bending moment diagram, and determine approximately the maximum positive bending moment.

8.4. A steel beam is continuous over five equal spans of 16 ft each; it carries a uniformly distributed live load of 1,000 lb per ft run, and a uniformly distributed dead load of 1,000 per ft run. Determine the bending moments at all supports, and hence sketch the entire bending moment diagram. What are the maximum negative and positive moments?

8.5. Determine the maximum negative moments for a beam continuous over three spans of 25 ft, 20 ft, and 30 ft (20 ft being the central span), and carrying a uniformly distributed dead load of 1,200 lb per foot run. Sketch the complete bending moment diagram and from it determine the approximate maximum positive moment.

8.6. Explain the difference between simply supported concrete beams and slabs, fixed-ended beams, and continuous beams, considering bending moment, deflection, and arrangement of reinforcement.

8.7. Derive the maximum positive and negative bending moments for a concrete slab, spanning 10 ft built-in at the supports, and carrying a uniformly distributed load of 200 lb per sq ft (including its own weight).

8.8. A primary reinforced concrete beam, built-in at the supports, carries two secondary beams at third points. Draw the bending moment diagram, neglecting the weight of the primary beam.

If the reaction of secondary beams on the primary beam is 5,000 lb, determine the maximum positive and negative bending moment in the primary beam, which has a span of 24 ft.

8.9. A rectangular steel portal frame spans 20 ft. Its columns are 10 ft high, built-in at the supports. The frame is built up from the same 6 in deep steel section ($I = 43.7$ in^4). Determine the flexural stress at the base of the portal, at the knee-junction, and at mid-span due to a uniformly distributed vertical load of 10,000 lb.

8.10. A reinforced concrete rectangular portal, built-in at both ends, has a span of 24 ft and a height of 12 ft. The beam carries a uniformly distributed load of 40,000 lb. The second moment of area (moment of inertia) of the beam is 15,000 in^4, and that of the column is 2,500 in^4. Determine the bending moment at the upper and lower ends of the column.

8.11. Explain and illustrate the characteristics of the following forms of steel construction, and consider where each may be suit-

†Similar problems in metric units are given in Appendix G.

able: (a) open-web steel joists; (b) a three-pin portal frame; and (c) a bracing system satisfactory for a medium-sized building.

8.12. Discuss and illustrate the various methods of ensuring the stability of medium-to-tall steel-framed buildings. Indicate the effect of each on the facade of the building.

8.13. At a certain point in a brick shear wall there is a compressive stress of 200 p.s.i. and vertical and horizontal shear stresses of 100 p.s.i. due to wind load. Calculate the magnitude and direction of the principal stresses.

8.14. An aluminum handrail extrusion has a cross-sectional area of 1 sq in; it is securely fixed on top of a steel handrail, the top member of which has a cross-sectional area of 2 sq in. Calculate the resulting stresses in the steel and the aluminum, if the temperature falls by 50°F. The moduli of elasticity of steel and aluminum are 30,000,000 and 10,000,000 p.s.i., respectively, and the coefficients of linear expansion are 6×10^{-6} and 12×10^{-6} per °F, respectively.

8.15. A 16-story office building is to be built with perimeter walls suspended by hangers at 12-ft centers from roof girders which cantilever 30 ft from a central service core. Determine the minimum size of the hangers, if the floor construction is to be as light as possible (say, light-gauge steel floor with lightweight concrete topping, steel beams with lightweight fireproofing, and a suspended metal ceiling). Conduct your own investigation on the loads, and on the quality of the steel suitable for the hangers.

Having determined the size of the hangers, calculate their elastic deformation under dead and live load, and their temperature movement (assuming that they are exposed to sunlight through glass walls in an air-conditioned interior).

Design of Curved Roofs $\quad$ **9**

An egg is always an adventure: it may be different.
Oscar Wilde

In Chapters 4 to 8 we examined the design of structures composed of linear elements; however, a large proportion of the structural types described in Chapter 1 was curved. We will now explain the mechanics of the simplest examples. Because of the mathematical complexities, largely caused by the geometry of these structures, we cannot go deeply into this subject.

9.1. Cables

A single cable is a statically determinate structure. Let us consider a cable carrying a load W uniformly distributed in plan over a span L (Fig. 9.1), with a sag s at mid-span. One-half of the cable is kept in equilibrium by the end reaction, R, the half-load, $\frac{1}{2} W$, acting at quarter-span, $\frac{1}{4} L$, and the cable tension at mid-span, T_0. The end reaction, R, has a horizontal component, R_H, and a vertical component, R_V.

For horizontal equilibrium:

$$R_\text{H} = T_0$$

For vertical equilibrium:

$$R_\text{V} = \tfrac{1}{2} W$$

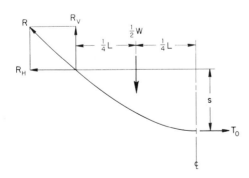

Fig. 9.1.
Tension in a suspension cable.

Taking moments about the cable support:

$$T_0 \times s = \tfrac{1}{2} W \times \tfrac{1}{4} L = \tfrac{1}{8} WL$$

Consequently the cable tension at mid-span:

$$T_0 = \frac{WL}{8s} \tag{9.1}$$

The load W, and the vertical reactions $R_V = \tfrac{1}{2} W$ set up a bending moment, just as in a simply supported beam, equal to $WL/8$, and this is resisted at mid-span by a tensile force, with the sag acting as resistance arm between T_0 and R_H. Because of the curvature of the structure we achieve this resistance arm with only a small expenditure of material, whereas in the simply supported beam we have to provide a depth of beam to accommodate it; this in turn increases the weight (see Fig. 1.11).

The cable tension, T, at the end of the cable equals R.

$$T = R = \sqrt{R_H{}^2 + R_V{}^2} = \sqrt{T_0{}^2 + \tfrac{1}{4} W^2} \tag{9.2}$$

[†]Example 9.1. *Two buildings, 40 ft apart, are to be joined by an unstiffened suspension bridge. Determine the cable size required, if each cable carries a load of 1,000 lb per foot run of span.*

Let us assume a ratio of span to sag of 8. This gives $L = 40$ ft, $s = 5$ ft, and $W = 40,000$ lb. From Eq. (9.1) the cable tension at mid-span

$$T_0 = \frac{40,000 \times 40}{8 \times 5} = 40,000 \text{ lb}$$

From Eq. (9.2) the maximum cable tension

$$T = \sqrt{40,000^2 + 20,000^2} = 44,720 \text{ lb}$$

If the maximum permissible cable stress is 60,000 p.s.i., the steel area is 0.745 sq in. A cable is normally twisted from strands, and its effective steel area is approximately $\tfrac{2}{3} \times \tfrac{1}{4} \pi d^2$, where d is the cable diameter. The steel area thus corresponds to a $1\tfrac{1}{4}$-in diameter cable.

Let us next consider a circular area roofed by cables suspended between an inner tension ring and an outer compression ring. This is best done with the aid of an example.

Example 9.2. *Determine the size of the cables and the size of the compression ring for a circular suspension roof over an arena, 200 ft in diameter (Figs. 1.17 and 9.2).*

The circumference of the compression ring is $\pi L = 629$ ft. If we use 62 cables, we obtain a spacing of almost exactly 10 ft around the circumference (Fig. 9.2).

[†]This and all following examples are worked in metric units in Appendix G.

Design of Curved Roofs

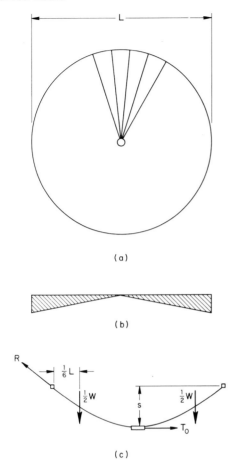

(a)

(b)

(c)

Fig. 9.2.
Design of a circular suspension roof (Example 9.2).
(a) Plan of roof.
(b) Distribution of load carried by each pair of cables.
(c) Tension in a pair of cables. The compression ring is on the outside and the tension ring in the center.

The roof needs a fairly large sag to be effective, and this also meets its architectural requirements (see Fig. 1.16). Let us make $s = L/5 = 40$ ft.

Let us take the weight per unit area of cable roof, allowing for the weight of the cables, the roofing material, and the live load, as 100 lb/sq ft. Since each cable carries a triangular slice, this varies linearly from a maximum at the compression ring to zero at the tension ring. The total load for a pair of cables (of which there are 31) is

$$W = \frac{100 \times \frac{1}{4}\pi \times 200^2}{31} = 101,340 \text{ lb}$$

Taking moments about the compression ring, we obtain for a single cable (i.e., from compression to tension ring)

$$T_0 \times s = \tfrac{1}{2} W \times \tfrac{1}{6} L$$

The cable tension at the tension ring

$$T_0 = \frac{101,340 \times 200}{12 \times 40} = 42,220 \text{ lb}$$

This is also the horizontal reaction of the cable at the compression ring. The vertical reaction is $\frac{1}{2} W = 50,670$ lb, and the combined reaction

$$R = \sqrt{42,220^2 + 50,670^2} = 66,000 \text{ lb}$$

For a maximum permissible steel stress of 60,000 p.s.i., the steel area is 1.10 sq in, and the approximate cable size is

$$\sqrt{1.5 \times 1.10 \times 4/\pi} = 1.45 \text{ in } \left(1\tfrac{1}{2}\text{-in diameter}\right)$$

Let us assume that the compression ring is vertically supported, so that the vertical reaction is absorbed by a wall or by columns. The horizontal reaction is 42,220 lb for each 10-ft run of ring, or 4,220 lb per ft. This is a condition like that in a thin pipe under suction, and we can solve it in the same way. We thus have to balance a pressure of 4,220 lb per ft over a diameter of 200 ft (Fig. 9.3), and this is resisted by the force in two cuts of the ring. Consequently the hoop compression in the ring is

$$\frac{4,220 \times 200}{2} = 422,000 \text{ lb}$$

Assuming a maximum permissible concrete stress of 1,500 p.s.i., we require a compression ring of 282 sq in, say 12 in wide by 24 in deep. In practice we would use a smaller ring, suitably reinforced in compression (see Example 7.10).

The diameter of the tension ring is so small that its dimensions are not primarily determined by structural considerations.

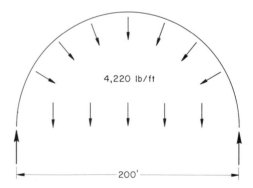

4,220 lb/ft

200'

Fig. 9.3.
The compression ring behaves like a thin tube under suction (Example 9.2).

A cable is the lightest of all structural types. A high-tensile cable can support its own weight over a span of about 20 miles, as compared with about 15 ft for a plain concrete beam.* It suffers, however, from lack of stability. Because of their flexibility, cables

*The corresponding metric dimensions are about 30 km and 5 m.

Design of Curved Roofs

change their shape easily with a change of loading. Another problem is thermal movement. Thus a steel cable 200 ft long extends an inch when the temperature increases by 70° F. (A steel cable 60 m long extends 28 mm when the temperature increases by 40° C.) Evidently cables are liable to move unless they are restrained. We can place a second prestressed cable system at right angles to the load-bearing cables to control temperature movement and instability of the load-bearing cables (see Fig. 1.18); or we can use two load-bearing cable systems joined together (Fig. 9.4).

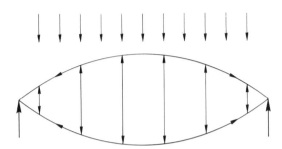

Fig. 9.4.
Interconnected cable structure for circular roof.

These interconnected cable structures are no longer statically determinate. Statically indeterminate cable structures are solved, like statically indeterminate frames, by ascertaining additional equations from the equality of the deflection of two cables at the joints where they are fixed together; but the solution is complicated by the nonlinearity of the equations, which results from the relatively large horizontal movements of the cables under load.

Although suspension roofs are generally neither as long nor as flexible as suspension bridges, aerodynamic instability can be a problem. The period of oscillation of a cable

$$t = \sqrt{\frac{4WL}{gn^2T}} \qquad (9.3)$$

where W is the total uniformly distributed load carried by the cable, L is its span, and T the cable tension; g is the acceleration due to gravity, and n is 1, 2, 3, etc. The number $n = 1$ gives the fundamental period for up and down motion; $n = 2$ gives the period for antisymmetrical oscillations which cause torsion in suspension bridges when two cables are out of phase (this caused the collapse of the Tacoma Narrows Bridge, near Seattle).

Quite small wind speeds may cause dangerous oscillations, but often it is possible to stop these by altering the natural frequency of the structure by ballasting the roof (which increases the load W) or by reducing the sag (which increases the cable tension T).

When the period of the structure is greater than the period of the wind gusts, the wind forces can be treated as static forces.

Rapid vibrations of small amplitude—known as flutter—which do not produce a rapid collapse, may nevertheless be dangerous because they can cause a fatigue failure of the materials in the roof structure if the vibrations continue over a long time.

Example 9.3. *Determine the period of the cable in* Example 9.1.

The acceleration due to gravity is 32.2 ft/sec/sec. From Eq. (9.3) the fundamental period $(n = 1)$ is

$$t = \sqrt{\frac{4 \times 40,000 \times 40}{32.2 \times 1 \times 44,800}} = 2.1 \text{ sec}$$

9.2. Arches

Since arches are rigid, they do not have the instability problems inherent in cable structures; however, because of the rigidity, the shape cannot adjust itself to the load system like a cable (see Figs. 1.19 to 1.21), and thus arches are almost invariably subject to some bending. Because the arch is much closer to the shape of the bending moment diagram than the portal frame, bending moments in arches are much lower than in portals (see Sections 5.7 to 5.9).

Arches can be made statically determinate by the insertion of two pins at the springings, and another pin at the crown. We have already discussed the design of three-pin arches (see Section 5.9 and Example 5.15).

If we do not insert a pin at the crown, we get a two-hinged arch which is statically indeterminate with one redundancy. The rigid arch, without pins, has three redundancies (see Fig. 8.13).

Statically indeterminate arches are most conveniently solved by the *strain energy method*, developed by A. Castigliano in 1870. As the arch (or any other structure) is loaded, it is strained and energy is stored in it. Due to the deflection of the arch, the loads drop slightly, and thus lose potential energy. Neglecting the small amount of energy converted into heat (and occasionally sound, when the structure deforms audibly), the loss of potential energy equals the strain energy stored in the elastic structure. The first derivative of the strain energy with respect to any of the redundancies must be zero, since each redundancy is there for the purpose of preventing displacement; this also follows from the Principle of Least Work. The method is, however, far from simple in application, since it involves integration to obtain the total strain energy, and differentiation of this integral with respect to each redundancy. In the case of a rigid arch, three simultaneous equations result. We cannot in this elementary text describe the strain energy method, and reference should be made to a more advanced text (e.g., S. Timoshenko and D. H. Young: *Theory of Structures*, McGraw-Hill, New York, 1945, Chapter 9, pp. 419–481).

Design of Curved Roofs

Fortunately it is found that the difference between the thrust and bending moment in a two-pin and a three-pin arch is small, and a preliminary design made statically determinate by the insertion of an extra pin is unlikely to differ greatly from the precise solution. Nor is the difference usually large for the rigid arch, except at the springings where the zero moment of the pins is replaced by a large restraining moment. For most arches, however, this is dissipated rapidly as we move away from the springings.

Thus a preliminary design based on a three-pin arch should be sufficient for most architectural problems.

*9.3. Domes

Shells and lattice shells form a large and complex field of design which we can only touch upon in this elementary text. We have already mentioned briefly the main classification of shell structures (see Section 1.11), and referred to geodesic domes (see Fig. 1.54). We will now examine the design of shell domes.

Domes have a long and successful history in masonry construction. In the age of reinforced concrete they suffer from the disadvantage that the surface cannot be formed by a series of straight lines, and thus the formwork, which is normally constructed from straight pieces of timber, is more expensive than for cylindrical shells or hypars (however, see footnote in Section 1.11.)

As a compensating factor, the forces in thin-shell domes are almost entirely membrane forces, i.e., direct and shear forces contained within the surface of the shell. A membrane can be likened to a balloon or a soap bubble. If we stretch a soap film or a rubber membrane between four boundaries (as shown in Fig. 9.5) we can, without destroying it, pull on it in two perpendicular directions, and we can distort it. Thus there are three distinct internal forces possible in a membrane: N_x and N_y are direct forces, assumed tensile; if the numerical result for either is negative, it represents a

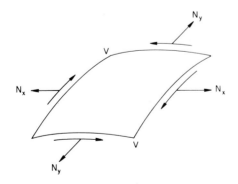

Fig. 9.5.
Direct and shear forces in a membrane.

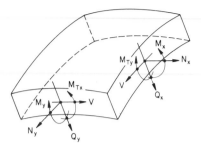

Fig. 9.6.
Forces and moments in a "thick" shell.
N_x and N_y = direct membrane forces; V
= membrane shear forces; Q_x and Q_y
= transverse shear forces; M_x and M_y
= bending moments; M_{Tx} and M_{Ty}
= torsional moments.

compressive force. V are shear forces acting within the surface of the membrane.

It is rarely possible to achieve equilibrium with membrane forces alone. A "thick" shell is one which has sufficient thickness to accommodate moments and transverse shear forces; but "thick" is a relative term since it may denote a few inches in a span of more than a hundred feet (relatively thinner than a eggshell).

The forces and moments in a thick shell are shown in Fig. 9.6 Eight distinct internal forces and moments are possible:

(a) the membrane forces N_x, N_y, and V
(b) shear forces Q_x and Q_y acting *across* the x- and y-sections
(c) bending moments M_x and M_y
(d) torsional moments M_{Tx} and M_{Ty}

The membrane forces only can be accommodated in a thin shell. The transverse shear Q requires thickness, and the moments require a resistance arm within the thickness of the shell (see Sections 6.4, 6.7, and 7.10).

The object of thin-shell design is to limit nonmembrane forces to a small region near the support of the shell, so that a thick shell is required only where its weight causes no significant bending moment (see Fig. 1.1). The appearance of a shell is also improved thereby.

Two conditions have to be satisfied:

(i) The internal forces (and moments, if necessary) must balance the external forces and moments at any point, in accordance with the normal laws of static equilibrium. This problem is statically determinate.

(ii) The deformation of the shell must be compatible with the restraints imposed on it by its own supports. These are called the *boundary conditions*, and since they involve determination of the elastic deformation of the shell, the problem is statically indeterminate.

Only the design of a shell floating in space, such as a rubber balloon, is fully statically determinate; but the boundary conditions of some shells are so simple that they can be designed by the membrane theory. The spherical dome is one of them.

Let us analyze the dome on the assumption that it is a statically determinate thin shell, and examine later the consequences of this assumption. The *membrane theory of shells* was developed by F. Dischinger and C. Bauersfeld in 1923 for the construction of the first planetarium in Jena, the home of the Zeiss optical works. Like every other statically determinate theory it depends only on equating the components of the external and the internal forces in each direction, and taking moments about suitable points; but the mathematics is complicated by the geometry of the shell. We will confine ourselves to a statement of the solution. Some references are

Design of Curved Roofs

given at the end of this chapter for the benefit of readers with a knowledge of differential geometry.

It is convenient to think of the spherical dome in terms of the coordinates used on the earth's surface. If the North pole is the crown of the dome, then the circles of latitude form horizontal hoops, and the meridians of longitude form vertical circles. We thus define the *hoop forces*, N_ϕ, as acting along horizontal circles of latitude, and the *meridianal forces*, N_θ, as acting along vertical circles of longitude. According to the membrane theory for thin spherical shells, the shear

$$V = 0 \tag{9.4}$$

The hoop force

$$N_\phi = wR \left(\frac{1}{1 + \cos\theta} - \cos\theta \right) \tag{9.5}$$

and the meridianal force

$$N_\theta = -wR \frac{1}{1 + \cos\theta} \tag{9.6}$$

where w is the load acting on the shell per unit area, measured on the shell surface; R is the radius of curvature of the dome, constant for a sphere; and θ is the angle subtended by the element under consideration with the crown (Fig. 9.7).

A positive sign denotes a tensile force (for which reinforcement must be provided), and a negative sign a compressive force (which can be resisted by the concrete).

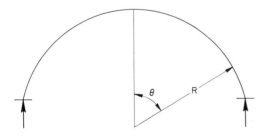

Fig. 9.7.
Coordinates of dome. R = radius of curvature; θ = angle to crown.

Example 9.4. *Determine the forces in a hemispherical concrete dome, 3 in thick, which spans 70 ft.*

The load consists of

weight of 3-in concrete shell	36
internal and external finishes	9
equivalent uniformly distributed live load	3
equivalent wind load	6
Total load per sq ft of shell surface	54 lb

The membrane shears are zero; the direct membrane forces (Fig. 9.7) are as follows:

Angle from crown of dome	N_ϕ (lb/ft)	N_θ (lb/ft)
0°	945 C[a]	945 C
30°	624 C	1,013 C
45°	229 C	1,107 C
60°	315 T	1,260 C
90°	1,890 T	1,890 C

[a]C = compression; T = tension.

The maximum compressive stress in the concrete is due to the meridianal force at the springings

$$f_c = \frac{1,890}{12 \times 3} = 52.5 \text{ p.s.i.}$$

The area of the reinforcement required for the hoop tension at the springings, for a maximum permissible steel stress of 20,000 p.s.i.

$$A_s = \frac{1,890}{20,000} = 0.095 \text{ sq in per ft}$$

This is so low (see Table C2) that the actual amount of reinforcement is determined by constructional considerations, both in the hoop and in the meridianal directions.

Evidently 3 in is a far greater thickness than the structure requires. Whether we can make it thinner depends on waterproofing. If a separate waterproofing membrane is to be used, or if the builder and his workmen have the skill to produce a thinner waterproof structure, we can greatly reduce the thickness. However, the main item in the cost of the dome is the cost of the formwork and the cost of placing concrete on a steeply sloping surface. The cost of the concrete materials and of the steel is of less importance.

Let us assume that the hemispherical dome is supported on a thin ring girder, and that the girder is in turn supported on flexible columns. The girder and the columns transmit the vertical compression of the meridianal forces to the ground. The hoop tension does not require a horizontal reaction, since it is self-anchoring. However, it causes the shell to expand by f_s/E_s (where f_s is the actual stress in the steel). In actual fact we are going to use a great deal more reinforcement than the above calculations indicate, so that the steel stress would be no more than 9,000 p.s.i. (60 MPa), and thus it is reasonable to assume that the concrete has not cracked. The stress, and strain, is thus further reduced. Since the expansion of the shell at its supports is small, the ring girder and columns should be sufficiently flexible to accommodate it. In that case the boundary conditions are satisfied. There is no bending or transverse shear in the shell, and the problem is purely statically determinate.

Design of Curved Roofs

Hemispherical domes are rarely used today, except for planetaria and other structures which require the shape for functional reasons. Classical domes were frequently hemispherical, and it was once thought that it is easier to build a dome in masonry as a hemisphere. As we have seen, there are large hoop forces. In classical domes these were countered by a great thickness of masonry to reduce the stresses; sometimes buttresses, not always visible, were used or chains were employed as reinforcement.

The present popularity of shallow domes is due to functional rather than structural reasons. There is less demand for the monumentality of a tall dome, and the great height adds to the cost of heating and air-conditioning and possibly of sound reinforcement, which received little consideration in classical designs.

The shallow dome is also subject to tension, and this creates problems for masonry construction, although these are different in character, as the following example shows.

Example 9.5. *Design a shallow spherical concrete dome, 3 in thick, which subtends an angle of 60° at the center of curvature, and has a span of 70 ft.*

Since the dome subtends an angle of 60° at the center, its radius of curvature is

$$\frac{\frac{1}{2} \times 70}{\sin 30°} = 70 \text{ ft}$$

We had a similar dome in Example 9.4, except that its radius of curvature was 35 ft. We again take the load as 54 lb per sq ft. Both the hoop and meridianal forces at the crown are 1,890 lb per ft compression, from Eqs. (9.5) and (9.6).

At $\theta = 30°$, i.e., at the springings of the dome, the hoop force is 1,248 lb per ft compression, and the meridianal force is 2,026 lb per ft compression.

The meridianal force is inclined at an angle of 60° to the vertical. The vertical component is absorbed by vertical supports (a wall or columns), but we must absorb the horizontal component $R_H = 2,026 \sin 60° = 1,755$ lb per ft with a tension ring (Fig. 9.8), which is designed in precisely the same way as the compression ring for the cable structure of Example 9.2. An alternative, but expensive, solution is to absorb the thrust transmitted by the meridianal forces with buttresses tangential to the shell (Fig. 9.9).

As we explained in Example 9.2, the tension in the ring is

$$\frac{1,755 \times 70}{2} = 61,400 \text{ lb}$$

This is resisted entirely by reinforcement, and since it is essential to avoid large cracks in the tension ring, we limit the stress to 15,000 p.s.i. Thus $A_s = 61,400/15,000 = 4.1$ sq in. From Table C1 we select six No. 8 bars.

Since the ring is in tension, it *expands*. However, the hoop forces in the dome are still compressive (1,248 lb per ft), so that the edge of the dome *contracts* (Fig. 9.10). The boundary conditions of the dome and the tension ring are incompatible, and thus both are put in bending. We can solve this statically indeterminate problem by a procedure somewhat like that in

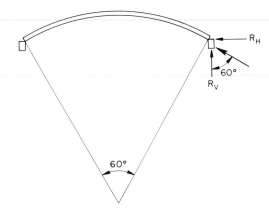

Fig. 9.8.
Tension ring to absorb the horizontal reaction (Example 9.5).

Fig. 9.9.
Buttresses to transmit the horizontal reaction to the ground. (Olympic Stadium, Rome, by P. L. Nervi.)

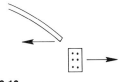

Fig. 9.10.
The tension ring expands elastically under load, while the shell contracts because the hoop force is compressive (Example 9.5).

Example 8.6, where we allowed for the differential stiffness of two joining members. However, this is a complicated and lengthy procedure, and we know from previous designs that the bending stresses are localized in the region of the tie. A proper junction with the tie requires local thickening of the shell, and we can introduce top and bottom steel to cope with the local bending stresses. Because bending stresses do not extend too far toward the crown, we can dimension the shell for bending by empirical rules.

*9.4. Cylindrical Shells

Cylindrical shells require ties for stability (Fig. 9.11). This is easily proved with a sheet of paper, which forms a useful shell structure if it is tied across the ends, but collapses immediately when the ties are removed. In practice the ties often take the form of edge beams, and additional ties are added along the straight edges; however, the latter are not essential.

The membrane forces in circular cylindrical shells (Fig. 9.12) are the longitudinal force

$$N_x = \frac{wx(L - x)\sin \phi}{R} \tag{9.7}$$

the tangential force

$$N\phi = - wR \sin \phi \tag{9.8}$$

Design of Curved Roofs

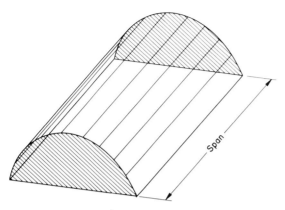

Fig. 9.11.
Cylindrical shells require ties for stability.

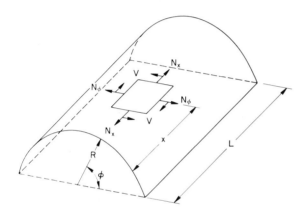

Fig. 9.12.
Coordinates of cylindrical shell: L = span; R = radius of curvature; x, ϕ = coordinates of element.

and the shear force

$$V = -w(L - 2x) \cos \phi \qquad (9.9)$$

where w is the load per unit area of shell, L and R are the length and the radius of curvature of the shell, and x and ϕ are the coordinates of the element under consideration

Cylindrical shells are frequently divided into "short" and "long" shells. Shells are "short" when $2R$ is much bigger than L, and the design is then dominated by the arch action of the tangential force N_ϕ. Shells are "long" when L is much larger than $2R$, and the shell is then dominated by the bending action of the longitudinal force N_x. In intermediate shells (L approximately equals $2R$) neither trend predominates, and it is better to use a noncylindrical shell over a square opening (Fig. 9.13).

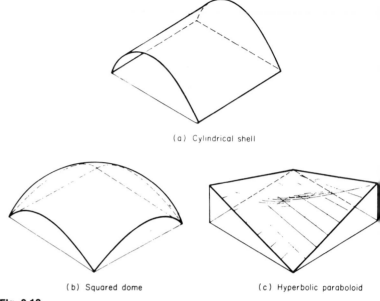

(a) Cylindrical shell

(b) Squared dome   (c) Hyperbolic paraboloid

Fig. 9.13.
Although there are no constructional problems, the mechanics of cylindrical shells makes them less suitable for covering a square opening (a). It is better to use a squared dome (b) or a hyperbolic paraboloid (c).

In fact "short" shells, so called because the straight lines run in the short direction, can be used over larger spans than "long" shells (in which the straight lines run in the long direction), because the stresses in the "short" shell are closer to the membrane condition. Since bending due to the N_x forces (compressive at the crown, and tensile at the bottom edges) is more significant, "long" shells behave rather like beams.

"Long" shells cannot be designed by the membrane theory with any accuracy; but they can be designed as beams spanning between the end frames (Fig. 9.11). We can then apply the methods of normal reinforced concrete design (see Sections 7.10 and 7.11). When shells are designed by the *beam method* it is useful to have edge beams which lower the neutral axis. Since the concrete shell contains a large amount of concrete, and since for horizontal equilibrium $T = C$ (Fig. 9.14a), a substantial quantity of steel has to be accommodated in the edge beams.

The depth of the neutral axis, kd, and the second moment of area (moment of inertia) I, are calculated from the geometry of the section; this involves additional work if the shell has a variable thickness. The maximum compressive stress occurs at the crown at mid-span

$$f_c = \frac{M \times kd}{I} \qquad (9.10)$$

Design of Curved Roofs

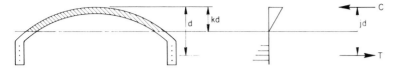

(a) Resistance to bending moment

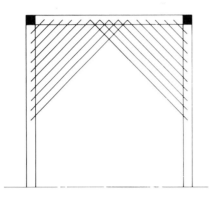

(b) Resistance to shear force

Fig. 9.14.
The "long" cylindrical shell acting as a beam.

and the maximum steel stress at mid-span

$$f_s = \frac{nM(d - kd)}{I} \tag{9.11}$$

where $n = E_s/E_c$ is the ratio of the moduli of elasticity of the steel and the concrete. When the shells are simply supported $M = WL/8$. However, the beam method makes it relatively simple to take account of the continuity of monolithic cylindrical shell roofs, since the shells can also be designed as continuous beams (see Section 8.3).

The shear forces, which have a maximum near the supports in both simply supported (see Section 5.4) and continuous beams, cause diagonal tension in the shell (see Section 6.9) and the main reinforcement is bent up at 45° from the edge of the shell.

Cylindrical shells are more frequently used than any other form of concrete shell, and their design for membrane and bending stresses has been tabulated in *Design of Cylindrical Shells*, American Society of Engineers, New York, 1952, and in *Circular Cylindrical Shells*, Teubner, Leipzig (German Democratic Republic) 1959. Specialized computer programs are also available.

*9.5. Saddle Shells

It may be helpful to coordinate the various teminologies in use for describing the geometry of shells (Table 9.1).

Cylindrical shells are singly curved, with straight horizontal lines in one direction. Domes are doubly curved, with convex curves in two directions at right angles. Saddle shells are doubly curved, and two cuts at right angles normally produce one convex and one concave curve.

Table 9.1
Terminologies for Description of the Geometry of Shells

Dome	Cylindrical Shell	Saddle Shell
Postive Gaussian curvature	Zero Gaussian curvature	Negative Gaussian curvature
Synclastic	Singly curved	Anticlastic
Mountain top	Ridge	Pass

If we think of a mountain range, a dome corresponds to a mountain top, a cylinder shell to a ridge, and a saddle shell to a pass between two mountains.

Domes and cylindrical shells have been used for more than 2,000 years, but they cannot be formed entirely from straight pieces of timber. Since shells are mostly poured in concrete on the site, those saddle shells which can be formed by straight lines have economic advantages. These are the *hyperbolic paraboloid* and the *hyperboloid*.

The hyperbolic paraboloid is formed by a straight line moving over two other straight lines inclined to one another (Fig. 9.15a), but if the shell is cut at 45° to the generators the same surface looks quite different (Fig. 9.15b). The hyperboloid is formed by rotating a straight line placed at an angle about a central axis, and another series of straight lines at an angle to the first can be drawn on the same surface. This is the familiar "cooling tower" shape (Fig. 9.15c). We can produce a different looking surface by cutting the shell at a right angle (Fig. 9.15d).

The hyperbolic paraboloid shell, generally abbreviated to *hypar shell*, has found the widest application of all the saddle shells, because it can be made to cover a square or rectangular plan or a series of rectangles. It can also be assembled into spectacular curved shapes (Fig. 9.16).

The membrane solution of the hypar shell is simple. The direct forces

$$N_x = N_y = 0 \qquad (9.12)$$

The shear force

$$V = \frac{w'ab}{2c} \qquad (9.13)$$

Design of Curved Roofs

(a) Hyperbolic paraboloid generated by a straight line moving over two other straight lines at an angle to one another.

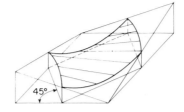

(b) The same shell cut at an angle of 45°.

(c) Hyperboloid generated by a straight line moving over a circle.

(d) The same shell cut at an angle of 90°.

Fig. 9.15.
Saddle shells formed by straight lines.

Fig. 9.16.
The hyperbolic paraboloid can be formed to cover square, rectangular, or circular plans. It can be used conservatively for industrial applications or be assembled into spectacular curved shapes (see also Fig. 1.66).

where w' is the load per unit area, measured on the *plan* of the shell (not its surface), and a, b, and c are the dimensions of the shell, as shown in Fig. 9.17).

Slightly more complicated equations need to be used if the hypar shell, instead of being relatively flat, as in Fig. 9.17, is strongly curved, as in Fig. 9.16d.

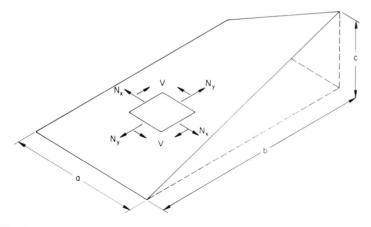

Fig. 9.17.
Coordinates of hyperbolic paraboloid: a, b = span; c = rise.

Example 9.6. *Determine the forces in a 3 in thick hypar shell, spanning 25 ft in both directions, with a rise of 5 ft.*

The load consists of:

weight of concrete, measured on plan	38
external finish (off-the-form inside)	3
equivalent uniformly distributed load	5
wind load (actually suction)	0
Total load per sq ft, on plan	46 lb

The direct forces are zero, and the uniform shear force in the shell

$$V = \frac{46 \times 25 \times 25}{2 \times 5} = 2,880 \text{ lb/ft}$$

The shear force produces numerically equal compressive and tensile forces at 45° to the sides of the shell (see Section 6.8). The tensile forces take the form of cable curves, and the compressive forces the shape of arches. Thus we can think of the hypar shell as a series of interconnected arches and suspension cables.

Since the maximum (diagonal) compressive force in the concrete is 2,880 lb/ft, the maximum compressive concrete stress

$$f_c = \frac{2,880}{12 \times 3} = 80 \text{ p.s.i.}$$

The area of reinforcement required

$$A_s = \frac{2,880}{20,000} = 0.144 \text{ sq in/ft}$$

It is not advisable to lay the steel in the direction of the principal tension, partly because every bar along the diagonals would have to be cut a different length, and partly because we require reinforcement in both directions, if only to control shrinkage and temperature stresses (see Section 8.9). If we use two sets of bars parallel to the sides, each contributes cos 45° to the tensile force, so that we have $2A_s \cos 45° = \sqrt{2} \; A_s$. Consequently the area of steel required parallel to each set of sides is

$$\frac{0.144}{\sqrt{2}} = 0.102 \text{ sq in}$$

This is too low (see Table C2), and the actual amount of reinforcement is determined by constructional considerations.

The membrane theory of hypar shells is deceptively simple; in fact, the bending stresses in most hypar shells are quite high. While the statically indeterminate theory of shells has been worked out for a wide range of domes and cylindrical shells, there are no specific solutions available for hypar shells at the time of writing. The results can be calculated from the general theory of shells, but this is laborious, even by computer.

The great majority of hypar shells built so far have had a much smaller span than the largest domes and cylindrical shells in exis-

Design of Curved Roofs

tence. Empirical design methods, based on the membrane theory, can be used so long as the spans are small.

From a structural point of view, we can make the shell thinner than 3 in; but the thickness depends on the skill of the builder and his workmen and on the extent to which an occasional leakage of water is acceptable.

*9.6. Folded Plates

Folded plates with simple support conditions can be designed by the beam theory. Thus the folded plates shown in Figs. 1.67 and 1.69 are simply supported "beams," whose cross-section is a folded plate (Fig. 9.18). The depth of the neutral axis, kd, and the second moment of area (moment of inertia), I, are calculated from the geometry of the section, and the maximum compressive stress in the concrete and the maximum steel stress then follow from Eqs. (9.10) and (9.11), as for cylindrical shells. Shear reinforcement is provided by bending up the main reinforcement toward the support, as for "long" cylindrical shells (see Fig. 9.14b).

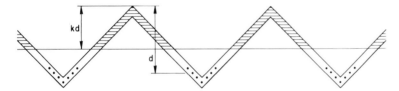

(a) Cross section of structure shown in Fig. 1.67

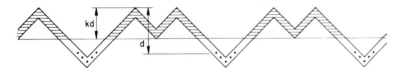

(b) Cross section of structure shown in Fig. 1.69
(the cross section varies along the span)

Fig. 9.18.
Folded plate structure acting as a beam.

Folded plates which form part of rigid frames (Fig. 1.70), or which act like plates or shells, can only be designed as statically indeterminate structures. The simplest method for the design of folded plates is by moment distribution (see Section 8.6). The plate is "cut apart" at each angle joint, and the straight pieces are rigidly restrained at their ends. Each joint is then released in turn, and the moments are distributed, until the residual differences are sufficiently small (Fig. 9.19).

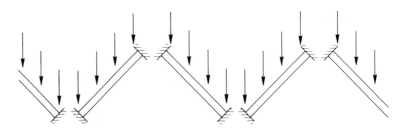

Fig. 9.19.
Solution of folded plate by moment distribution. The structure is "dissected" at each joint, which is firmly clamped. The joints are then released in turn, and the moments are distributed until the residual differences are sufficiently small.

Suggestions for Further Reading

This chapter deals only superficially with the design of curved roofs. A more thorough treatment requires a better knowledge of mathematics than the author has assumed throughout this book.

The best book on cable roofs is:
F. Otto: *Tensile Structures*. M.I.T. Press, Cambridge, Massachusetts. Vol. 1, 1967, 320 pp., Vol. 2, 1969, 171 pp.

There is a vast literature on shell structures, but a great deal of it is directed toward the problems of the aircraft industry. The best book on the elementary membrane theory is:
A. Pfluger: *Elementary Statics of Shell*. Dodge, New York, 1961, 122 pp.

The most architecturally oriented books on the general theory of shells are:
D. P. Billington: *Thin Shell Concrete Structures*. McGraw-Hill, New York, 1965, 322 pp.
A. M. Haas: *Thin Concrete Shells*. Wiley, New York, Vol. I, 1962, 129 pp.; Vol. II, 1967, 242 pp.

Although several books have been published on the design of folded-plate roofs, none can be recommended to architects. The original paper on the moment distribution method is:
G. Winter and M. Pei: Hipped plate construction. *Proc. American Concrete Institute*, Vol. 43 (1947), p. 505.

A formulation of folded-plate design particularly suitable for computer analysis is:
A. G. Scordelis: Matrix formulation of folded plate equations. *J.*

Design of Curved Roofs

Structural Division, American Society of Civil Engineers, Vol. 86, No. ST10 (October 1960), p. 1.

Problems

9.1. Explain the principles which make suspension structures particularly suitable for single-cell buildings of great span. What are the principal limitations?

9.2. Compare the relative advantages and limitations of suspension structures and space frames for long-span roofs. Sketch two alternative designs each for a roof for (a) a covered market, and (b a covered sports stadium, using either of these methods. In each case give reasons for your choice, and approximate dimensions for the principal members.

†**9.3.** A steel cable spans 360 ft between supports, with a sag of 25 ft at mid-span. It carries a uniformly distributed load of 50 lb per foot run. Determine the cable tension at mid-span, the cable tension at the supports, and the reactions transmitted to the supports. If the maximum permissible steel stress is 120,000 p.s.i., calculate the cross-sectional area required for the cable.

9.4. Figure 9.20 shows a mast for the support of a suspension cable. Calculate the forces in the members AB and AC. Hence

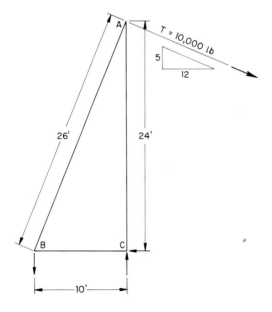

Fig. 9.20.
Problem 9.4.

†Similar problems are given in Appendix G.

calculate the cross-sectional area required for AB and AC, using permissible stresses of 6,500 p.s.i. in compression, and 20,000 p.s.i. in tension.

9.5. Why are three-pin arches statically determinate, while arches with fewer pins are statically indeterminate? How are statically determinate arches designed?

9.6. A laminated-timber three-pin arch of semicircular shape is loaded by two beams as shown in Fig. 9.21, each beam transmitting a load of 2,000 lb to the arch. The cross-section of the arch is 6 in wide by 12 in deep. Determine the reactions at the support of the arch, and sketch the bending moment diagram. From measurements on this diagram, calculate the approximate maximum tensile and compressive stresses, and indicate which is on the outside of the arch.

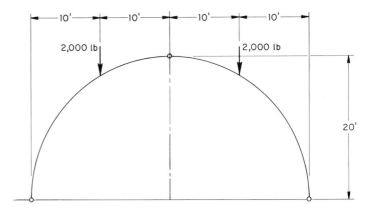

Fig. 9.21.
Problem 9.6.

9.7. With the aid of diagrams, discuss the basic assumptions made in the membrane theory of shells, and the requirements which the structure must meet so that this theory can be used as a basis for design. Describe the principal shell forms in present use, and indicate which can be designed by the membrane theory, giving your reasons.

9.8. Explain how thin-shell domes carry vertical loads across a horizontal gap by membrane action (a) if the dome is hemispherical, and (b) if it is shallow. Define the boundary conditions required in each case.

With the aid of modern examples, describe two methods of absorbing the horizontal reaction in shallow domes, and explain how each affects the design of the shell.

9.9. Distinguish between "short" and "long" cylindrical shells, and explain what effect the distinction has (a) on structural design and (b) on architectural design.

Design of Curved Roofs

9.10. With the aid of sketches describe several different ways in which hypar and hyperboloid shells can be used singly or in combination.

9.11. A residential college requires a covered spectators' stand for its sports ground to accommodate approximately 1,000 persons. This stand is to be built on level ground and is to be situated at a prominent position within the college campus. Only the minimum of ancillary service rooms are required to be incorporated, i.e., changing rooms and toilets.

Accepting these broad requirements, give a suitable form for the primary structure of this stand. Any suitable materials may be used, and no detailed planning is necessary. Justify the major structural sizes nominated by means of approximate calculations or, where this is too complex, a critical written appraisal.

9.12. A new university building includes a lecture theatre located in front of the main building. The theatre capacity is 250, and full artificial lighting and air conditioning is required.

A proposal has been put forward for a circular theatre with a shallow domical roof, the whole theatre being supported on columns. The access is from the main building via a bridge linking the two blocks.

Accepting the above decisions, present a possible structural solution in any structural material appropriate. Consideration should be given to the problems of acoustics, noise insulation, and the thermal properties of the theatre.

9.13. It is proposed to erect a wholesale market hall as part of an area for the marketing of primary produce. The broad requirements for this building are as follows:

(a) overall area—approximately 80,000 sq ft (8,000 m^2) on one level contained within a site of 400 ft (120 m) by 250 ft (75 m) size;

(b) functional requirements—good natural lighting and ventilation, with a minimum of maintenance;

(c) structural requirements—material either steel or concrete; the structure should have the minimum number of internal columns consistent with economy.

Design a suitable structural system, considering all relevant aspects of the functional efficiency of the building and taking account of the fact that the site is in a prominent position on the perimeter of the city.

Aids to Structural Teaching   **A**

This Appendix is for the Benefit of Instructors

A.1. The Value of Visual Aids

Teaching science to architects presents problems unlike those encountered in any other part of the university curriculum. The architect is not a scientist, and he has rarely the basic training expected from a student in a science or engineering department. At the same time he is required to have a knowledge of scientific principles of building that goes far beyond the scope of "popular science."

The investigation of the technical problems of building on an extensive scale by scientific methods is comparatively new, and our educational methods have generally failed to keep in step with developments in architectural practice and with the ambitions of our better students.

These remarks apply to the entire field of architectural science, and it is convenient to arrange visual aids as a unified service to all subjects. However, we will confine ourselves here to structural teaching.

We find today students introducing into their design structural forms so complex that some are not included in the undergraduate engineering curricula, but are reserved for graduate courses. Clearly no conceivable revision of the mathematics prerequisites would enable us to teach these structures in the same way as the engineers do. Yet students are entitled to expect the same type of guidance from the teaching staff on the use of complex forms, as they would expect if they used the older and simpler statically determinate structures.

It may be assumed that the architect calls on a structural consultant for all but the simplest work, so that he needs to understand structural principles and sizes only in general terms; the determination of detailed structural dimensions is the task of the consultant. It may also be assumed that the average architecture student is more receptive to visual demonstrations than to mathematical treatment.

The exclusive use of computations in structural courses for architecture students is therefore unhelpful. Structure is in fact a physical concept, which can be explained in terms of mathematics; but some aspects can be explained even better with models. An "architectural model," i.e., something just to look at, is of limited value, particularly if it is not to scale. On the other hand, a working model need not be accurate, since we are not concerned with precise numerical answers; we want the model to demonstrate principles.

It is desirable to have a store of models which can be used during lecture demonstrations and for demonstrations in the studio to explain a design under discussion. It is even better to provide a proper laboratory in which students can perform experiments for themselves. A model demonstration is roughly equivalent to looking at design drawings. This may be preferable to just talking about design; but it is probably still more instructive for the student to work on the drawing board. In the same way students can discover the solutions for themselves in the laboratory.

A laboratory used purely for teaching need not be large, nor need the equipment be expensive. Indeed, simplicity is a merit in both demonstration models and laboratory experiments. Research in architectural science, which requires precision equipment, is a different matter.

A.2. Techniques for Demonstration Models

Demonstration models should be properly made. It is irritating when a demonstration performed before a class fails to work, and it wastes much valuable time. Models made by students during a class are therefore not satisfactory. Even if they work when first made, they deteriorate without proper maintenance. There is, moreover, no educational value in making structural models; it involves skills which bear no relation to an understanding of structures and of architecture. The author thus regards the services of a technician who has access to a workshop as essential. The workshop requires only the tools normally used for architectural model-making, and the operation can be combined with the facilities which many schools already possess for making models for the architectural studio.

Structural models are mainly of five kinds:

(i)   Models which demonstrate forces with the aid of spring balances. Compression is demonstrated by crossing the arms of a tension balance (Fig. A.1).

(ii)   Models which demonstrate elastic deformation by means of

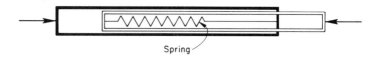

Spring

Fig. A.1.
Compression balances jam unless their loading is perfectly concentric. It is therefore better to use tension balances in models and to reverse the direction of the force.

a flexible material, such as rubber. The deformation is shown by the distortion of a grid drawn on the rubber.

(iii)   Models which demonstrate tension, compression, or shear by means of highly deformable materials, such as foamed plastic. These materials are also obtainable with a grid of regularly spaced holes.

(iv)   Models which demonstrate strains by the opening up of slots. These can be made from plywood or plastic. The slots terminate in a drilled hole to prevent propagation of the crack. Ordinary timber is liable to split after one or two demonstrations, and metal is not sufficiently flexible.

(v)   Models which demonstrate slope and deflection. Strips of spring steel, and strips of Plexiglas are suitable for making frames and arches.

A.3.   Some Demonstration Models for Use With This Book

Chapter 1.   (a)   We pointed out in Section 1.2 that bending causes measurable deflections in beams. We can load a beam (which·is too thick to buckle) and show that a load which produces a visible deflection in bending produces no visible deflection in compression or tension.

(b)   A curved structure requires horizontal as well as vertical reactions. This is shown with a cable set up as in Fig. 1.10. A string of beads loaded with lead sinkers is suitable.

(c)   The varying shape of the cable under concentrated and distributed load can be shown as in Fig. 1.19.

(d)   The various methods of supporting a floor slab (Section 1.9) may be illustrated with a mat of foamed plastic. This is laid directly on the columns, or on beams ($\frac{1}{2}$-in-sq timber), arranged one-way or two-way on the columns. The load, by lead sinkers, can be arranged in various ways. The shape of the curvature indicates the distribution of tension and compression in the slab; the deflection is also observed.

(e)   A pneumatic membrane (Section 1.10) is demonstrated with a toy balloon, or with soap bubbles.

(f)   We can illustrate with a sphere (or an orange) and a ruler that a dome cannot be formed from straight pieces of timber (Section 1.11).

(g)   The formation of a hyperbolic paraboloid and a hyperboloid from straight lines merely requires a frame and some string. The difference in the appearance of the hypar shell at 90° and 45° is also instructive (see Fig. 9.15).

(h)   The need for a tie in a cylindrical shell is best illustrated with thin sheet brass; paper tears too easily. String is sufficient for the ties (Fig. 1.62).

(i)   The strength of shell structures is illustrated by an egg, loaded through two rubber caps to provide approximately uniformly distributed loading. The rubber caps can be cut from the rubber bungs used in large bottles. The egg is blown, since the fracture of a full egg is too messy. If concentrically loaded, it carries at least 100 lb (45 kg); the author once reached 180 lb (82 kg) with an ordinary hen's egg.

(j)   The experiment on folded plates (Fig. 1.67) can be repeated for other folded plates (Figs. 1.69, 1.70, 1.71, etc.). The variation of the cross section, required for solving Eqs. (9.10) and (9.11) in Section 9.6, is demonstrated by cutting the section with scissors.

Chapter 2.   We will discuss buckling models under Chapter 7, and a creep model under Chapter 8.

(a)   The collapse of the masonry arch (Fig. 2.1) illustrates that structures can fail without overstressing the material. The blocks need to be heavy, but they should be protected from damage when the arch falls. This rules out lead and timber by themselves. The author recommends hardwood voussoirs with a large hole drilled through the center, which is then filled with lead. It is advisable to suspend a net under the arch, to catch the blocks when they fall. If the corners are damaged during collapse, the experiment does not work the next time.

(b)   Temperature movement is demonstrated by heating a steel rod. The end of the rod has a hole, which is connected to a cold bar with a glass rod. Students with some knowledge of physics may regard this as obvious.

(c)   Wind-pressure patterns can be demonstrated two-dimensionally with liquid which contains small pellets of unexpanded polystyrene. The image can be projected on a screen with a slide projector.

Chapter 3.   (a)   The experiments illustrated in Figs. 3.5 and 3.13 are so simple that students may regard their performance as superfluous.

(b)   The middle-third rule (Fig. 3.31) is demonstrated with a block of foamed plastic, divided into three parts by thick lines. The same technique can be used for retaining walls.

(c)   Many students have difficulty in visualizing moments, or at least in distinguishing between positive and negative moments. It would be helpful for students to spend some time exercising with models which illustrate moments in the laboratory (if there is one) or else in the studio (see Section A.5).

Chapter 4.   (a)   The forces in trusses are illustrated with tension and compression (Fig. A.1) spring balances. Only two balances may be used at any one time (otherwise the frame distorts too much); but the balances can be changed over quickly when the members are joined by pins pushed through holes.

Aids to Structural Teaching

(b)   The $n = 2j - 3$ rule is demonstrated with a set of bars and screws. It is desirable to show that the length of any member of a statically determinate truss ABCDF (Fig. A.2) can be changed without changing the character of the truss. If we introduce a statically indeterminate member BE, it is no longer possible to change the length of the members, unless we use force. If we remove a member from the statically determinate truss, we get a mechanism.

(c)   Architectural models help to illustrate the shape and composition of space frames, and to show the essential difference between geodesic domes ($n = 2j - 3$) and three-dimensional space frames ($n = 3j - 6$).

(d)   If there is a laboratory, an experiment on space frames is recommended, to show the nature of the forces and the end restraints (see Section A.5).

Chapter 5.   (a)   The Gerber beam becomes clearer with a pin-jointed model (see Fig. A.2).

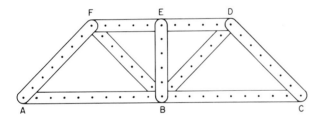

Fig. A.2.
Statically determinate frame. The frame is assembled from members with holes along their length (obtainable in a toy shop). It is thus possible to vary their length to show that the length of members in an statically determinate frame ABCDF can be varied without producing stresses. If a statically indeterminate member BE is introduced, the length of BD, BE, BF, and DF can no longer be varied without "prestress." If any one member is removed from the frame ABCDF, it becomes a mechanism. The same pieces can be used to assemble a Gerber beam.

(b)   Bending moment and shear force are difficult concepts. Even engineering students, who know exactly how to calculate them, sometimes fail to understand why they balance the external loads. Students should be given an opportunity of handling the model shown in Fig. 5.4. In a laboratory the same experiment can be performed quantitatively (see Section A.5).

(c)   The horizontal reaction of the three-pin portal and the action of the top pin can be shown in a pinned model tied with a spring balance and a turnbuckle (to bring the portal back to its original span when the spring extends).

(d)   The superiority of the arch over the portal is shown with two models made of slotted plywood, which indicate the nature of the strains. The arch is seen to have much lower tensile strains.

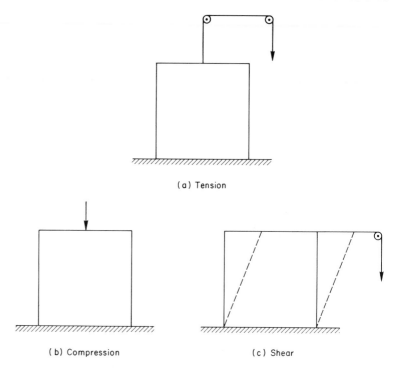

(a) Tension

(b) Compression　　　　　　　　(c) Shear

Fig. A.3.
Model demonstration of strain.

Chapter 6. (a) Strain is another difficult concept. It can be illustrated with a cube of foam rubber which is tensioned, compressed, and sheared (Fig. A.3).

(b) Some specimens of steel and concrete (before and after testing) help to illustrate the difference between ductile and brittle failure, and between shear failure and tension failure. Where there is a laboratory, the student should subsequently perform the tests for himself.

(c) Navier's assumption is illustrated with a rubber beam on which a grid has been drawn (Fig. 6.5).

(d) Torsion (Section 6.7) also can be illustrated with a rubber beam.

(e) Another simple torsion experiment illustrates the superiority of the tube. We place a piece of flat sheet metal in a torsion balance, and measure the rotation. We then test a channel section bent from a piece of the same dimensions, and a tube bent from a piece of the same dimensions, but with the joint open. All have the same torsional rotation. If we now take an identical tube with the joint welded, the torsional stiffness is greatly increased.

(f) Quite simple photoelastic apparatus is adequate for demonstration (Fig. 6.30). The patterns are easier to see if they are projected on a screen.

Aids to Structural Teaching

Chapter 7. (a) The laws of buckling become clearer with models (Figs. 7.1, 7.2, and 7.3). It is important to demonstrate that the pure buckling failure is an elastic phenomenon, and the strut returns to its original shape when the load is removed. This can be compared with a piece of thick, short steel permanently deformed in a compression test.

(b) Joints (Section 7.4) in structural steel, aluminum, and timber are more readily explained on samples, preferably on the model scale. Manufacturers may offer to supply these.

(c) The layout of reinforcing cages also becomes clearer with models. A transparent resin, such as Araldite, can be cast around a model brass cage.

(d) Model (d) cited under Chapter 1 in this Section is useful for Examples 7.11 and 7.12.

(e) The model illustrated in Fig. 3.31 can be used again for eccentrically loaded columns.

(f) Prestressing is illustrated with a model consisting of wood blocks, rubber, and a screw (Fig. 7.25). The rubber should be about 1/16th in thick for blocks about 2 in wide (1.5 mm by 50 mm).

Chapter 8. (a) The slope-basis (Section 8.2) of the Theorem of Three Moments is illustrated with a slotted plywood beam (Fig. 8.4). Whichever way the load is applied, the slope over any continuous support is the same on either side (up on one side, down on the other).

(b) The superior strength of rigid frames is illustrated with a simply supported beam and a rigid frame made from the same spring-steel. Both are supported over the same span, and both carry an obviously identical load. The simply supported beam has a much greater deflection.

(c) The pinned-frame concept for wind loading can be demonstrated with a frame made from spring steel or Plexiglas (Fig. 8.9a).

(d) Moment distribution (Section 8.6) can be illustrated with a model devised by Professor A. J. S. Pippard (*The Experimental Study of Structures*, Arnold, London, 1947, p. 74). Each joint is released in turn by the operation of a spring-loaded mechanism.

(e) Sidesway (Fig. 8.16) is illustrated with a spring-steel portal, preferably with a height greater than the span. Even symmetrical loading is seen to induce large bending moments in the columns.

(f) Architectural models illustrating diagonal wind-bracing (Section 8.8) can be made from cardboard and string. If the diagonals are left slightly loose, they come visibly into action.

(g) Creep (Section 8.9) is best explained in terms of rheology. Every motor car and motor cycle contains a spring and dashpot in parallel, for damping, and this is the rheological model for creep. The motor-car assembly is too large for a lecture room, but a small model can be made (Fig. A.4). A dashpot is obtainable from any store which supplies school science laboratories.

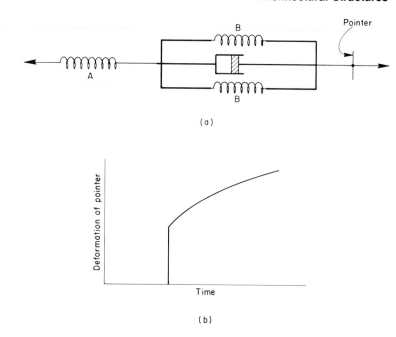

(a)

(b)

Fig. A.4.

Rheological creep model. When the model is loaded, there is an instant elastic deformation due to the spring A. The system B deforms only with time, because of the delaying action of the dashpot. The movement of the pointer with time is shown in (b).

Chapter 9. Some of the models mentioned under Chapter 1 are worth showing a second time.

If a laboratory is available, students can gain insight into the behavior of shells by personally conducting model tests; otherwise they may be able to do so in the studio.

The minimal surface between a given set of boundaries is easily established. We form the outline from soft copper wire and dip it into concentrated detergent solution. The surface remains long enough to make a sketch. A more permanent result is obtained with rubber solution, or with canvas treated with the dope used for model aeroplanes.

If the school has facilities for making precision models (see Chapter 4, H. J. Cowan *et al.*, *Models in Architecture*, Elsevier, London, 1968, 228 pp.), some strain gauge experiments on models are most illuminating.

The author regrets that it is not practicable to describe the models in detail or reproduce more illustrations, since this would increase the price of the book. However, he will be pleased to supply additional information to any instructor who wishes to use them in a course.

A.4. A Structural Laboratory Program for Architecture Students

Students gain a much clearer understanding of structural mechanics if they can carry out experiments for themselves. It is necessary to lay down precise rules in the junior years and to provide supervision and instruction sheets. In the senior years students should be able to frame their own programs, preferably in conjunction with their studio work (see Section A.6).

Although it is desirable to have the experiments in the same week as the lectures, it is impracticable to provide every student with equipment for every experiment at the same time. A rota must be adopted; for example, if a laboratory program includes 10 experiments, then 10 laboratory periods are required for every group to get a turn; and some students will perform the experiments early in the course, and some toward the end. It is expensive in terms of space and equipment to allow students to work singly, and some experiments need four hands. However, groups should be as small as possible; two or three is a good number. The structural experiments can be integrated with laboratory work on other subjects, to provide simpler scheduling.

Some of the experiments require only small apparatus on a table. For breaking actual structural materials, we require testing machines. A portable concrete cylinder-press may, with packing, also be used for timber and metal specimens tested in compression. A testing machine for beams and shells can be made with standard hydraulic jacks pressing against a welded steel frame (Fig. A.5); it is not as accurate as a proper testing machine, but it is much cheaper.

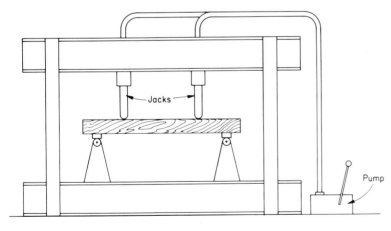

Fig. A.5.
Inexpensive beam testing machine. The testing machine consists of a welded frame, two pin bearings, two hydraulic jacks, and a pump. The machine can also be used for testing shells, which are loaded through a load-distributing bed of sand.

We do not require the accuracy, since we only wish to demonstrate principles, and the homemade machine has a clearer action then the fully enclosed commercial tester.

Since concrete is the dominant structural material, it is desirable to be able to make and break reinforced and prestressed members. This requires a platform for mixing concrete and some bins for material storage. A small mixer is helpful.

The author found only one real difficulty in setting up an architectural laboratory in an old building never intended for the purpose, and that was the disposal of surplus wet concrete. The cost of the hardware for a small laboratory is a minor consideration; it corresponds roughly to the salary of one professor for one year. The real cost lies in staffing the operation after the initial enthusiasm has worn off.

Although the laboratory program depends on the school's timetable, the reader may find it convenient to have experiments arranged by chapter. The author will be pleased to give additional information to interested instructors.

A.5. Some Laboratory Experiments for Use With This Book

Chapter 2. a) The difference between the failure of weak joints and the overstressing of the material is demonstrated by a comparison of the failure of a voussoir arch (Fig. 2.1) and a plaster arch of the same dimensions. The plaster arch fails in tension. If time permits, another plaster arch may be cast with reinforcement to fail in compression.

(b) If the school has access to a wind tunnel, smoke patterns produced by various building shapes can be studied, and the total wind pressure measured.

Chapter 3. (a) The coefficient of friction for various building materials is determined from the force required to move a heavy weight over a horizontal platform. Although the experiment is very elementary, it illustrates that rough concrete may have a lower coefficient of friction than smooth concrete, because particles broken off the surface form rollers.

(b) The center of gravity of standard steel angles, a gravity retaining wall, and a free-shape table is determined by hanging cut-outs from several points around the circumference and finding the intersection of the vertical lines. The result is checked by calculation. Many students will find this too elementary.

(c) The principle of moment is verified with levers and wheels, variously loaded. Although some students may argue that they know this already, the experiment will probably prove them wrong.

Aids to Structural Teaching

Chapter 4. (a) The forces in a plane frame are measured with spring balances and checked by means of a stress diagram or some other method.

(b) The behavior of a plane frame at failure differs from that under working loads. Students design a truss in balsa wood (which buckles in compression) and copper wire (which breaks in tension) and test it to destruction. The weakest member is strengthened, and the test is repeated.

(c) The forces in a space frame are determined with spring balances, and the result is checked by calculation. The movement of the members at the supports are observed.

Chapter 5. (a) The shear force and bending moment in a cantilever are checked experimentally by inserting spring balances which record the component forces (Fig. 5.4 with spring balances).

(b) The reactions of a simply supported beam are calculated and measured with two kitchen balances. The loads at which the mid-span bending moment changes sign is then calculated, and this is checked with a hinged beam (Fig. A.6).

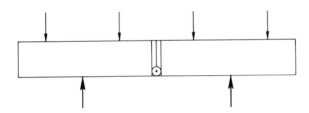

Fig. A.6.
Demonstration of bending moment reversal. The beam is hinged on one side, so that it is stable as long as this is the tension side. The inner loads are applied first, and the outer loads later. When the bending moment reverses, the joint opens up.

(c) The horizontal reactions in a three-pin arch are calculated for various loading conditions and then checked with a spring balance. The model must include a turnbuckle to restore the original span. The experiment could usefully be duplicated with a north-light portal frame.

Chapter 6. (a) The modulus of elasticity of steel, copper, and aluminum wires is determined as described in Section 6.2 (Fig. 6.4).

(b) This is a good time to introduce students to mechanical and electrical strain gauges with some simple tests, which can be checked by calculation.

(c) Navier's assumption that plane sections remain plane is verified with mechanical or electrical resistance strain gauges (Fig. 6.6).

(d) The central deflection of a beam is measured with a dial gauge, and the end slope with a long pointer moving over a scale.

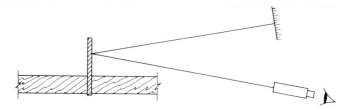

Fig. A.7.
Slope can be measured with a mechanical pointer or a light "pointer." If the angle is small, it is more accurate to fix a mirror to the beam and observe in it the reflection of a lit-up scale through a telescope (which has a diaphragm as a reference line).

Some may find it easier to measure the slope with a mirror, which reflects the reading of a lit-up scale (Fig. A.7).

(e)   Photoelastic equipment adequate for quantitative work is quite expensive, but it is a simple matter to study the general pattern of the trajectories for various structural sections, and the effect of stress concentrations.

Chapter 7.   (a)   If the school has a press, it is instructive to test structural materials in compression (tension tests require better facilities): steel, aluminum, cast iron, or hard brass for a brittle metal failure, hardwood, softwood, laminated timber, brick, and concrete.

(b)   Concrete is such an important and easily misused material that it is desirable for students to make their own cylinders and see how the strength and the workability vary with the water:cement ratio.

(c)   If facilities permit, students should cast and test reinforced and prestressed concrete beams. Each group could cast a different type of beam. Some should be designed to fail in tension, some in compression, and some in shear. Although this is a long experiment, most students enjoy it.

(d)   Tests on joints, particularly in timber, are worthwhile. Students should make the specimens, as well as test them in the press.

Chapter 8.   If the school has strain measuring equipment, some portals and rigid frames should be designed, and model-analyzed. A model is easily made by gluing strips of Plexiglas.

Chapter 9.   Experiments on minimal membrane surfaces are recommended (see Section A.3).

Experiments on model suspension structures and model shells depend on the availability of the necessary equipment. Electric resistance gauges are most suitable for strain measurement. The simplest terminal box is a battery-operated, transistorized one employing a Wheatstone bridge circuit.

It is simple to make model suspension structures, but attaching strain gauges to the wires requires great skill.

Aids to Structural Teaching

It is simple to attach strain gauges to models made from brass or Plexiglas, but forming shell models requires special facilities.

Model domes can often be cut from spherical shapes commercially available, and cylindrical shells may similarly be obtainable from tubes. It is only necessary to locate a well-made piece with a reasonable ratio of thickness to span. For other shell types it is necessary to develop suitable casting or forming techniques (using heat and pressure or vacuum).

Jointing the parts of a model is not difficult. Brass can be brazed, and Plexiglas welded with the glue provided for the purpose.

The parts of a folded plate can easily be cut from a flat sheet.

A.6. Approximate Structural Design in the Architecture Studio

The most important aid to the successful teaching of structures is its integration with architectural design. In practice the architect produces the original design concept, which includes the form of the structure; the engineer undertakes the structural calculations.

No significantly useful purpose is served if architecture students work out some reinforced concrete slabs or some steel connections in the same way as they might draw up a detail for a nonstandard door. The structural decisions made by the architect lie mainly at the preliminary design stage, and it is there that structure should be considered in a university design course.

To accomplish this successfully, the instructors for both structural and architectural design must accept some restrictions. The design program must allow time for structural decisions at the right moment, i.e., while the student is still forming his ideas on the solution of the problem. The structural instructor must be available at this critical time.

Ultimately the decisions must be made by the student himself, and if the structural course has not taught him to do so, it has not been successful. The instructor is there to advise, not to act as the consulting engineer.

The student should be able to state clearly—if need be, in writing —why he has chosen a particular structural solution in preference to others. He should understand the manner in which the loads are transferred to the foundations and how they are absorbed.

He should make small models from cardboard and string, to check that diagonal wind-bracing or shear panels have been correctly provided.

He should pick out the critical members, and make rough calculations of their sizes. To do this effectively, he should make appropriate simplifications.

A statically indeterminate arch or frame should be converted into one which is statically determinate as a reasonable approximation. This is conservative, and the sizes will normally be smaller for the statically indeterminate structure. The calculation should take less than half an hour.

Similarly a Warren truss or a Vierendeel truss can be roughly sized by considering it as a beam consisting of two flanges separated by a resistance arm D (Fig. A.8).

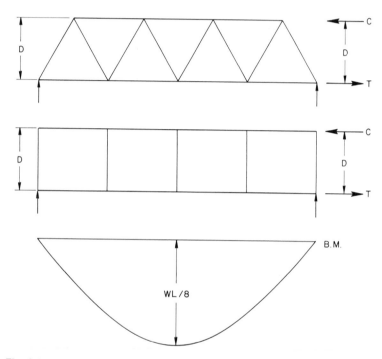

Fig. A.8.
The feasability of a Warren truss and of a Vierendeel truss can be estimated by considering each as a beam, consisting of two flanges separated by a resistance arm.

The size of the columns of a building a few stories high depends mainly on the load carried by the building, W. The column load is approximately W/n, where n is the number of symmetrically spaced columns. The weight of a building consists of its own weight and the live load. The weight of brick, concrete, and concrete-encased steel is approximately 144 lb per cu ft, or 1 lb for each 1 ft length per sq in of section, and in most buildings this is by far the heaviest load. In metric units the force due to the weight of brick, concrete, and concrete-encased steel may be taken as approximately 0.02 N per m length per mm^2 of section.

It is not necessary to be too accurate at the conceptual design stage. Even precise structural calculations are rarely correct within 1%, and preliminary calculations are quite satisfactory if they are within 20%. Even a 100% error is not disastrous, although the structure may look heavier than the architect intended. An error of 1,000%, however, makes a complete revision of the concept inevitable.

The structural instructor should not insist on unreasonable precision in architectural design, but should, on the contrary, encourage simplifications which enable students to form their own estimates with a minimum of calculation. In turn, the architectural design instructor should not encourage the use of structures that neither he nor the student can understand.

Unfortunately the structures used in the architectural design problems of many schools bear little relation to the structural teaching or, indeed, to the realities of building. Many architectural instructors believe that the use of a rectangular frame shows a lack of imagination, even though it is the obviously sensible solution to the problem, and they suggest exotic structures in the early years of the course.

The use of unusual structures should be encouraged in the final year; but students should also be given facilities for model-analyzing the critical parts of their structures, if they are unable to determine their safety by theoretical arguments.

It would be idle to pretend that all is well with the teaching of architectural structures, or that we could not achieve far better integration between the design of architecture and structure. It is a challenge to be met by the instructors in both subjects.

Preliminary Structure Design Charts*

B

Contributed by Philip A. Corkill*

B.1. Horizontal Support Charts

The architectural designer is aware that the thickness, depth, or height of any structural system is closely related to the span of the system and to such variables as the spacing of structural elements, loads and loading conditions, continuity of system, and cantilevering. He is also aware that the structure should be considered at an early stage in his design synthesis because of the influence it will have upon his design.

These charts have been developed to provide the architectural designer with a quick and easy method of obtaining this basic structural information without the necessity of detailed mathematical analysis of the many possible structural solutions that might logically be integrated with his preliminary design.

Each chart indicates the range of thickness, depth, or height to span normally required for each of the systems indicated. This normal range is a composite of analytical solutions, structural design tables, and many constructed architectural examples.

The few structures that may exceed the range of these charts are generally composed of double systems or the combination of two or more integrated systems. Sometimes one system may be an extension of another system, and in these cases the span and height should be considered for only the primary system. These charts then consider only the normal use of a single system and do not consider the extreme possibilities for either depth or span.

To use the charts effectively, a designer must determine the approximate span required for his design, then choose a system appropriate to the design requirements, and read vertically from the appropriate span to the center of the range, then horizontally to the left of the chart to determine the normal thickness, depth, or height. If, however, greater-than-normal loads are anticipated, or a wider-than-normal spacing of members is desirable, the upper portion of

*Professor of Architecture, University of Nebraska, Lincoln, Nebraska

305

the range should be used. If light loads or closer-than-normal spacing of members is anticipated, the lower portion of the range should be used.

Structures such as frames, arches, or suspension systems can be used to cover or enclose both rectangular or circular spaces. In these cases, the upper portion of the range is more appropriate for rectangular or vaulted areas, the lower portion for circular or domed areas.

Thicknesses or depths when indicated across the top of these charts reflect the averages for the spans indicated. These figures may, however, need some adjustment. For example, domed areas would require somewhat less thickness or depth of material than vaulted areas, or the thickness indicated for folded plates should be increased somewhat if the lower portion of the range is used, and decreased if the upper portion is used.

The use of cantilevers extended from normal spans or a continuous beam system would generally result in less thickness or depth of a system for a given span and would indicate the use of the lower portion of the range, or even below in some cases. On the other hand, pure cantilevers require a much greater thickness or depth and go beyond the range of these charts.

The masonry vault-and-dome charts below have been included for comparative use only. However, if their use is anticipated with contemporary materials and methods of construction, the lower portion of the range should be used.

B.2 Vertical Support Charts

Each chart indicates the normal range of thickness or depth of independent or continuous vertical members. This range is a composite of analytical solutions, structural design tables and codes, and many constructed architectural examples.

These charts are presented in two parts:

1. Structural supports which have an unsupported height of from 1 to 50 ft and primarily support only a roof or one story load.
2. Structural supports which have a normal one story unsupported height and must support from 1 to 50 stories.

For single story structures the designer must assume an unsupported height for his vertical supports and choose an appropriate chart designated single story design. From the unsupported height shown across the bottom of the chart read vertically to the center of the range then horizontally to the left of the chart to find the required depth or thickness of the support.

For multistory design the same procedure is followed except the charts designated multistory should be used. Normal unsupported

Preliminary Structure Design Charts

story height for these charts is from 8 to 12 ft.

Loads and spacing of supports would determine whether the designer should use the upper, middle, or lower portion of the chart range. If, for example, greater than normal loads are anticipated or column spacings are great the upper portion of the range should be used. If, however, loads are light or the spacing of columns is closer than the normal the lower portion of the range should be used. The middle portion of the range would generally be used for most design.

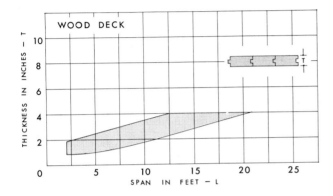

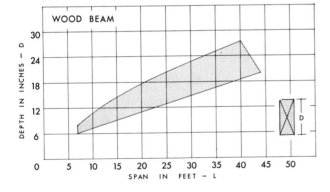

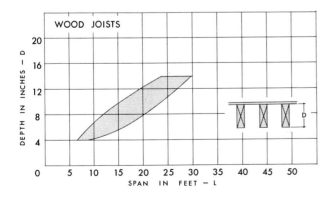

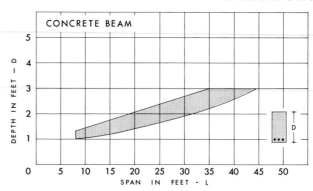

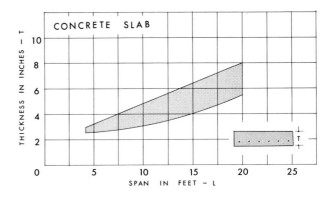

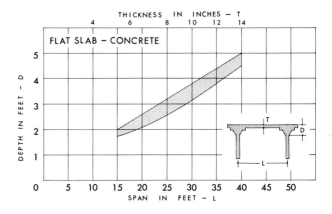

Preliminary Structure Design Charts

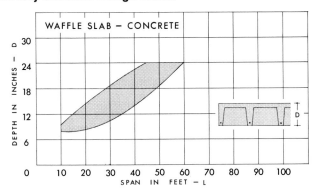

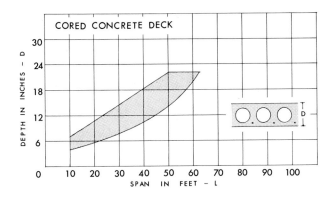

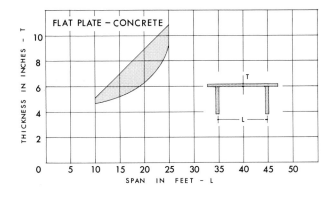

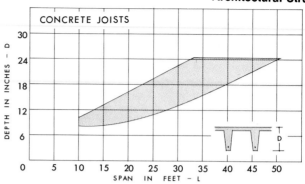

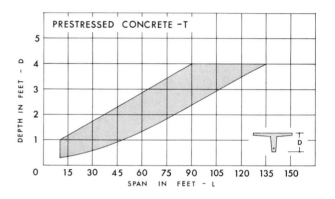

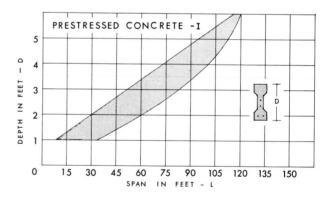

Preliminary Structure Design Charts

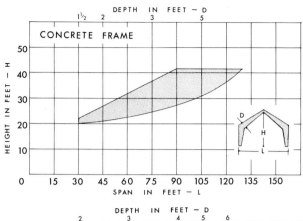

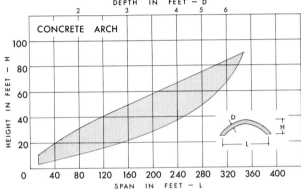

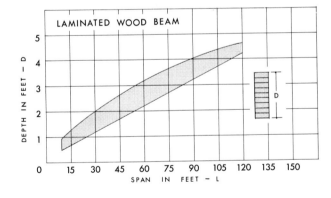

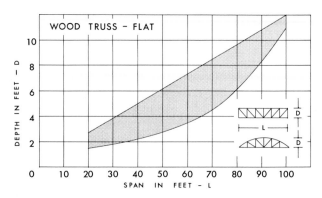

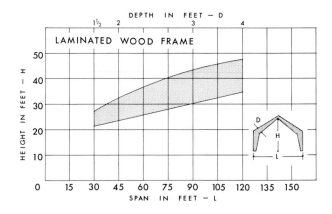

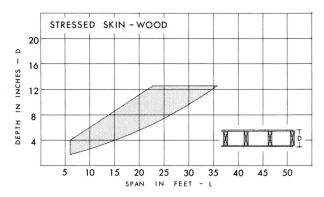

Preliminary Structure Design Charts

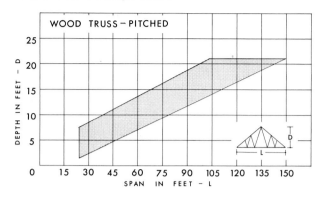

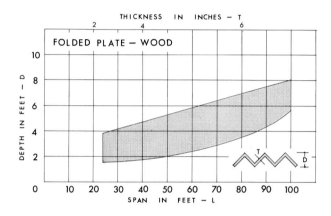

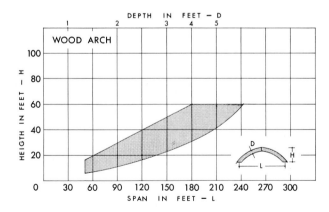

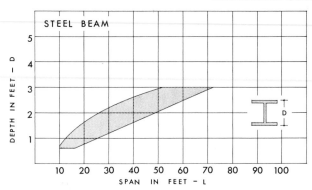

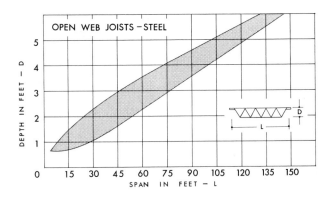

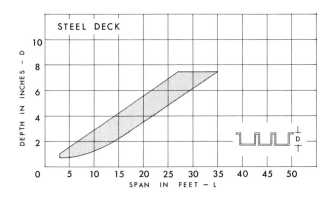

Preliminary Structure Design Charts

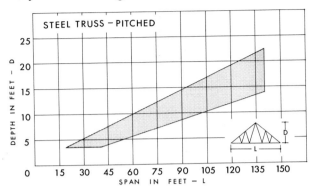

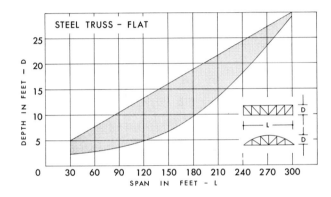

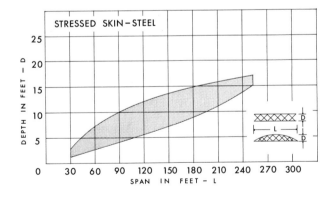

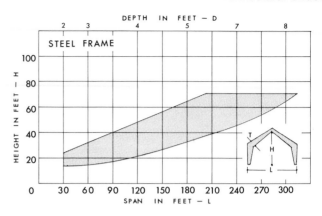

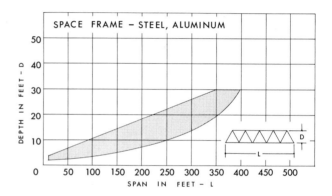

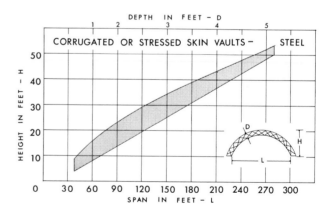

Preliminary Structure Design Charts

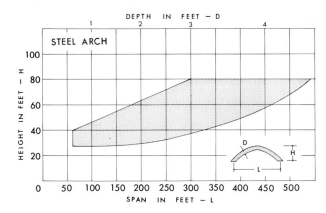

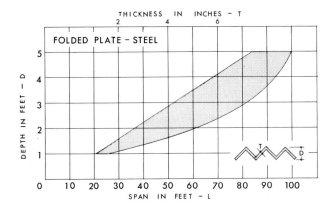

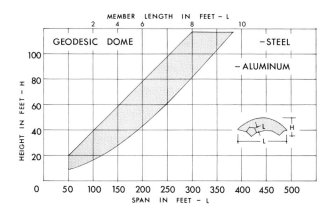

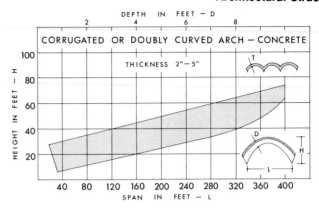

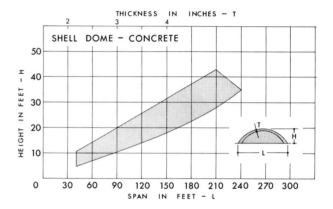

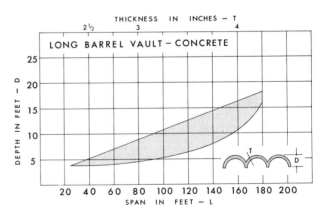

Preliminary Structure Design Charts

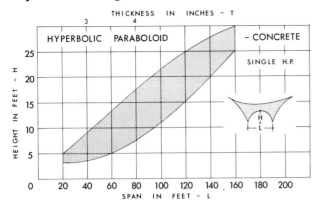

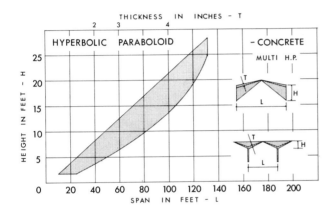

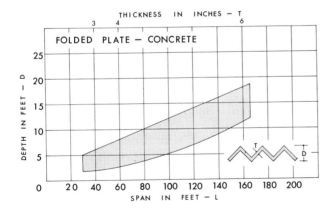

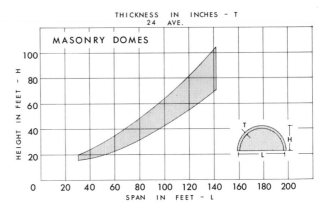

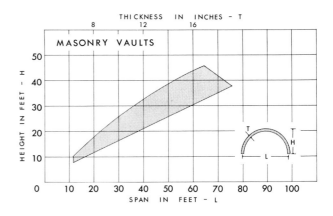

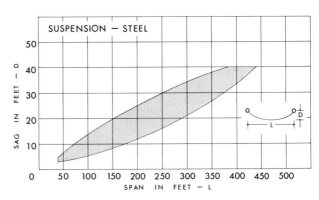

Preliminary Structure Design Charts

single story

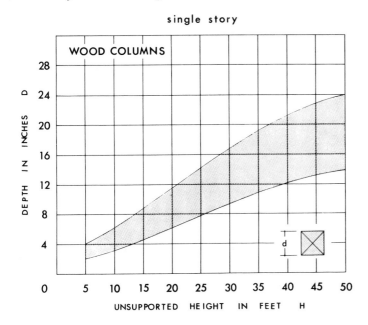

single story

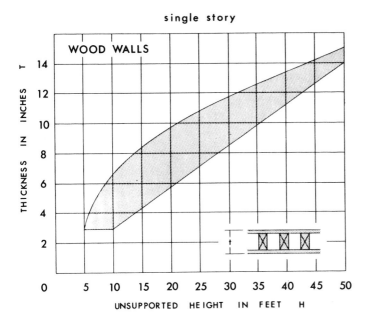

single story

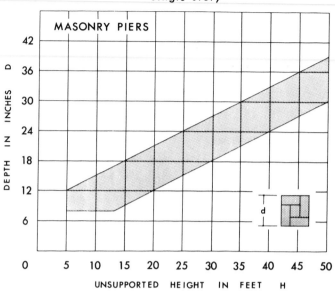

single story

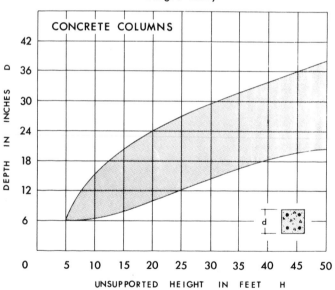

Preliminary Structure Design Charts

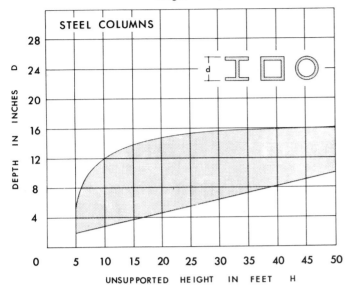

single story

STEEL COLUMNS

DEPTH IN INCHES D

UNSUPPORTED HEIGHT IN FEET H

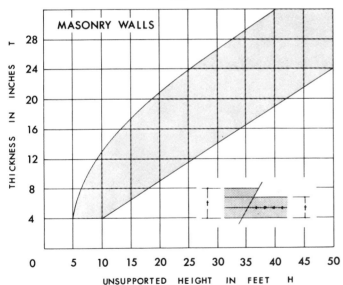

single story

MASONRY WALLS

THICKNESS IN INCHES t

UNSUPPORTED HEIGHT IN FEET H

single story

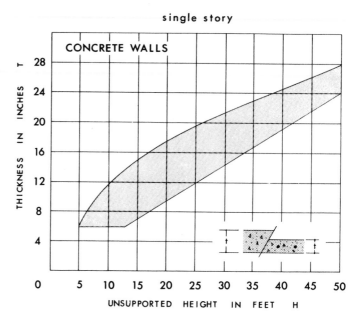

CONCRETE WALLS

THICKNESS IN INCHES T

UNSUPPORTED HEIGHT IN FEET H

single story

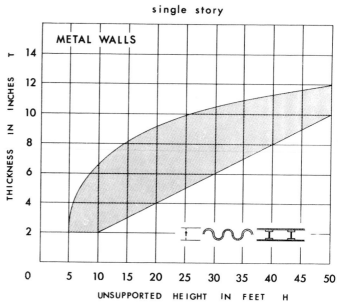

METAL WALLS

THICKNESS IN INCHES T

UNSUPPORTED HEIGHT IN FEET H

Preliminary Structure Design Charts

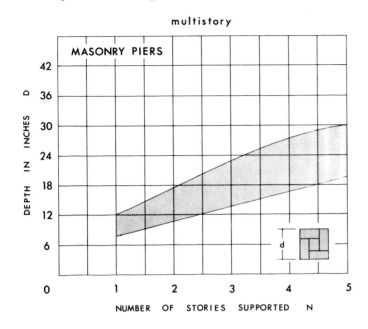

multistory

MASONRY PIERS

DEPTH IN INCHES D

NUMBER OF STORIES SUPPORTED N

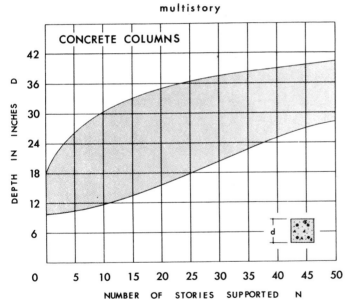

multistory

CONCRETE COLUMNS

DEPTH IN INCHES D

NUMBER OF STORIES SUPPORTED N

multistory

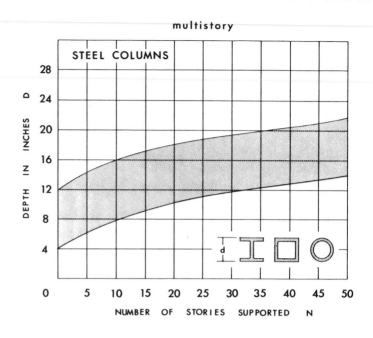

STEEL COLUMNS

DEPTH IN INCHES D

NUMBER OF STORIES SUPPORTED N

multistory

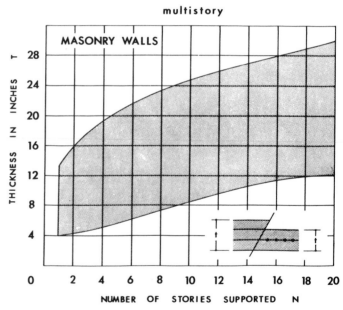

MASONRY WALLS

THICKNESS IN INCHES T

NUMBER OF STORIES SUPPORTED N

Preliminary Structure Design Charts

multistory

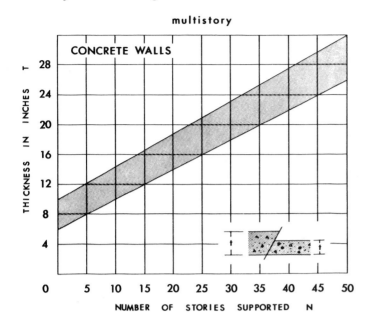

multistory

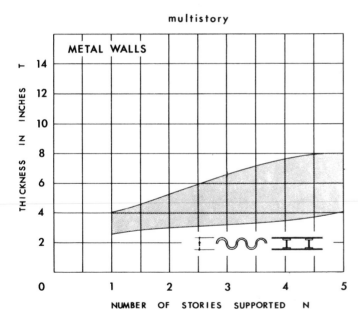

multistory

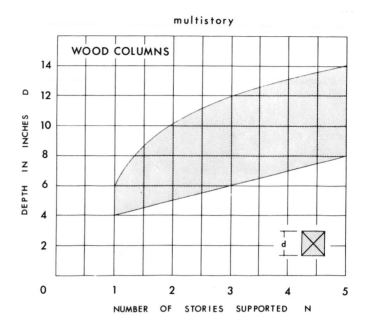

WOOD COLUMNS

DEPTH IN INCHES D

NUMBER OF STORIES SUPPORTED N

multistory

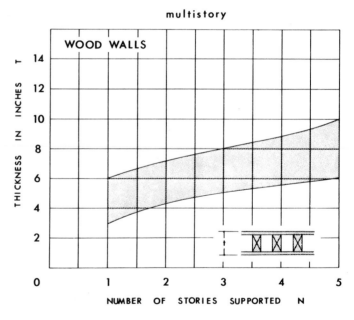

WOOD WALLS

THICKNESS IN INCHES T

NUMBER OF STORIES SUPPORTED N

Tables C

C.1. List of Reference Tables Given in Chapters 1–9

Table 6.1 Formulas for the second moment of area (momenet of inertia) and for the section modulus p. 164.

Table 6.2 Coefficients for the maximum values of the shear force, the bending moment, the slope, and the deflection of statically determinate beams pp. 168–9.

Table 8.1 Continuous beams: bending moment coefficients for dead loads p. 236.

Table 8.2 Continuous beams: bending moment coefficients for live loads p. 237.

Table 8.3 End-moments in built-in beams p. 239.

C.2. Conversion Table for American Customary, Conventional Metric, and Metric S. I. Units

Note: The *Système International d'Unités* has recently been introduced in Australia, Canada, Ireland, New Zealand, South Africa, and the United Kingdom. Although it has been adopted by the International Organization for Standardization, it is not yet used in other countries that conform to the metric system.

American Customary Units	Conventional Metric Units	Metric S. I. Units
Length		
1 foot	0.3048 metre	0.3048 metre
1 inch	2.540 centimetres	25.40 millimetres
Area		
1 sq ft	929.0 cm^2	0.09290 m^2
1 sq in	6.452 cm^2	645.2 mm^2
Section Modulus		
1 ft^3	28.32 × 10^3 cm^3	28.32 × 10^{-3}m^3
1 in^3	16.39 cm^3	16.39 × 10^3 mm^3

	American Customary Units	Conventional Metric Units	Metric S. I. Units
		Second Moment of Area (Moment of Inertia)	
	$1\ \text{ft}^4$	$863 \times 10^3\ \text{cm}^4$	$8.63 \times 10^{-3}\ \text{m}^4$
	$1\ \text{in}^4$	$41.62\ \text{cm}^4$	$416.2 \times 10^3\ \text{mm}^4$
		Velocity	
	1 mile per hour	1.609 kilometres per hour	1.609 km/h
		Density	
	1 lb/cu ft	$16.019\ \text{kg/m}^3$	$16.019\ \text{kg/m}^3$
		Force	
	1 lb	0.4536 kilogram	4.448 newtons
		Moment	
	1 lb ft	0.1383 kg m	1.356 N m
	1 lb in	0.01152 kg m	0.1130 N m
		Stress	
	1 lb/sq ft (p.s.f.)	$4.882\ \text{kg/m}^2$	$47.88\ \text{N/m}^2 = 47.88\ \text{pascals}$
	1 lb/sq in (p.s.i.)	$0.07031\ \text{kg/cm}^2$	$6895\ \text{N/m}^2 = 6.895\ \text{kPa}$

Note:
$1\ \text{Pa} = 1\ \text{N/m}^2$

$1\ \text{kPa} = 1 \times 10^3\ \text{Pa} = 1 \times 10^3\ \text{N/m}^2 = 1 \times 10^{-3}\ \text{N/mm}^2$

$1\ \text{MPa} = 1 \times 10^3\ \text{kPa} = 1 \times 10^6\ \text{Pa}$
$= 1 \times 10^6\ \text{N/m}^2 = 1\ \text{N/mm}^2$

C.3. Bar Area Tables

Table C1
Cross-Sectional Area of Groups of Round Bars and Deformed Bars (square inches)

Designation No. of Deformed Bar	Diameter of Round Bar (in)	Number of Bars in Group									
		1	2	3	4	5	6	7	8	9	10
2	$\frac{1}{4}$	0.049	0.098	0.147	0.196	0.245	0.294	0.343	0.392	0.441	0.491
3	$\frac{3}{8}$	0.110	0.220	0.331	0.441	0.552	0.662	0.772	0.883	0.993	1.104
4	$\frac{1}{2}$	0.196	0.392	0.588	0.785	0.981	1.177	1.374	1.570	1.766	1.963
5	$\frac{5}{8}$	0.306	0.613	0.920	1.227	1.534	1.840	2.147	2.454	2.761	3.068
6	$\frac{3}{4}$	0.441	0.883	1.325	1.767	2.209	2.650	3.092	3.534	3.976	4.418
7	$\frac{7}{8}$	0.601	1.202	1.803	2.405	3.006	3.607	4.209	4.810	5.411	6.013
8	1	0.785	1.570	2.356	3.141	3.927	4.712	5.497	6.285	7.068	7.854
9	$1\frac{1}{8}$	0.994	1.988	2.982	3.976	4.970	5.964	6.958	7.952	8.946	9.940
10	$1\frac{1}{4}$	1.227	2.454	3.681	4.908	6.136	7.363	8.590	9.817	11.044	12.272
11	$1\frac{3}{8}$	1.484	2.969	4.454	5.939	7.424	8.909	10.394	11.879	13.364	14.849
12	$1\frac{1}{2}$	1.767	3.534	5.301	7.068	8.835	10.602	12.369	14.136	15.903	17.671

Table C2
Cross-Sectional Area of Evenly Spaced Round Bars and Deformed Bars
(square inches per foot width)

Def. Bar No.	Diameter Round Bar (in)	Pitch (inches)															
		3	$3\frac{1}{2}$	4	$4\frac{1}{2}$	5	$5\frac{1}{2}$	6	$6\frac{1}{2}$	7	$7\frac{1}{2}$	8	9	10	12	14	18
2	$\frac{1}{4}$	0.196	0.168	0.147	0.130	0.117	0.107	0.098	0.095	0.084	0.078	0.074	0.065	0.059	0.049	0.042	0.033
3	$\frac{3}{8}$	0.440	0.377	0.330	0.294	0.264	0.240	0.220	0.203	0.188	0.176	0.165	0.147	0.132	0.110	0.094	0.074
4	$\frac{1}{2}$	0.784	0.673	0.588	0.522	0.470	0.428	0.392	0.362	0.336	0.313	0.294	0.261	0.235	0.196	0.168	0.131
5	$\frac{5}{8}$	1.224	1.040	0.918	0.816	0.734	0.667	0.612	0.565	0.525	0.489	0.459	0.408	0.367	0.306	0.262	0.204
6	$\frac{3}{4}$	1.768	1.511	1.326	1.179	1.061	0.964	0.884	0.816	0.758	0.707	0.663	0.589	0.530	0.442	0.379	0.295
7	$\frac{7}{8}$	2.404	2.060	1.803	1.603	1.442	1.311	1.202	1.110	1.030	0.962	0.901	0.801	0.721	0.601	0.515	0.401
8	1	3.141	2.691	2.355	2.094	1.884	1.713	1.570	1.449	1.346	1.256	1.177	1.047	0.942	0.785	0.674	0.532

Table C3
Cross-sectional Area of Groups of Round and Deformed Bars (mm ^{2})

Diameter or Nominal Diameter of Bar (mm)	Number of Bars in Group									
	1	2	3	4	5	6	7	8	9	10
6	28	57	85	113	141	170	198	226	255	283
8	50	101	151	201	251	302	352	402	452	503
10	80	160	240	320	400	480	560	640	720	800
12	110	220	330	440	550	660	770	880	990	1,100
16	200	400	600	800	1,000	1,200	1,400	1,600	1,800	2,000
20	310	620	930	1,240	1,550	1,860	2,170	2,480	2,790	3,100
24	450	900	1,350	1,800	2,250	2,700	3,150	3,600	4,050	4,500
25	491	982	1,473	1,963	2,454	2,945	3,436	3,927	4,418	4,909
28	620	1,240	1,860	2,480	3,100	3,720	4,340	4,960	5,580	6,200
32	800	1,600	2,400	3,200	4,000	4,800	5,600	6,400	7,200	8,000
36	1,020	2,040	3,060	4,080	5,100	6,120	7,140	8,160	9,180	10,200
40	1,260	2,520	3,780	5,040	6,300	7,560	8,820	10,080	11,340	12,600
50	1,960	3,920	5,880	7,840	9,800	11,760	13,720	15,680	17,640	19,600

There are slight differences between British and Australian bars of the same nominal diameter.

Table C4
Cross-sectional Area of Evenly Spaced Round and Deformed Bars (mm ^{2}/m)

Diameter or Nominal Diameter of Bar (mm)	Pitch (mm)							
	75	100	125	150	175	200	250	300
6	377	283	226	188	162	141	113	94
8	670	503	402	335	287	251	201	168
10	1,067	800	640	533	457	400	320	267
12	1,467	1,100	880	733	629	550	440	367
16	2,667	2,000	1,600	1,333	1,143	1,000	800	667
20	4,133	3,100	2,480	2,067	1,771	1,550	1,240	1,033
24	6,000	4,500	3,600	3,000	2,571	2,250	1,800	1,500
25	6,545	4,909	3,297	3,272	2,805	2.454	1,963	1,636
28	8,266	6,200	4,960	4,133	3,543	3,100	2,480	2,067

There are slight differences between British and Australian bars of the same nominal diameter.

Symbols and Abbreviations

D

D.1. Notation

a	resistance arm
A	area; cross-sectional area
A_s	cross-sectional area of reinforcing steel
b, B	width
C	compressive force
d	depth; effective depth; diameter
D	overall depth; diameter
e	direct strain
e_y	yield strain of steel
e_c'	crushing strain of concrete
E	direct modulus of elasticity
E_c, E_s	modulus of elasticity of concrete, steel
f	direct stress
f_c'	crushing strength of concrete
f_y	yield stress of steel
F	force
G	shear modulus of elasticity
h, H	height
H	horizontal force
I	second moment of area (moment of inertia)
j	number of joints in truss
jd	lever arm of reinforced concrete beam at the service load
$j_u d$	lever arm of reinforced concrete beam at the ultimate load
kd	depth of neutral axis in reinforced concrete beam at the service load
K	stiffness
l, L	span
M	moment, bending moment, resistance moment
M'	ultimate bending moment
M_u	factored ultimate bending moment $= \phi M'$
M_T	torsional moment
M_+, M_-	positive, negative moment

n	number
N	direct membrane force
p	pressure
P	direct (tensile or compressive) force; thrust in arch
P'	ultimate column load
P_u	factored ultimate column load $= \phi P'$
r	rise of arch; *also* radius of gyration $= \sqrt{I/A}$
R	reaction; *also* radius of curvature
R_H, R_V	horizontal, vertical reaction
R_L, R_R	left-hand, right-hand reaction
s	sag of suspension cable
S	section modulus
S_f	factor of safety
t	time; *also* flange thickness; *also* tension coefficient
T	tensile force
v	shear stress; *also* velocity
V	shear force
w	load per unit area or per unit length
W	total load
x, y, z	distances along the x-, y-, and z-axes
$\bar{x}, \bar{y}$	coordinates of the center of gravity
y	deflection
α	angle
γ	shear strain
δ	sidesway
ϵ	eccentricity
θ	angle; slope of beam
π	circular constant $= 3.1416$
ρ	reinforcement ratio $= A_s/bd$
ϕ	capacity reduction factor.

D.2. Mathematical Symbols

$\geqslant$	greater than or equal to
$\leqslant$	smaller than or equal to
Σ	sum of
$\int$	integral of (geometrically the area contained under the curve)
dy/dx	differential coefficient (first derivative) of y with respect to x (geometrically the slope of the curve).

D.3. Abbreviations

ACI	American Concrete Institute
act.	actual
AISC	American Institute of Steel Construction

Symbols and Abbreviations

AITC	American Institute of Timber Construction
B.M.	bending moment
C	compression
C.G.	center of gravity
cm	centimetres
cm^2	square centimetres
cos	cosine
cu ft	cubic feet
ft	feet
GPa	gigapascals; 1 GPa = 1,000 MPa = 1,000,000,000 Pa
in	inches
kip	kilopounds (= 1,000 lb)
lb	pounds
kg	kilograms
kN	kilonewtons; 1 kN = 1,000 N
m	metres
mm	millimetres; 1 mm = 0.001 m
m^2	square metres
m^3	cubic metres
max.	maximum
min.	minimum
MN	meganewtons; 1 MN = 1,000 kN = 1,000,000 N
MPa	megapascals; 1 MPa = 1,000,000 Pa
N	newtons
Pa	pascals
PCA	Portland Cement Association
perm.	permissible
p.s.i.	pounds per square inch
sec.	second
S.F.	shear force
sin	sine
sq in	square inches
sq ft	square feet
T	tension
tan	tangent
udl	uniformly distributed load
2-D, 3-D	two-dimensional, three-dimensional
', "	inches, feet
°	degree
°F, °C	degree Fahrenheit, degree Celsius

Glossary E

Words in italics are further defined in alphabetical order.

anchorage. Device, frequently patented, for permanently anchoring the *tendons* at the end of a *post-tensioned* beam, or for temporarily anchoring *pre-tensioned* tendons while the concrete gains strength.

anti-clockwise. Opposite to *clockwise*.

arch. A structure designed to carry loads across a gap mainly by compression. See also *cable*.

balanced reinforcement ratio (ρ_b). In reinforced concrete, the *reinforcement ratio* at which a *primary compression failure* would occur simultaneously with a *primary tension failure*. The ACI Code limits the reinforcement ratio to 0.75 of the balanced reinforcement ratio.

beam. A structural member to carry loads across a horizontal gap by bending.

beam-and-slab floor. A floor system, particularly in reinforced concrete, whose floor slab is supported by beams; see also *flat slab*.

bending moment. The *moment* at any section of a *beam* or other flexural member of all the forces which act on the beam on one side of that section.

bending moment diagram. Diagram showing the variation of *bending moment* along the span.

bent. A two-dimensional frame capable of supporting horizontal as well as vertical loads.

buckling. A *column* or *strut* is said to buckle if it loses its stability and bends sideways. It is a type of failure associated with high *slenderness ratios*. The material of the column or strut is not necessarily damaged by buckling.

built-in. A condition of support which prevents the ends from rotating.

butt joint. A joint between two pieces of material which are in line, butting against each other. See also *lap joint*.

cable. A structural member which is flexible and can therefore resist only tension, but not compression or flexure. See also *arch*.

cantilever. A beam, truss, or slab, supported only at one end.

capacity reduction factor (ϕ). A factor used in *ultimate strength design* to provide a margin of safety against collapse or serious structural damage. The column load or the resistance moment is multiplied by the capacity reduction factor($P_u = \phi P'$ and $M_u = \phi M'$). See also *factor of safety* and *load factor*.

Cartesian coordinates. The coordinates conventionally used for plotting a curve, measured perpendicularly from two axes, called x and y, at right angles to one another. In three dimensions, Cartesian coordinates are denoted by x, y, and z.

catenary. The curve assumed by a freely hanging *cable* of uniform section due to its own weight. Its mathematical equation is $y = a \cosh (x/a)$, where a is a constant.

center of gravity (C.G.). A point in a plane figure at which the centroidal axes intersect. A centroidal axis is one about which the first moment of the area is zero. The total weight of a body acts at its center of gravity.

chord. A horizontal member in a truss.

clockwise. Rotation in the same direction as the hands of a clock. The movement of a screwdriver driving a right-handed screw into its hole is clockwise.

column. A vertical member of a frame carrying a compressive load.

column capital. Enlargement of the top of a column, built as an integral unit with the column and the *flat slab*. It is designed to increase the shearing resistance of the flat slab.

column head. *Column capital.*

concrete. Artificial stone made from aggregate (gravel, broken stone or other suitable material about $\frac{3}{4}$ in in size), sand, and portland cement.

continuous beam. A beam which is continuous over intermediate supports, and thus *statically indeterminate*. See also *simply supported beam* and *Gerber beam*.

counterclockwise. Opposite to *clockwise*.

couple. A pair of equal and parallel forces, not in line and oppositely directed. The *moment* of a couple equals the magnitude of one of the forces times the perpendicular distance between them.

cover. In reinforced concrete, the thickness of concrete overlying the steel bars nearest to the surface. An adequate cover is needed to protect the reinforcement from rusting.

creep. Time-dependent deformation due to load. In concrete, a sustained load squeezes water from the cement gel at ordinary temperatures, and this produces deformation, which may eventually be two to three times as great as the *elastic* deformation.

creep deflection. Deflection of a beam due to *creep*. Elastic deflection occurs instantly, while creep deflection requires time to develop, and it may be several months before it becomes noticeable.

dead load. A load which is permanently applied to the structure (for example, its own weight) and acting at all times, as opposed to a *live load*.

deflection. The flexural deformation of a structural member, particularly the vertical deformation of a horizontal beam. Although elastic deflection is recovered when the load is removed, it may damage brittle finishes, such as plaster, if it is excessive. In concrete and timber there is an additional deflection due to *creep*.

direct strain. Elongation or shortening caused by tensile and compressive stresses respectively. See also *shear strain*.

dome. A vault of double curvature, both curves being convex upward. Most domes are spherical, i.e., they are portions of a sphere.

draped tendon. A *tendon* whose eccentricity varies along the length of the beam. The object is to set up a bending moment which opposes the bending moment due to the imposed load.

dropped panel. The portion of a *flat slab* which is thickened throughout the area surrounding the *column capital* to reduce the magnitude of the shear stress.

effective depth. In reinforced concrete, the distance of the center of the reinforcement from the compression face of the concrete.

eccentric load. A load which does not act through the *center of gravity*. The eccentricity is the distance of the line of action of the load from the center of gravity.

elastic design. Design based on the assumption that structural materials behave elastically. The stresses due to the *service loads* should be as close as possible to, but not greater than, the *maximum permissible stresses*. See also *ultimate strength design*.

elasticity. The ability of a material to deform instantly under load and recover its original shape instantly when the load is removed. See also *Hooke's Law*.

extrados. The outer curve of an arch.

factor of safety. Factor used in *elastic design* to provide a margin of safety against collapse and serious structural damage. See also *load factor* and *maximum permissible stress*.

fillet weld. A weld of approximately triangular cross-section joining two surfaces at approximately right angles to one another.

flat plate. A concrete slab reinforced in two directions at right angles, and supported directly on the columns without *column capitals*, *dropped panels*, beams or girders.

flat slab. A concrete slab reinforced in two or more directions and supported directly on *column capitals*, without beams or girders. It may have *dropped panels* surrounding the column capitals.

flexure. Bending.

flexural member. A structural member which resists loads acting at right angles to it by flexure.

flutter. Rapid vibrations of small amplitude

folded plate. A structure formed by flat plates, usually of reinforced concrete, joined at various angles.

force (F). Defined as anything that changes or tends to change the state of rest of a body, or its motion in a straight line. The external forces most commonly acting on buildings are the weight of the materials from which they are built, the weight of the contents (including people), and the forces due to snow, wind, and earthquakes. These external forces produce internal forces in the structural members.

Gerber beam. A *continuous beam* that has a sufficient number of *pin joints* to make it *statically determinate*.

gigapascal (GPa). One-thousand million *pascals*.

girder. A large beam.

hinge. A joint which allows rotation, i.e., it transmits the *shear force*, but not the *bending moment*.

Hooke's Law. "Stress is proportional to strain, or $f = Ee$." The constant of proportionality, E, is called the *modulus of elasticity*.

hypar. *Hyperbolic paraboloid.*

hyperbolic paraboloid. A geometric surface which has the equation $z = kxy$, where k is a constant, and x, y, and z are the *Cartesian coordinates*. It is generated by moving one straight line over two other straight lines.

hyperboloid. A surface generated by a straight line at an angle to a vertical axis. It has the familiar "water-tower" shape.

hyperstatic reaction. A reaction that renders a structure *statically indeterminate*.

intrados. The inner curve of an arch.

isostatic line. A line tangential to the direction of one of the *principal stresses* at every point through which it passes. Also called a stress trajectory.

isostatic structure. A structure which is *statically determinate*.

joist. A *secondary beam*.

lap joint. A joint between two pieces of material which overlap. See also *butt joint*.

leeward. On the side sheltered from the wind, as opposed to *windward*.

lever arm. The distance between the resultant compressive and the resultant tensile force in a section. These two forces and the lever arm form a *couple*, which is the *resistance moment*.

live load. A load which is not permanently applied to the structure, as opposed to the *dead load*.

Glossary

load-deformation diagram. A diagram that shows the variation of the deformation with increase in load. The diagram shows whether a material is ductile or brittle.

load factor. Factor used in *ultimate strength design* to provide a margin of safety against collapse or serious structural damage. The *service loads* are multiplied by the appropriate load factors. See also *factor of safety* and *capacity reduction factor*.

maximum permissible stress. The greatest stress permissible in a structural member under the action of the *service loads*. It is usually defined as (limiting strength of material)/(*factor of safety*).

mechanism. A frame that is capable of movement. In structural mechanics, it is defined as a *statically determinate* truss from which one or more members have been removed.

megapascal (MPa). One million *pascals*.

modulus of elasticity (E). The ratio of direct stress to strain in an elastic material obeying *Hooke's Law*.

moment (M). The moment of a *force* about a given point is the turning effect, measured by the product of the force and its perpendicular distance from this point.

moment of inertia. Used frequently, but incorrectly, in structural mechanics as a synonym for the *second moment of area* (I).

monitor. A series of windows on both sides of single-story factory roof which admit daylight and sometimes also provide ventilation. They are intended to exclude direct radiation from the sun.

neutral axis. The line in the cross-section of a beam where the flexural stress changes from tension to compression, i.e., a line where the direct stress is zero. It passes through the *center of gravity*.

newton (N). The unit of force in the SI metric system. It is the force exerted by a mass of 1 kilogram due to an acceleration of 1 m per second per second. In the conventional metric system a force of 1 kilogram is the force exerted by a mass of 1 kg due to the acceleration of the earth's gravity, which is 9.8 m per second per second. Similarly, in conventional American units, a force of 1 lb is the force exerted by a mass of 1 lb due to the acceleration of the earth's gravity (32.2 ft per second per second). The newton is a small unit, and forces are usually given in kilonewtons (kN).

north light truss. A truss used in the northern hemisphere to limit the admission of direct radiation from the sun. The glazing faces north.

pascal (Pa). The unit of stress in the SI metric system. It is the stress due to a uniform load of 1 *newton* acting on 1 m^2. The pascal is a very small unit. Stresses are usually stated in *megapascals* (MPa), and moduli of elasticity in *gigapascals* (GPa).

pin joint. A joint which allows rotation and effectively forms a hinge. Pin joints are so called because in the mid-nineteenth century they frequently consisted of pins pushed through holes in each member to be joined. Today true pin joints are rare, and the term is commonly used for a joint which transmits the *shear force*, but not the *bending moment*.

plane frame. A frame whose members all lie in one plane, as opposed to a space frame.

plastic deformation. Continuous permanent deformation in metals, which occurs above a critical stress, called the *yield stress*.

polygon of forces. A figure analogous to the *triangle of forces*, representing the statical equilibrium of more than three forces.

post-tensioning. *Prestressing* the *tendons* after the concrete has gained sufficient strength, as opposed to *pretensioning*.

Pratt truss. A statically determinate truss, consisting of top and bottom chords, regularly spaces vertical compression members and diagonal tension members.

prestressed concrete. Concrete which is precompressed in the zone where tension occurs under load. Prestressing may be done by *pre-tensioning* or *post-tensioning*.

pre-tensioning. Tensioning the *tendons* before the concrete is cast, as opposed to *post-tensioning*.

primary compression failure. In reinforced concrete, a failure initiated by crushing of the concrete.

primary girder. A girder supported directly on the columns, as opposed to a *secondary beam*.

primary tension failure. In reinforced concrete, a failure initiated by yielding of the steel.

radian. The angle between two radii of a circle, cut off on the circumference of an arc equal in length to the radius. It is the absolute angular measure. Slope is obtained in radians when it is calculated by the methods used in this book. The degree is an artificial (if more convenient) unit. 1 radian = 57.3 degrees. 1 degree = 0.01745 radians.

radius of gyration (r). Convenient shorthand notation for $\sqrt{I/A}$, where I is the *second moment of area* (moment of inertia) and A is the cross-sectional area.

reaction (R). The opposition to an action, e.g., the downward pressure of a beam or frame produces an equal upward reaction from its support.

redundancy. A structural member or support in excess of those required for a *statically determinate structure*.

reinforced concrete. Concrete which contains reinforcement, usually of steel, to improve its flexural resistance. If the steel is prestressed, the material is called *prestressed concrete*.

reinforcement ratio (ρ). The ratio of the cross-sectional area of the reinforcement to the product of width and *effective depth*

($\rho = A_s/bd$). It is not the ratio of the area of reinforcement to the area of concrete, because the effective depth is less than the overall depth of the concrete.

resistance arm. *Lever arm.*

resistance moment. The internal moment which resists the bending moment due to the loads. It is exactly equal to the *bending moment*, but of opposite sign. See also *reaction*.

resultant. The *vector* sum of two or more forces.

rigid frame. A frame with *rigid joints*.

rigid joint. A joint which is capable of transmitting the full moment at the end of the member to other members framing into the joint. See also *semi-rigid joint* and *pin joint*.

riveting. Joining two pieces of metal with red-hot rivets which contract on cooling and press the two pieces together.

secondary beam. A beam supported on *primary girders*, not directly on the columns. Also called a joist.

second moment of area (I). The sum of the products obtained by multiplying each element of an area, dA, by the square of its distance from the *neutral axis*. In mathematical terms, $I = \int y^2 dA$. The second moment of area is used in the theory of bending and in calculations involving slope and deflection. It is frequently, but mistakenly, called moment of inertia.

section modulus (S). A convenient shorthand notation for I/y, where I is the *second moment of area* (moment of inertia), and y is the greatest distance of any point in the cross section from the *neutral axis*.

semi-rigid joint. A joint that is capable of transmitting some bending moment, but not the full moment. It is intermediate between a *pin joint* and a *rigid joint*.

service core. A vertical unit in a high-rise building, which contains the vertical runs of most of the services, particularly the elevators. It is usually built in reinforced concrete, and it can be used as a load-bearing element.

service load. The actual *dead* and *live* load that the structure is designed to carry. It is synonymous with working load. See also *ultimate load* and *elastic design*.

shear force. The resultant of all the vertical forces acting at any section of a horizontal beam or slab on one side of the section; or the resultant of the horizontal forces acting at any section of a vertical column in a frame.

shear force diagram. Diagram showing the variation of *shear force* with span.

shear strain. Distortion caused by shear forces. See also *direct strain*.

shear wall. A wall which in its own plane resists shear forces due to wind, earthquake forces, etc..

shrinkage. Contraction of concrete and timber due to drying of the materials.

simply supported beam. A beam freely resting on two supports, and not joined to the adjacent structure, as opposed to built-in or *continuous beams*. It is *statically determinate*.

slenderness ratio. The ratio of the effective length of a column or strut to its least *radius of gyration*.

slope (θ). In structural mechanics, the rotation of a structural member due to its deformation. Slope is measured in *radians*.

south light truss. The southern hemisphere equivalent of the *north light truss*.

space frame. See *plane frame*.

span. The distance between the supports.

spandrel. The part of the wall between the head of a window and the sill of the window above it. The term is frequently used as a synonym for a spandrel beam placed within the spandrel, or a structural beam at the edge of the building.

specified compressive strength of concrete (f_c'). The strength of the concrete used in ultimate strength design.

specified yield stress (f_y). The steel stress used in the ultimate strength design of reinforced concrete.

statically determinate structure. A structure whose internal forces or moments can be determined by the use of *statics* alone. Also called an isostatic structure.

statically indeterminate structure. A structure that has more members or restraints than can be determined by *statics* alone. It is necessary to derive additional equations, e.g., by considering the elastic deformation of the structure. Also called a hyperstatic structure.

statics. The branch of the science of mechanics which deals with forces in equilibrium.

stiffness (K). In structural mechanics, the ratio EI/L, where E is the *modulus of elasticity*, I is the *second moment of area* (moment of inertia), and L is the span.

strain. A *direct strain* is deformation per unit length. A *shear strain* measures distortion.

stress. Internal force per unit area, considering an infinitesimally small element of the structure.

stress-strain diagram. A diagram that shows the variation of strain with increase in stress. The diagram shows whether a material is ductile or brittle.

stress trajectory. *Isostatic line.*

strut. A member in compression, as opposed to a *tie*.

tendon. A bar, wire, strand, or cable of high-tensile steel, used to impart *prestress* to concrete when the element is tensioned.

tension coefficient (t). A convenient shorthand notation for (force)/(length).

tie. A member in tension, as opposed to a *strut*.

three-pinned arch (portal). An arch (portal) with two pins at each of the supports and a third pin at the crown. The structure is *statically determinate*.

thrust. Compressive force, particularly in an arch or a portal frame.

triangle of forces. A graphical representation of three forces in equilibrium. If the forces are not in equilibrium, their *vectors* do not form a triangle.

triangulated structure. A structure composed of members arranged in triangles. It is *statically determinate*.

trussed rafter. A *triangulated* rafter in a roof truss.

ultimate load. The greatest load which a structure can actually support before it fails. To obtain the *service load*, we must apply *load factors* and a *capacity reduction factor*.

ultimate strength design. Design based on the *ultimate load*. See also *elastic design*.

vector. A quantity which has direction, as well as magnitude.

Vierendeel truss. An open-frame girder without diagonals. It consists of two horizonal chords connected to vertical members with rigid joints. It is *statically indeterminate*.

welding. Uniting two pieces of metal by raising the temperature of the metal surfaces to the plastic or molten condition, with or without the addition of additional welding metal, and with or without pressure.

wind bracing. Structural members, usually diagonal, specially designed to resist the wind forces.

windward. The opposite to *leeward*.

working load. *Service load*.

yield strength. *Yield stress*.

yield stress. The stress at which there is a great increase in strain without an increase in stress. The *elastic* limit of steel is only slightly less than its yield stress. In structural design it is generally assumed that steel is elastic up to the yield stress, and then becomes *plastic*.

Answers to Problems   **F**

C = compression; T = tension; BM = bending moment;
SF = shear force; LH = left hand; RH = right hand;
S = section modulus.

Chapter 3

3.1. The force needed is 543.6 lb (horizontal component = 171.9 lb pointing towards the left; vertical component = 515.7 lb pointing upwards).

3.2. 74.43 lb; the shortest distance from A is 0.345 in.

3.3. The LH reaction is 700 lb; the RH reaction is 600 lb.

3.4. The vertical reaction is 8,000 lb. The moment reaction is 43,400 lb ft.

3.5. The RH reaction is 1,559 lb, acting vertically; the LH reaction is 2,381 lb, inclined at an angle of 49°6′ to the vertical (the vertical component is 1,559 lb, and the horizontal component is 1,800 lb).

3.6. The center of gravity is 7.34 in to the right and 2.56 in above the bottom LH corner.

3.7. The answer is 765.6 lb; it is most conveniently obtained by taking moments about E.

3.8. The answer is most conveniently obtained by taking moments about the leeward edge. The block will overturn, since the resultant falls 0.4 in outside the base.

3.9. The building will overturn if the resultant falls outside the base. This requires a wind pressure of 19.1 lb per sq ft.

3.10. The resultant must fall within the middle third. This requires a minimum density of 150 lb per cu ft.

Chapter 4

4.4. AB: 12,500 lb; AE: 77,500 lb; A3: 2,500 lb C (compression); A1: 12,500 lb T (tension); B1: 17,700 lb C; 12: 10,600 lb C; C2: 5,000 lb C; 23: 10,600 lb T; 34: 53,000 lb C; D4: 40,000 lb T; E4: 56,600 lb C (see Fig. F1).

4.5. The stress diagram is drawn in Fig. F.2. The forces, scaled

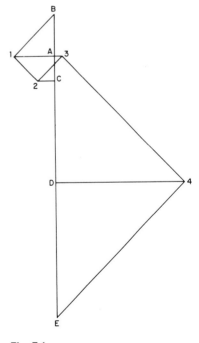

Fig. F.1.
Answer to problem 4.4.

347

from the diagram to the nearest 5 kips (5,000 lb) are:
AL 125; DE 115; A1 180 C; A3 185 C; B5 195 C; C7 185 C;
D8 185 C; D-10 180 C; D-12 165 C; D-14 160 C; E-14 115
T; F-13 115 T; G-11 155 T; H9 175 T; I6 190 T; J4 175 T;
K2 125 T; L1 125 T; 1-2 20 T; 3-4 10 C; 5-6 20 T; 7-8 45 T;
9-10 5 T; 11-12 10 C; 13-14 20 T; 2-3 70 T; 4-5 25 T; 6-7 15
C; 8-9 10 T; 10-11 35 T; 12-13 60 T.

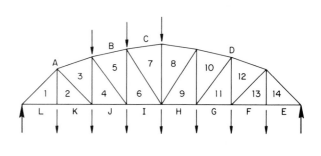

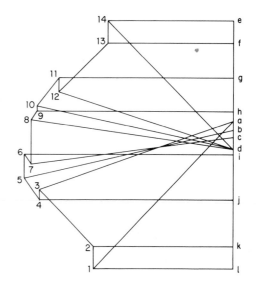

Fig. F.2.
Answer to problem 4.5.

4.6. The force in the horizontal member is 126,250 lb T, the force
in the inclined member is 178,540 lb C.

4.7. 20,000 lb T.

4.8. The reactions are 2,500 lb each. By resolution at the joints:
joint A: AO 4,167 lb T; joint O: BO 4,167 lb C.
By method of sections: moments about M: CD 11,333 lb C;
moments about D: ML 12,000 lb T.

4.9. Moments about C: BD (and DF by symmetry) 2,000 lb C;
moments about D: CE 2,500 lb T; moments about B: CD
(and DE by symmetry) 707 lb C.

Answers to Problems

4.10. The force in AB increases by 2,121 lb.
4.11. The force in XY is 42.43 lb C. The reaction at Z is 58.31 lb; it makes an angle of 31° to the vertical, and its horizontal component is 30 lb.
4.12. AB: 1,800 lb C; BC and BD: 975 lb T each.
4.13. The reactions are, at A: 10,700 lb (down); at B: 10,000 lb (up); at C: 10,700 lb (up). The forces are: AB 8,000 lb C; AD 13,333 lb T; BC 8,000 lb C; BD 5,000 lb T; BE 18,788 lb C; CD 13,333 lb C; DE 12,370 lb T.

Chapter 5

5.1. The max. SF occurs at the support $= +8,000$ lb; the max. BM also occurs at the support $= -43,400$ lb ft.
5.2. The max. SF occurs at the support $= +12,000$ lb; the max. BM also occurs at the support $= -42,000$ lb ft.
5.3. The max. SF occurs at the RH support $= -7,600$ lb; the max. BM occurs under the 8,000 load $= +56,800$ lb.
5.4. The BM varies linearly from 0 at the supports to $+\frac{3}{8}$ NL at the quarter points, and then linearly to $+\frac{1}{2}$ NL at mid-span.
5.5. The SF varies linearly from $+7,000$ lb (max.) at the LH support to $-1,000$ lb at mid-span, and then linearly to $-5,000$ lb at the RH support. The BM varies parabolically, with a discontinuity at mid-span. The maximum occurs where $dM/dx = 0$, which is 3.5 ft from the LH support. The max. BM is $+12,250$ lb ft.
5.6. The SF varies linearly from 0 at the LH end to $-6,000$ lb at the LH support; then changes to $+7,825$ lb; remains constant to the concentrated load; changes to $-10,125$ lb; remains constant to RH support; changes to $+12,000$ lb (max.) and reduces linearly to 0 at the RH end.
   The BM varies parabolically from 0 at the LH end to $-18,000$ lb ft at the LH support; changes linearly to $+45,000$ lb ft (max.) under the concentrated load; changes linearly to $-36,000$ lb ft at the RH support; and reduces to 0 at the RH end.
5.7. This problem is basically the same as Example 5.9, worked out in the text. The smallest max. BM is obtained when the columns are 4.142 ft from the ends of the beam.
5.8. The horizontal reactions are 75 lb, the vertical reaction is 200 lb, and the max. BM is 750 lb ft.
5.9. The beam reactions are 190 lb at A and 72.5 lb at C. The forces in the cables are 83.1 lb T (AF and AE) and 72.5 lb T (CD).
5.10. The BM and SF vary in the same way in the cantilevered and the supported spans. The SF varies linearly from $+10,000$ lb at the LH support to $-10,000$ lb at the RH

support. The BM varies parabolically from $-25,500$ lb ft at the supports to $+24,500$ lb ft at mid-span.

5.11. The vertical reactions are 10,000 lb each, and the horizontal reactions are 5,000 lb each. The BM diagram is shown in Fig. F.3; max. BM $= -75,000$ lb ft at the rigid joint.

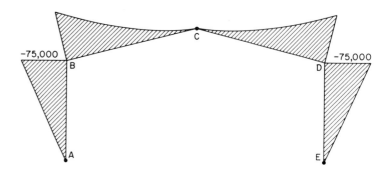

Fig. F.3.
Answer to problem 5.11.

5.12. $R_A = -250$ lb (uplift); $R_D = +250$ lb; $H_A = 1,000$ lb; $H_D = 0$. The BM diagram is shown in Fig. F.4; max. BM $= +5,000$ lb ft at the rigid joint.

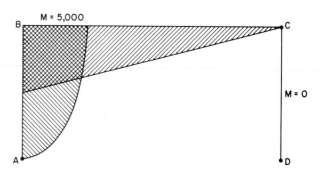

Fig. F.4.
Answer to problem 5.12.

5.13. The vertical reactions are 20,000 lb (LH) and 10,000 lb (RH); the horizontal reactions are 10,000 lb. The BM diagram is shown in Fig. F.5; max. BM $= +250,000$ lb ft under the concentrated load.

5.14. Vertical reactions: 50,000 lb; horizontal reactions 25,000 lb. The BM diagram is shown in Fig. F.6. The cross-sectional area of the cable is 0.25 sq in.

Answers to Problems

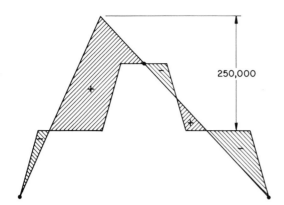

250,000

Fig. F.5.
Answer to problem 5.13.

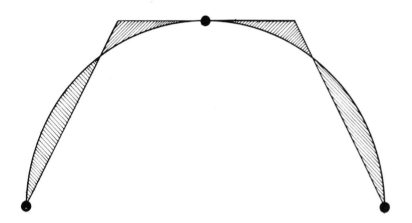

Fig. F.6.
Answer to problem 5.14.

Chapter 6

6.2. Max. SF = 1,130 lb; max. BM = 6,300 lb ft; S required = 75.6 in^3. Say 4 × 12 nominal ($3\frac{5}{8}$ × $11\frac{1}{2}$ actual, S = 79.90 in^3).

6.3. Max. SF = 4,778 lb; max. BM = 38,222 lb ft. S = 382.222 in^3. Depth required = 13.824 in, say 14 in.

6.4. I = 238,464 in^4; S = 9,936 in^3.

6.5. Max. SF = 8,000 lb; max. BM = 24,000 lb ft (see Fig. F.7). S = 115.2 in^3. Say 7 × 12 nominal ($6\frac{1}{2}$ × $11\frac{1}{2}$ actual, S = 143.27 in^3).

6.6. Either above or below the existing beam, not adjacent to it, and connected to it by sufficient bolts to resist the horizontal shear. If appearance is no objection, a greater advantage can

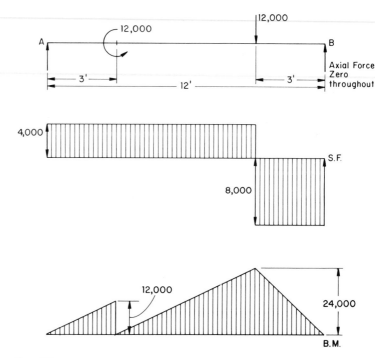

Fig. F.7.
Answer to problem 6.5.

be obtained by cutting the length of timber in half, and fixing one 6-ft length above and another below the existing beam covering the range of maximum bending moment.

6.7. The deflection is -2.308 in (downwards). The camber required to eliminate it is 2.308 in upwards.

6.8. The horizontal shear stress and both principal stresses are 156.26 p.s.i..

6.9. The principal stresses are 28,000 p.s.i. T and 7,000 p.s.i. C.

6.10. The principal stresses are 356.16 p.s.i. C and 56.16 p.s.i. T. The angle of inclination is $38°0'$.

6.11. The maximum forces are 4,217 lb C and 5,217 lb T. The max. compressive stress in the concrete is 100.4 p.s.i.; the area of reinforcement required is 0.261 sq in per ft width.

Chapter 7

7.1. 0.45 sq in.

7.4. Using the same maximum permissible stresses as in Section 7.4; (a) a 4-in long $\frac{3}{8}$-in fillet weld on each side of the angle (3.58 in required); (b) five $\frac{3}{4}$-in rivets (4.52 rivets required);

(c) seven $\frac{3}{4}$-in black bolts (6.80 bolts required); (d) five $\frac{3}{4}$-in high-strength bolts (4.52 bolts required).

7.5. By resolution at the joints, the most highly stressed diagonal tension member is the diagonal closest to the support (2,828 lb T); the most highly stressed diagonal compression member is the adjacent diagonal (2,828 lb C).

By the method of sections, or by taking the bending moment (120,000 lb in) and dividing by the resistance arm (10 in), the most highly stressed portion of the bottom chord is at mid-span (12,000 lb T).

7.8. The (factored) ultimate column load is 2,225,250 lb. The area of reinforcement required is 13.111 sq in. Say 12 No. 10 (nominal $1\frac{1}{4}$ in) bars (14.726 sq in).

7.9. Assuming No. 5 (nominal $\frac{5}{8}$ in) bars, the effective depth is 2.937 in and the area of reinforcement required is 0.7005 sq in per ft. No. 5 at 5-in centers (0.734 sq in per ft) on the top of the slab.

7.11. The flat plate requires bars in both directions, and one set of bars lies inside the other. Considering the lower (interior) set, and assuming No. 6 (nominal $\frac{3}{4}$ in) bars, the effective depth is 3.125 in. The area of reinforcement required is 0.7901 sq in per ft; No. 6 at $6\frac{1}{2}$ in centers (0.816 sq in per ft).

7.12. From Fig. 6.8: A = 8 sq ft; I = 2.167 ft^4; S_t (wall side) = 2.889 ft^3; S_b (pier side) = 1.773 ft^3. Vertical load = 25,200 lb; BM = 4,500 lb ft; stress on pier side = 5,747 lb per sq ft; stress on wall side = 1,592 lb per sq ft, both compressive.

7.13. Vertical reactions = 4,000 lb each; horizontal reactions = 2,500 lb each; max. BM = 300,000 lb in; stress in column varies from 1,378 p.s.i. C to 1,289 p.s.i. T.

7.14. After prestressing, the concrete stress varies from 0 (top) to 2,000 p.s.i. (bottom). Under full load it varies from 2,000 p.s.i. C (top) to 0 (bottom).

Chapter 8

8.3. Using the Theorem of Three Moments, or Table 8.2, the negative BM at the central support is 50,000 lb ft. This is the max. negative BM. The moments M_1 and M_2 are +50,000 lb ft, and the BM diagram is sketched in Fig. F.8. Scaling from this diagram, or by calculation from Table 8.2, the max. positive BM is 28,000 lb ft.

8.4. Using Table 8.2, the positive BM in the first span = 45,568 lb ft (max.); second span = 28,928 lb ft; center span = 33,792 lb ft. The negative BM at the outer support = 0; at the first interior support = 56,576 lb ft (max.); at the support next to the center span = 47,616 lb ft.

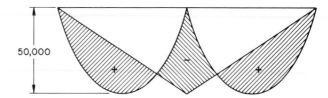

Fig. F.8.
Answer to problem 8.3

8.5. The problem is similar to Example 8.1. Using the notation of Fig. 8.6, $M_A = 0$; $M_B = -57,994$ lb ft; $M_C = -93,401$ lb ft (max. negative BM); $M_D = 0$; $M_1 = +93,750$ lb ft; $M_2 = +60,000$ lb ft; $M_3 = +135,000$ lb ft. The net BM diagram is similar to Fig. 8.6, and the max. positive BM can be scaled.

8.7. The max. negative BM can be calculated from theory, or using Table 8.3; it is 1,667 lb ft per ft. The max. positive BM is the difference between this and WL/8; this is 833 lb ft per ft.

8.8. The problem is similar to Example 8.3. Max. negative BM = 26,667 lb ft; max. positive BM = 13,333 lb ft. The BM diagram is sketched in Fig. F.9.

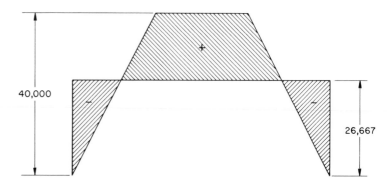

Fig. F.9.
Answer to problem 8.8

8.9. The stiffness of the beam is half the stiffness of the column because it is twice as long. After three moment distributions, the BM at the base of the portal = +6,615 lb ft, and at the knee-junction = -13,280 lb ft. From the latter, the BM at mid-span is obtained = +11,720 lb ft. S = 14.57 in³. The flexural stresses at the base = 5,450 p.s.i.; at the knee-junction = 10,940 p.s.i.; at mid-span = 9,655 p.s.i..

8.10. The BM at the upper end of the column = -31,300 lb ft; at the lower end of the column = +15,700 lb ft.

Answers to Problems

8.13. The angle is 22°30′, and the principal stresses are 241.4 p.s.i. (compression) and 41.4 p.s.i. (tension).

8.14. The difference in strain due to temperature change is 300×10^{-6}, and this is the sum of the tensile strain in the aluminum and the compressive strain in the steel, since they are fixed to one another. Furthermore, the tensile force in the aluminum is equal to the compressive force in the steel for equilibrium. The tensile stress in the aluminum is 2,571 p.s.i., and the compressive stress in the steel is 1,286 p.s.i..

Chapter 9.

9.3. Cable tension at mid-span = 32,400 lb; reaction at support = 33,627 (horizontal component = 32,400 lb, vertical component = 9,000 lb); max. cable tension (at support) = 33,627 lb; cross-sectional area of cable = 0.280 sq in.

9.4. Resolving horizontally and vertically at A: force in AB = 24,000 lb T; force in AC = 26,000 lb C. Cross-sectional areas required for AB: 1.2 sq in; for AC 4.0 sq in.

9.6. Vertical reactions = 2,000 lb; horizontal reactions = 1,000 lb. BM diagram is sketched in Fig. F.10. Max. positive BM occurs under concentrated load. Max. negative BM (and the absolute max. BM) occurs approx. 3 ft 9 in from supports and is approx. 59,000 lb in (scaled from Fig. F.10). At this point, the angle of inclination of the arch to the horizontal is approx. 54°. Hence the thrust in the arch is 1000 cos 54° + 2000 sin 54° = 2,206 lb. The maximum stresses are 440 p.s.i. (compression, on the inside of the arch) and 380 p.s.i. (tension, outside).

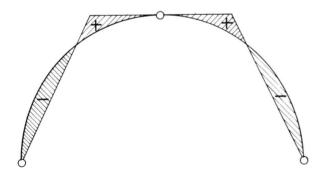

Fig. F.10.
Answer to problem 9.6.

Examples and Problems in Metric Units

G

This appendix is for the benefit of students and instructors in countries which have converted to the SI metric system, and for those American students and instructors who wish to anticipate the conversion. All numerical examples and problems are re-stated in appropriate metric units, and the examples are solved in exactly the same way as the corresponding examples in the previous chapters. The solutions to the metric problems are given in Appendix H.

Some of the examples and problems in Chapters 6 and 7 require the geometric properties of standard steel and timber sections. There are, at the time of writing this appendix, no metric U.S. or Canadian steel and timber sections, so Australian standard metric sections are used. British standard sections are also given in Examples 6.1 and 6.2. In the other examples, the additional use of British sections would require the complete repetition of the calculations. However, the differences between the British and Australian standard sections are slight, and readers or instructors can easily make the necessary substitution.

Examples 7.10, 7.11, and 7.12 are solved in metric dimensions using British and Australian metric standard bars. The examples are, however, solved in accordance with the (American) ACI Code. The British Concrete Code CP 110-1972 and the Australian Concrete Code AS 1480-1974 are based on the same principles, but they differ slightly from the ACI Code and from one another.

Chapter 3

Example 3.1. *A force of 10 N is inclined to the horizontal at an angle of 30°. Determine the horizontal and vertical components.*

Horizontal component $= 10 \cos 30° = 10 \times 0.866 = 8.66$ N

Vertical component $= 10 \sin 30° = 10 \times 0.500 = 5.00$ N

Example 3.2. *Determine the reactions of the simply supported beam shown in Fig. 3.14m.*

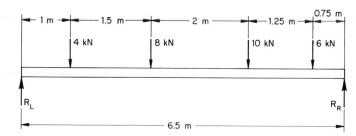

Fig. 3.14m.
Determination of reactions (see Example 3.2).

Let us take moments about the left-hand support. Since $\Sigma M = 0$, the sum of the moments of the individual loads about the left-hand support equals the moment of the right-hand reaction about the left-hand support.

$$4 \text{ kN} \times 1 \text{ m} + 8 \text{ kN} \times (1 + 1.5) \text{ m} + 10 \text{ kN} (1 + 1.5 + 2) \text{ m}$$

$$+ 6 \text{ kN} (1 + 1.5 + 2 + 1.25) \text{ m} = R_R \times 6.5 \text{ m}$$

Solving for the right-hand reaction

$$R_R = \frac{4 \text{ kN m} + 20 \text{ kN m} + 45 \text{ kN m} + 34.5 \text{ kN m}}{6.5 \text{ m}} = 15.92 \text{ kN}$$

Let us now take moment about the right-hand support:

$$R_L \times 6.5 \text{ m} = 4 \text{ kN} \times 5.5 \text{ m} + 8 \text{ kN} \times 4 \text{ m} + 10 \text{ kN} \times 2 \text{ m}$$

$$+ 6 \text{ kN} \times 0.75 \text{ m}$$

Consequently the left-hand reaction $R_L = 12.08$ kN.

For vertical equilibrium, $\Sigma F_y = 0$, or the sum-total of the loads must equal the sum of the reactions. We may thus check the values determined for R_L and R_R:

$$4 \text{ kN} + 8 \text{ kN} + 10 \text{ kN} + 6 \text{ kN} = 15.92 \text{ kN} + 12.08 \text{ kN}$$

Example 3.3. *Determine the moment reaction required at the end of the cantilever shown in Fig. 3.15m.*

The anticlockwise restraining moment must equal the clockwise moments about the support produced by the forces due to the loads acting on the cantilever:

$$M = 4 \times 1 + 8 \times 2.5 + 10 \times 4.5 + 6 \times 5.75 = 103.5 \text{ kN m}$$

Examples and Problems in Metric Units

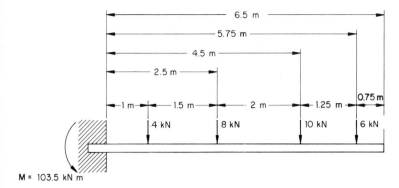

Fig. 3.15m.
Determination of moment reaction (see Example 3.3).

Example 3.4. *Determine the moment reaction required at the end of a cantilever, 8 m long, carrying a uniformly distributed load of 24 kN.*

The load equals 3 kN per m run of beam. If only the restraining moment is required, it can be replaced by a single load of 24 kN halfway along the beam (Fig. 3.16m). Consequently

$$M = 24 \times 4 = 96 \text{ kN m}$$

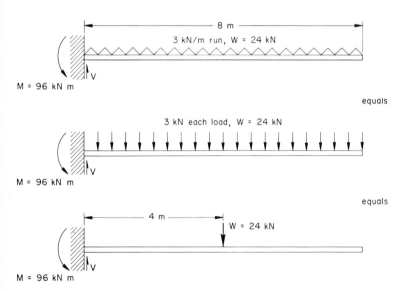

Fig. 3.16m.
Determination of moment reaction (see Example 3.4).

Example 3.5. *Determine the reactions of the roof truss shown in Fig. 3.20m.*

359

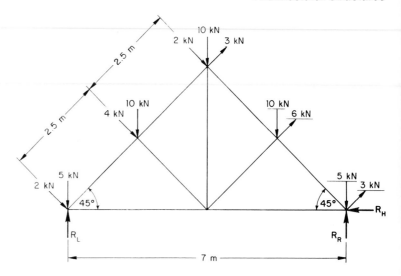

Fig. 3.20m.
Determination of reactions (see Example 3.5).

The truss carries vertical dead loads at all panel points, and in addition there is wind pressure on the windward face, and wind suction on the leeward face. We will assume that the columns supporting the truss are sufficiently flexible to allow slight horizontal expansion of the truss under load; but one horizontal reaction is needed for equilibrium, to keep the entire building from being pushed over by the wind. In addition there are vertical reactions at both columns.

Resolving horizontally:

$$(2 + 4 + 2) \cos 45° + (3 + 6 + 3) \cos 45° = R_{\mathrm{H}}$$

This gives $R_{\mathrm{H}} = 14.14$ kN.

Resolving vertically:

$$(3 \times 10 + 2 \times 5) + 8 \sin 45° - 12 \sin 45° = R_{\mathrm{L}} + R_{\mathrm{R}}$$

This gives $R_{\mathrm{L}} + R_{\mathrm{R}} = 40 - 4 \times 0.7071 = 37.17$ kN.

Taking moment about the right-hand support:

$$5 \times 7 + 10 \times 5.25 + 10 \times 3.5 + 10 \times 1.75 + 2 \times 5$$

$$+ 4 \times 2.5 - 3 \times 5 - 6 \times 2.5 = R_{\mathrm{L}} \times 7$$

This gives $R_{\mathrm{L}} = 18.58$ kN.

Since $R_{\mathrm{L}} + R_{\mathrm{R}} = 37.17$ kN, $R_{\mathrm{R}} = R_{\mathrm{L}}$; however, this result is not obvious.

Example 3.6. *Determine the vertical and horizontal reactions of the retaining wall shown in Fig. 3.21m.*

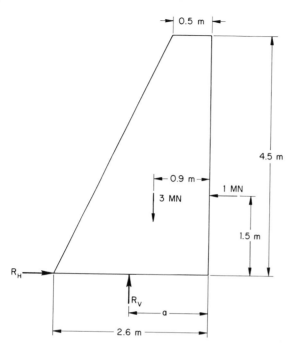

Fig. 3.21m.
Determination of reactions (see Example 3.6).

The wall retains soil, which exerts a horizontal pressure of 1 MN at two-thirds of the depth. The wall itself weighs 3 MN. Let us summarize the reactions as a horizontal force R_H, which must be provided by friction or mechanical anchorage under the wall, and a vertical reaction R_V, which must be provided by the upward pressure of the soil under the wall.

Resolving horizontally: $R_H = 1$ MN

Resolving vertically: $R_V = 3$ MN

Since the horizontal reaction acts at the base of the wall, the foundation pressure is unlikely to be uniform; and to determine its distribution, we must determine its line of action, defined by the distance a from the vertical wall face.

Taking moments about the bottom of the vertical face, we obtain

$$3 \times 0.9 + 1 \times 1.5 = R_H \times 0 + R_V \times a$$

which gives $a = 4.2/3 = 1.4$ m.

Example 3.7. *Determine the center of gravity of the retaining wall shown in Fig. 3.23m.*

If the wall is made of the same material throughout (say, concrete) the weight of any part of the section is proportional to the

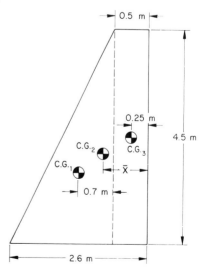

Fig. 3.23m.
Determination of center of gravity (C.G.) (see Example 3.7).

361

area of that section. For convenience, we therefore divide the wall into a rectangle, whose center of gravity lies at half-height and half-width; and a triangle, whose center of gravity lies at one-third height and one-third width.

We are concerned with determining the location of the center of gravity through which the weight of the combined wall acts, and since this is a vertical force, we are interested only in the location of the center of gravity along the x-axis. The component rectangle has a weight proportional to the area of 0.5×4.5, and its center of gravity (C.G.$_3$) is 0.25 m from the right-hand (vertical) face. The moment about that face is $0.5 \times 4.5 \times 0.25$. The triangle has an area of $\frac{1}{2} \times 2.1 \times 4.5$, and its center of gravity (C.G.$_1$) is $0.5 + 0.7$ from the vertical face. The moment about it is $\frac{1}{2} \times 2.1 \times 4.5 \times 1.2$.

If the distance of the center of gravity of the entire wall (C.G.$_2$) from the vertical face is $\bar{x}$, then taking moments about the vertical face

$$\left(0.5 \times 4.5 + \tfrac{1}{2} \times 2.1 \times 4.5\right)\bar{x} =$$

$$0.5 \times 4.5 \times 0.25 + \tfrac{1}{2} \times 2.1 \times 4.5 \times 1.2$$

so that $\bar{x} = 6.2325/6.975 = 0.894$ m

Example 3.8. *Determine the location of the center of gravity of the prestressed concrete section shown in Fig. 3.24m.*

Since the section is vertically symmetrical, the center of gravity is evidently situated centrally, 300 mm from either side. To determine its vertical location, let us take moments about the top face:

$$(600 \times 50 + 300 \times 75)\,\bar{y} = 600 \times 50 \times 25 + 300 \times 75(50 + 150)$$

$$\bar{y} = \frac{5{,}250{,}000}{52{,}500} = 100 \text{ mm}$$

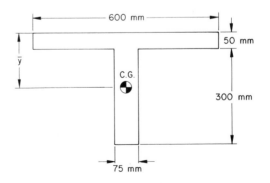

Fig. 3.24m.
Determination of center of gravity (see Example 3.8).

Example 3.9 *Determine the depth of the center of gravity of a composite steel section consisting of an Australian 250 universal beam (37 kg/m) and a 250 mm × 25 mm plate, as shown in Fig. 3.25, p. 70.*

From section tables, the universal beam is 146 mm wide and 256 mm deep; its cross-sectional area is 4,750 mm². Its center of gravity is placed symmetrically, i.e., 128 mm from the edge of the flange. The plate has an area of 250 × 25 = 6,250 mm².

Taking moments about the top face of the compound section

$$(4,750 + 6,250)\,\bar{y} = 6,250 \times 12.5 + 4,750(128 + 25)$$

$$\bar{y} = \frac{804,875}{11,000} = 73.17 \text{ mm}$$

The Australian metric section is lighter than the American section used in inch units in Chapter 3. The nearest British universal beam is very similar to the Australian beam. It has a nominal size of 254 × 146 (actual size 256 mm × 146.4 mm), weighs 37 kg/m, and has a cross-sectional area of 4,740 mm².

Example 3.10. *Determine the width of the foundation required for a wall carrying a load of 150 kN/m run (including the weight of the wall and the footing) if the maximum permissible bearing pressure of the soil is 180 kPa.*

The width required is

$$\frac{150}{180} = 0.834 \text{ m} \qquad (\text{Say 1 m.})$$

Example 3.11. *Determine the size of the footing required for a reinforced concrete column, 600 mm square, carrying a load of 1.8 MN, if the maximum permissible bearing pressure is 180 kPa.*

Let us provide the footing with a reinforced concrete slab, 600 mm thick, and assume that the mass of the reinforced concrete is 2,400 kg/m³.

In American units and conventional metric units, the mass is numerically equal to the weight. When SI metric units are used, the mass must be multiplied by the acceleration due to gravity (9.81 m per sec per sec) The footing slab therefore has a mass of 1,440 kg/m² and produces a vertical pressure of

$$1,440 \times 9.807 \text{ N/m}^2 = 14.122 \text{ kN/m}^2 = 14.122 \text{ kPa}$$

We must subtract this from the maximum permissible bearing pressure which leaves

$$180 - 14.1 = 165.9 \text{ kPa}$$

Let us assume a square footing, of area B^2 (Fig. 3.27m) so that

$$B = \sqrt{\frac{1800}{165.9}} = 3.29 \text{ m} \qquad (\text{Say 3.5 m square.})$$

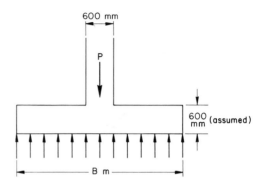

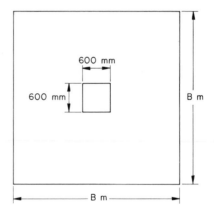

Fig. 3.27m.
Foundation pressure under column footing (see Example 3.11).

Example 3.12. *Determine the water pressure at the bottom of a water tower if the depth of the water is 2 m.*

The mass of water is $1,000 \text{ kg/m}^3$, and the pressure due to this mass is $1,000 \times 9.807 \text{ Pa} = 9.807 \text{ kPa}$ per metre depth. The water pressure at the bottom of the tower is therefore 19.61 kPa.

Example 3.13. *Determine the resultant horizontal force on a wall, retaining soil 4 m deep. The equivalent fluid mass of the soil is 480 kg/m^3.*

The horizontal pressure varies uniformly from zero at the surface to a maximum at the bottom of the wall. The equivalent fluid mass of the soil at a depth of 4 m is $4 \times 480 = 1{,}920$ kg/m³, which produces a pressure of $1{,}920 \times 9.807$ Pa $= 18.83$ kPa/m width of wall.

The average soil pressure is $\frac{1}{2} \times 18.83$ kPa, and the area on which this pressure acts is a depth of 4 m and a unit length of 1 m. The resultant horizontal force per metre run of wall is $H = \frac{1}{2} \times 18.83 \times 4 \times 1 = 37.66$ kN per metre run of wall.

This resultant force H acts at $\frac{1}{3} h = \frac{1}{3} \times 4 = 1.33$ m above the base of the wall (Fig. 3.28, p. 73).

Problems

3.1. Determine the magnitude of the forces needed to restore equilibrium in the system of forces shown in Fig. 3.33m.

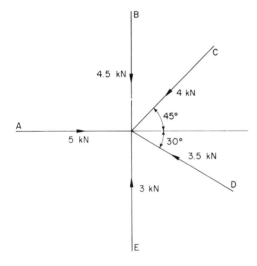

Fig. 3.33m.
Problem 3.1.

3.2. Find the resultant of all the forces acting on the bracket shown in Fig. 3.34m. Calculate the shortest distance from point A to the line of action of the resultant.

3.3. Determine the reactions of the cantilevered beam, shown in Fig. 3.35m, which carries a total load of 6.5 kN.

3.4. The cantilever shown in Fig. 3.36m carries a uniformly distributed load of 5 kN/m run, as well as concentrated loads totalling 15 kN. Determine the vertical and moment reactions at the support.

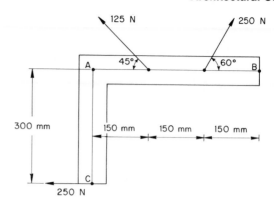

Fig. 3.34m.
Problem 3.2.

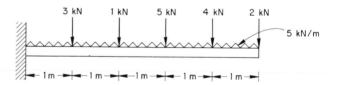

Fig. 3.35m.
Problem 3.3.

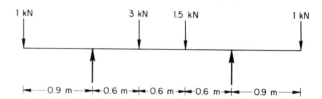

Fig. 3.36m.
Problem 3.4.

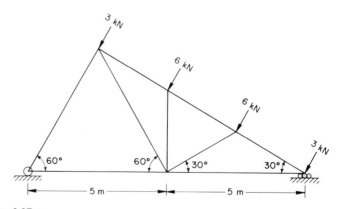

Fig. 3.37m.
Problem 3.5.

Examples and Problems in Metric Units

3.5. The roof truss shown in Fig. 3.37m is pinned at the left-hand support and able to slide at the right-hand support. It carries a wind load of 18 kN. Determine the magnitude of the support reactions.

3.6. Fig. 3.38m shows a steel plate with two holes in it. Determine the distance of the center of gravity from the bottom-left corner.

3.7. Figure 3.39m shows a large rectangular steel plate with a circular hole in it. The plate is freely suspended by a cable at point B. There is a weight hanging from one corner, point E. The force due to the mass of the plate is 0.5 kN/m² of plate. If the magnitude of the weight W is such that the plate hangs with the side AC exactly perpendicular to the cable BD, calculate the force in the cable.

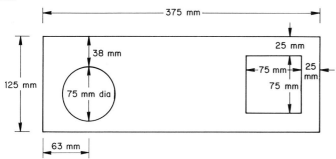

Fig. 3.38m.
Problem 3.6.

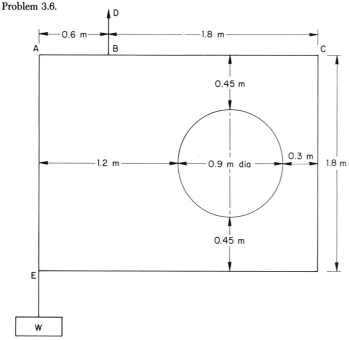

Fig. 3.39m.
Problem 3.7.

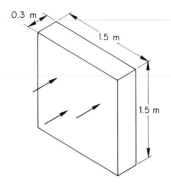

Wind Pressure = 1.5 kPa

Fig. 3.40m.
Problem 3.8.

3.8. A concrete block of density 2,000 kg/m³, measures 1.5 m by 1.5 m by 0.3 m. There is a horizontal pressure of 1.5 kPa on one side of the block as shown in Fig. 3.40m. Calculate whether the block will overturn.

3.9. A tall, narrow building has the following dimensions: plan 30 m by 9 m, height 100 m. If the mass of the material is approximately 100 kg/m³, determine the horizontal wind pressure required to overturn the building. Assume the wind velocity profile to be uniform.

3.10. A masonry wall is 1.5 m high and 0.3 m thick. There is a horizontal pressure of 0.5 kPa, acting uniformly on one side of the wall. Calculate the minimum average density of the material from which the wall needs to be constructed if no tension is to develop in any part of the structure.

Chapter 4

The solution of the unnumbered Example in Section 4.2 is shown in Fig. 4.4m.

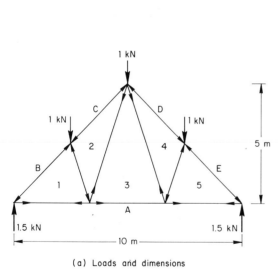

(a) Loads and dimensions

(b) Stress diagram

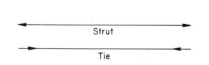

Strut

Tie

(c) Distinction between struts (compression member = C) and ties (tension members = T)

Member	Force (kN)	Strut or Tie
A1 A5	1.50	T
A3	1.00	T
B1 E5	2.12	C
C2 D4	1.82	C
12 45	0.82	C
23 34	0.82	T

(d) Magnitude of forces scaled from stress diagram

Fig. 4.4m.
Stress diagram for timber roof truss.

Examples and Problems in Metric Units

Example 4.1. *By graphic statics determine the forces in the king-post truss shown in Fig. 4.5m.a.*

The stress diagram is drawn in Fig. 4.5m.b, and the arrows are marked on Fig. 4.5m.a. The results measured off the stress diagram are tabulated in Fig. 4.5m.c. The force BC goes straight to the reaction AB, and does not contribute to the stress diagram. Note that the king post is not in compression, but in tension.

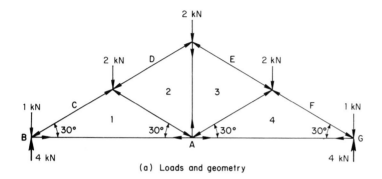

(a) Loads and geometry

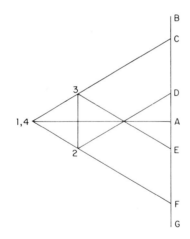

(b) Stress diagram

Member	Force (kN)	Strut or Tie
A1 A4	5.20	T
C1 F4	6.00	C
D2 E3	4.00	C
12 34	2.00	C
23	2.00	T

(c) Magnitude of forces scaled from stress diagram

Fig. 4.5m.
Stress diagram for king post truss (Example 4.1)

Example 4.2. *By graphic statics determine the forces in the parallel chord truss shown in Fig. 4.6m.a.*

The stress diagram is drawn in Fig. 4.6m.b. The arrows have been placed on Fig. 4.6m.a to distinguish between struts and ties, and the results are tabulated in Fig. 4.6m.c.

It may be noted that the reactions AB and GA are transmitted along the verticals B1 and G8, so that the bottom chords A1 and A8 have no internal forces and, theoretically, could be removed. As a matter of common sense, these members are required nevertheless; they become important as soon as the smallest horizontal force is placed against the truss. Unstressed members do occur under certain conditions of loading, particularly in space frames, where even a statically determinate frame has a large number of members.

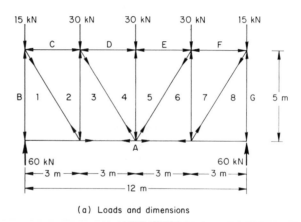

(a) Loads and dimensions

Member	Force (kN)	Strut or Tie
A1 A8	0	—
A3 A6	27.0	T
C2 F7	27.0	C
D4 E5	36.0	C
B1 G8	60.0	C
23 67	45.0	C
45	30.0	C
12 78	52.5	T
34 56	17.5	T

(c) Magnitude of forces scaled from stress diagram

(b) Stress diagram

Fig. 4.6m.
Stress diagram for Pratt truss (Example 4.2)

Examples and Problems in Metric Units

The solution of the unnumbered example (Fig. 4.8m) is as follows:

$$BC \times 4 = 7 \times 8 - 2 \times 6 - 2 \times 4 - 2 \times 2$$

which gives BC = $7 \times 2 - 2(1.5 + 1 + 0.5) = 8$ kN

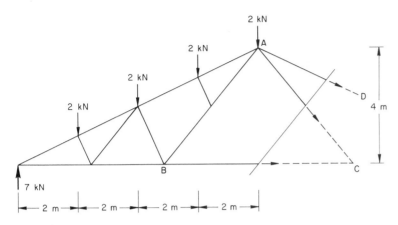

Fig. 4.8m.
The method of sections. The force in the tie BC is obtained by making a cut, and taking moments about A. (The horizontal distance between panel points is 2m).

Example 4.3. *Determine the magnitude of the greatest tensile and compressive forces in the horizontal chords of the truss shown in Fig. 4.6m.*

Since the truss is simply supported at its ends, the bending moment is a maximum at the center of the truss (see Sections 1.1 and 5.4). Since the truss has parallel chords, the resistance arm of the moment is the same throughout, and the highest tensile and compressive forces therefore occur in the panels nearest to the center.

Let us therefore cut the truss as shown in Fig. 4.9m, and replace the internal forces in AD, AC, and BC by imaginary external forces.

AD and AC have no moments about A, and the moments of the remaining "external" forces are:

$$BC \times 5 = 60 \times 3 - 15 \times 3$$

which gives BC = 27 kN.

Since we have drawn the force as tensile, and the answer is positive, the force is, in fact, tensile.

Let us now take moments about C:

$$AD \times 5 = -60 \times 6 + 15 \times 6 + 30 \times 3$$

which gives AD = -36 kN.

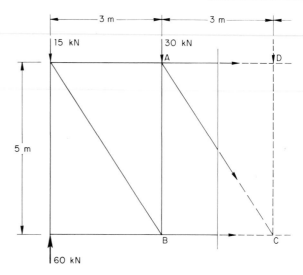

Fig. 4.9m.
Approximate design of Pratt truss by the method of sections (Example 4.3). The internal forces in the horizontal chords are obtained by making a suitable cut. The other members are of lesser importance, and they need not be considered at the preliminary design stage.

Since we have drawn the force as tensile and the answer is negative, the force is, in fact, compressive.

Example 4.4. *Determine the forces in the king-post truss shown in Fig. 4.10m.*

Bow's notation is unsuitable for resolving at the joints, and we letter the joints instead. Since the truss is obviously symmetrical, we can save ourselves some equations by lettering statically identical joints with the same letter.

In writing down our equations, we will assume that all internal forces in the truss are tensile. If the result comes out negative, the internal force is compressive.

Taking joint A first and resolving horizontally (see Fig. 3.6 and 4.10m.b)

$$T_{AB} \cos 30° + T_{AD} = 0 \tag{i}$$

Resolving vertically at joint A:

$$1 - T_{AB} \sin 30° - 4 = 0 \tag{ii}$$

Equation (ii) gives

$$T_{AB} = \frac{1 - 4}{\sin 30°} = \frac{-3}{0.5} = -6 \text{ kN (compression)}$$

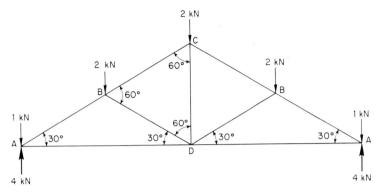

(a) Loads and geometry

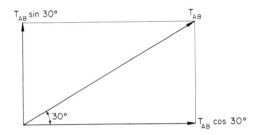

(b) Horizontal and vertical components of internal force T_{AB}

Fig. 4.10m.
The method of resolving at the joints (Example 4.4).

We can now substitute this value of T_{AB} into Eq. (i)

$$-6 \times 0.866 + T_{AD} = 0$$

which gives

$$T_{AD} = +5.20 \text{ kN (tension)}$$

Taking the next joint B and resolving horizontally:

$$T_{AB} \cos 30° + T_{BC} \cos 30° + T_{BD} \cos 30° = 0$$

which simplifies to

$$T_{BC} + T_{BD} = -6 \qquad \text{(iii)}$$

Resolving vertically at B:

$$2 - T_{BC} \sin 30° + T_{AB} \sin 30° + T_{BD} \sin 30° = 0$$

which simplifies to

$$2 - T_{BC} \times 0.5 - 6 \times 0.5 + T_{BD} \times 0.5 = 0$$

$$T_{BC} - T_{BD} = -2 \qquad \text{(iv)}$$

Adding Eq. (iii) to Eq. (iv)

$$2T_{BC} = -8 \quad \text{and} \quad T_{BC} = -4 \text{ kN (compression)}$$

From Eq. (iii) $T_{BD} = -2$ kN (compression).
Resolving horizontally at C we obtain:

$$T_{BC} = T_{BC}$$

which is one of our check equations.
Resolving vertically at C

$$2 + T_{BC} \sin 30° + T_{BC} \sin 30° + T_{CD} = 0$$

This gives

$$T_{CD} = -2 + 4 \times 0.5 + 4 \times 0.5 = +2 \text{ kN (tension)}$$

Resolving horizontally at D we obtain another obvious check equation:

$$T_{AD} + T_{BD} \cos 30° = T_{BD} \cos 30° + T_{AD}$$

Resolving vertically at D, we obtain a more useful check on our calculations:

$$T_{CD} + 2T_{BD} \sin 30° = 0$$

Substituting numbers:

$$2 - 2 \times 2 \times 0.5 = 0$$

which checks.

Example 4.5. *Using tension coefficients, determine the forces in the truss shown in Fig. 4.12m.*

The frame being symmetrical, A, B, C, and D are repeated with the suffix 2 on the right-hand side.

Since the solution is largely mechanical, errors in sign are not so easily discovered as in the previous methods, where they are often evident by inspection. The sign convention must therefore be strictly observed.

As with ordinary Cartesian coordinates, x-distances are positive to the right and negative to the left, and y-distances are positive up and negative down. If we take the joint C_1, the x-distances of E and of F are $+3$, and the x-distances of A and B are -3; the x-distance of D is 0. The y-distances of A and E are 0, and the y-distances of B, D and F are -5.

Similarly the loads A, C, and E are negative, because they are

Examples and Problems in Metric Units

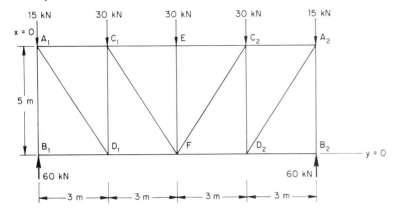

(a) Loads and dimensions

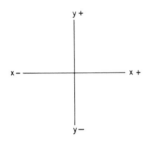

(b) Sign convention. Distances *and* loads positive up and right,
negative down and left.

Fig. 4.12m.
Solution of Pratt truss by tension coefficients (Example 4.5).

downward; and the reaction at A is positive, because it acts upward.

We will now write down the equations for each joint in turn, starting with the condition of horizontal equilibrium at the joint A, and followed by the condition of vertical equilibrium (y):

Joint	Direction	Equation	Eq. No.
A	x	$0t_{AB} + 3t_{AC} + 3t_{AD} = 0$	(i)
	y	$-5t_{AB} + 0t_{AC} - 5t_{AD} - 15 = 0$	(ii)
B	x	$+3t_{BD} = 0$	(iii)
	y	$+5t_{AB} + 60 = 0$	(iv)
C	x	$-3t_{AC} + 3t_{CE} + 3t_{CF} = 0$	(v)
	y	$-5t_{CD} - 5t_{CF} - 30 = 0$	(vi)
D	x	$-3t_{BD} - 3t_{AD} + 3t_{DF} = 0$	(vii)
	y	$+5t_{AD} + 5t_{CD} = 0$	(viii)
E	x	$-3t_{CE} + 3t_{CE} = 0$	(ix)
	y	$-5t_{EF} - 30 = 0$	(x)
F	x	$-3t_{DF} + 3t_{DF} - 3t_{CF} + 3t_{CF} = 0$	(xi)
	y	$+5t_{CF} + 5t_{CF} + 5t_{EF} = 0$	(xii)

Most of these equations are very easily solved:

(ix) and (xi) are check equations only and give $0 = 0$

(iii) $t_{BD} = 0$

(iv) $t_{AB} = -12$

(x) $t_{EF} = -6$

(ii) $t_{AD} = -t_{AB} - 3 = +9$

(i) $t_{AC} = -t_{AD} = -9$

(vii) $t_{DF} = t_{AD} + t_{BD} = +9$

(viii) $t_{CD} = -t_{AD} = -9$

(xii) $t_{CF} = -\frac{1}{2} t_{EF} = +3$

(v) $t_{CE} = t_{AC} - t_{CF} = -12$

This leaves one equation as a numerical check:

(vi) $-5(-9) - 5(+3) - 30 = 0$, which is correct.

We now tabulate the tension coefficients and the lengths of the members, and thus obtain the forces. Positive answers denote tension, and negative answers compression.

Member	Tension Coefficient (kN/m)	Length (m)	Force (kN)
AB	− 12	5	60.0 C
AC	− 9	3	27.0 C
AD	+ 9	5.83	52.5 T
BD	0	3	0
CD	− 9	5	45.0 C
CE	− 12	3	36.0 C
CF	+ 3	5.83	17.5 T
DF	+ 9	3	27.0 T
EF	− 6	5	30.0 C

This solution may be compared with that of Example 4.2.

Example 4.6. *Determine the forces in the bracket shown in Fig. 4.15m. The bracket is fixed to a pin in the wall at* A, *and it is sliding against the wall at* C_1 *and* C_2.

The frame being symmetrical, C_1 and C_2 are identical. There are four joints and six members, so that the frame is statically determinate. We obtain the reactions from Eq. (3.6).
From $\Sigma F_x = 0$ we obtain $R_V = 6$ kN
From $\Sigma F_y = 0$ we obtain $R_{HA} = 2 R_{HC}$

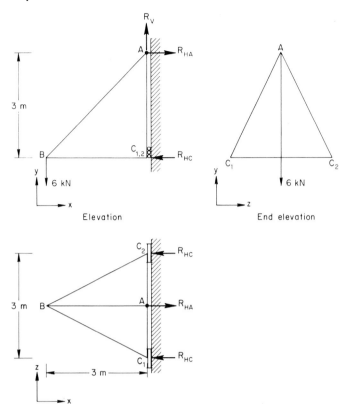

Fig. 4.15m.
Solution of space frame by tension coefficients (Example 4.6).

From $\Sigma M_{xy} = 0$ we obtain, by taking moments about A

$$2\,R_{\text{HC}} \times 3 = 6 \times 3$$

which gives $R_{\text{HC}} = 3$ kN and $R_{\text{HA}} = 6$ kN.

The remaining equilibrium conditions give $0 = 0$.

We will now write down the equations for each joint in turn, as in Example 4.5.

Joint	Direction	Equation	Eq. No.
A	x	$-3t_{\text{AB}} + 0t_{\text{AC}} + 6 = 0$	(i)
	y	$-3t_{\text{AB}} - 3t_{\text{AC}} - 3t_{\text{AC}} + 6 = 0$	(ii)
	z	$0t_{\text{AB}} - 1.5t_{\text{AC}} + 1.5t_{\text{AC}} = 0$	(iii)
B	x	$+3t_{\text{AB}} + 3t_{\text{BC}} + 3t_{\text{BC}} = 0$	(iv)
	y	$+3t_{\text{AB}} + 0t_{\text{BC}} - 6 = 0$	(v)
	z	$0t_{\text{AB}} - 1.5t_{\text{BC}} + 1.5t_{\text{BC}} = 0$	(vi)
C_1	x	$0t_{\text{AC}} - 3t_{\text{BC}} + 0t_{\text{CC}} - 3 = 0$	(vii)
	y	$+3t_{\text{AC}} + 0t_{\text{BC}} + 0t_{\text{CC}} = 0$	(viii)
	z	$+1.5t_{\text{AC}} + 1.5t_{\text{BC}} + 3t_{\text{CC}} = 0$	(ix)

As in Example 4.5, most of these equations solve very easily:

(i) $\quad t_{AB} = +2$

(viii) $\quad t_{AC} = 0$

(iv) $\quad t_{BC} = -\frac{1}{2} t_{AB} = -1$

(ix) $\quad t_{CC} = -\frac{1}{2} t_{BC} = +0.5$

The remainder are check equations only:

(ii) $\quad -6 + 0 + 6 = 0$

(iii) $\quad 0 = 0$

(v) $\quad 6 - 6 = 0$

(vi) $\quad 0 = 0$

(vii) $\quad +3 - 3 = 0$

It may seem at first sight that most of this operation is a waste of time if so many equations are superfluous; however, it is not always obvious which equations are redundant until they are written out.

We now tabulate the tension coefficients and the lengths of the members, and thus obtain the forces.

Member	Tension Coefficient (kN/m)	Length (m)	Force (kN)
AB	+ 2	4.24	8.48 T
AC	0	3.35	0
BC	− 1	3.35	3.35 C
CC	+ 0.5	3.00	1.50 T

Problems

4.4. Draw the stress diagram for the truss shown in Fig. 4.16m, and find the magnitude of the forces in the members D4, 34, and A3. State whether the forces are tensile or compressive.

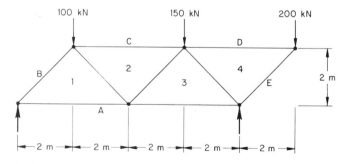

Fig. 4.16m.
Problem 4.4.

Examples and Problems in Metric Units

4.5. Draw the stress diagram, and determine the forces in all the members of the Pratt truss shown in Fig. 4.17m, indicating whether it is tensile or compressive.

4.6. Check the forces in the two members next to the left-hand support of the truss in Fig. 4.17m by the method of resolving at the joints.

4.7. Check the force in the vertical member nearest to the left-hand support of the truss shown in Fig. 4.17m by the method of sections.

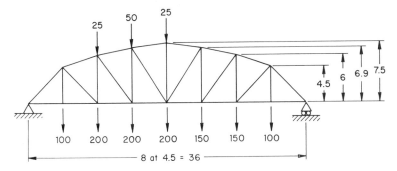

Fig. 4.17m.
Problems 4.5, 4.6, and 4.7.

4.8. Fig. 4.18m shows a truss of 7.2-m span and 450-mm depth, carrying five loads of 5 kN each. Calculate the forces in the members AO, OB, CD, and ML.

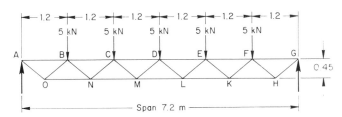

Fig. 4.18m.
Problem 4.8.

4.9. Using the method of sections, determine the magnitude of the forces in the members BD, CD, CE, DE, and DF of the truss shown in Fig. 4.19m. State whether they are tensile or compressive.

4.10. If the load at point D in Fig. 4.19m is increased by 15 kN, find the change in the magnitude of the force in the member AB by drawing the triangle of forces.

4.11. Figure 4.20m shows a bracket on a pole carrying electric cables. Assuming that the joints are free to rotate and the self-weight of the brackets is negligible, calculate:

(i) The force in the member XY; state whether it is compressive or tensile.

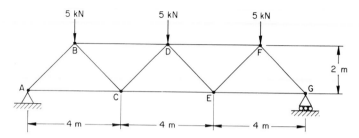

Fig. 4.19m.
Problems 4.9 and 4.10.

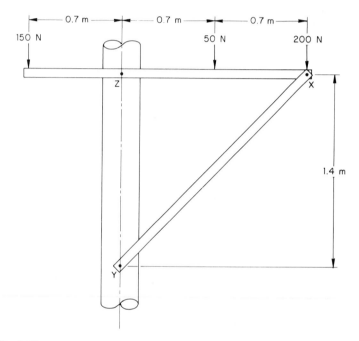

Fig. 4.20m.
Problem 4.11.

(ii) The angle which the reaction Z makes with the vertical center line of the pole; what is the horizontal component of that reaction?

4.12. An advertising sign whose weight equals a vertical force of 3 kN is suspended from a shop front by two oblique ties and a horizontal strut as shown in Fig. 4.21m. Using the method of tension coefficients, determine the forces in each of the three members, and state whether they are in tension or compression. It is advisable to start at point B.

4.13. Determine the forces in each of the members of the crane shown in Fig. 4.22m, and also the reactions at A, B, and C. Distinguish between tension and compression members. It is advisable to start at point E.

Examples and Problems in Metric Units

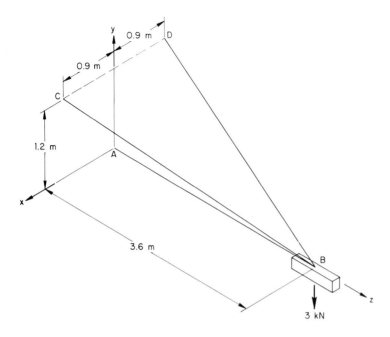

Fig. 4.21m.
Problem 4.12.

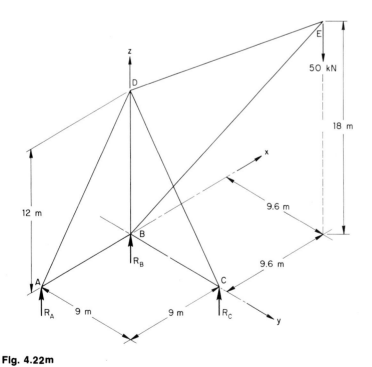

Fig. 4.22m
Problem 4.13.

Chapter 5

Example 5.1. *Draw the bending moment and shear force diagrams for a cantilever, spanning 4 m, (a) carrying a load of 5 kN at the end, and (b) 3 m from the end.*

The solution is shown in Fig. 5.6m.

Example 5.2. *Determine the maximum bending moment and shear force for a cantilever, 4 m long, carrying a load of 5 kN (a) if it is uniformly distributed with the whole of its length, and (b) if it is uniformly distributed over a length of 3 m from the support.*

The solution is shown in Fig. 5.8m.

Example 5.3. *A balcony slab, cantilevering 4 m, carries at the free end a concentrated load of 2.5 kN (due to a dwarf brick wall), a load of 5 kN uniformly distributed over the whole span (due to its own weight), and a load of 2 kN uniformly distributed over a length of 3 m from the support (due to the live load, caused by people and furniture on the balcony). Determine the maximum bending moment and the maximum shear force.*

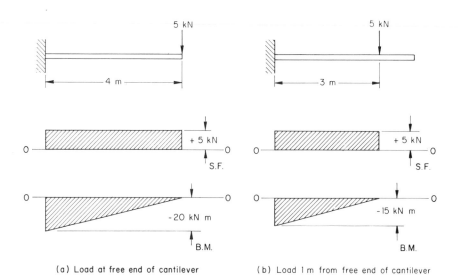

(a) Load at free end of cantilever   (b) Load 1 m from free end of cantilever

Fig. 5.6m.
Cantilever carrying a concentrated load (Example 5.1).

The maximum shear force

$$V = 2.5 + 5 + 2 = 9.5 \text{ kN}$$

The maximum bending moment

$$M = -2.5 \times 4 - \tfrac{1}{2} \times 5 \times 4 - \tfrac{1}{2} \times 2 \times 3 = -23 \text{ kN m}$$

Example 5.4. *Determine the maximum bending moment and shear force due to a central concentrated load of 5 kN, carried on a simply supported span of 4 m.*

The solution is $V = \pm 2.5$ kN and $M = 5$ kN m.

A comparison with Example 5.1 shows the structural superiority of the simply supported beam over the cantilever. The shear force is half that of the cantilever carrying the same load on the same span, and the maximum bending moment (with the load in the worst location in each case) is a quarter that of the cantilever.

Example 5.5. *Determine the bending moment and the shear force due to a concentrated load of 5 kN on a simply supported span*

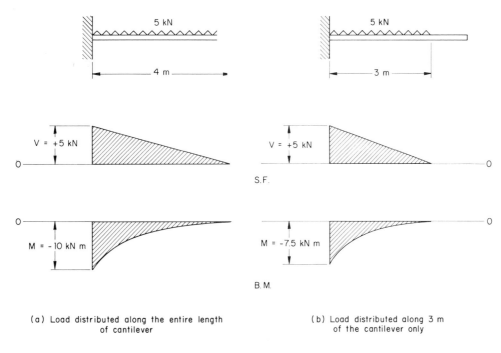

(a) Load distributed along the entire length
of cantilever

(b) Load distributed along 3 m
of the cantilever only

Fig. 5.8m.
Cantilever carrying a uniformly distributed load (Example 5.2).

of 4 m if the load is located (i) at $\frac{1}{10}$ of the span, (ii) at $\frac{1}{4}$ span, and (iii) at mid-span.

For $\frac{1}{10}$ span, $L_1 = 0.4$ m and $L_2 = 3.6$ m and $L_2/L = 0.9$
For $\frac{1}{4}$ span, $L_1 = 1$ m and $L_2 = 3$ m and $L_2/L = 0.75$
For mid-span $L_1 = L_2 = 2$ m and $L_1/L = L_2/L = 0.5$.
Consequently the maximum shear force is

(i) $V = 0.9 \times 5 = 4.5$ kN
(ii) $V = 3.75$ kN
(iii) $V = 2.5$ kN

The maximum bending moment, which occurs directly under the load is

(i) $M = 5 \times 0.4 \times 3.6/4 = 1.8$ kN m
(ii) $M = 5 \times 1 \times 3/4 = 3.75$ kN m
(iii) $M = 5 \times 2 \times 2/4 = 5$ kN m

The maximum bending moment increases as the load moves toward the center of the span; the maximum shear force increases as the load moves toward either support.

Assuming that we wished to determine the design condition for a heavy movable load, for example, a steel safe or a computing machine, we should consider (a) mid-span which gives the highest bending moment $M = \frac{1}{4}WL$, and (b) a location just off either support which gives the highest shear force, namely $V = W$. Since the bending moment is usually critical for the design, it may be sufficient to consider the mid-span location only.

Example 5.6. *A primary girder, spanning 12 m, carries two secondary beams at the third points of the span, each imposing a reaction of 500 kN. Determine the resulting maximum bending moment in the primary girder.*

$$M = \frac{1}{6}WL = \frac{1}{6} \times 2 \times 500 \times 12 = 2{,}000 \text{ kN m}$$

Example 5.7. *Determine the distribution of shear force and bending moment for the beam shown in Fig. 5.16m.*

The beam carries a total load of 5 kN so that

$$R_1 + R_2 = 5 \text{ kN}$$

Taking moments about R_2

$$R_1 = \frac{1.5 \times 3.4 + 0.5 \times 2.4 + 1 \times 1.6 + 2 \times 1}{4} = 2.475 \text{ kN}$$

which gives $R_2 = 2.525$ kN.

Examples and Problems in Metric Units

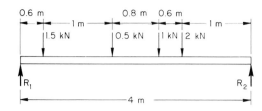

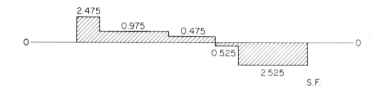

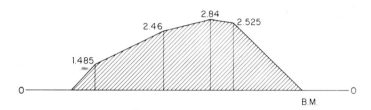

Fig. 5.16m.
Simply supported beam carrying several concentrated loads (Example 5.7).

The shear force at the left hand support is therefore 2.475 kN. This is reduced by 1.5 kN at the first concentrated load, by 0.5 kN at the next, and so forth (Fig. 5.16m.b.).

The bending moment at the first concentrated load is $2.475 \times 0.6 = 1.485$ kN m; at the second concentrated load it is $M = 2.475 \times 1.6 - 1.5 \times 1 = 2.46$ kN m.

The bending moment at the fourth concentrated load, proceeding from the right-hand support, is $2.525 \times 1 = 2.525$ kN m; and at the third concentrated load it is

$$M = 2.525 \times 1.6 - 2 \times 0.6 = 2.84 \text{ kN m}$$

The maximum bending moment is 2.84 kN m, 2.4 m from the left-hand and 1.6 m from the right-hand support (Fig. 5.16m).

Example 5.8. *Determine the maximum shear force and bending moment for a beam carrying a uniformly distributed load of 5 kN over a simply supported span of 4 m.*

The answer is

$$V = \tfrac{1}{2} \times 5 = 2.5 \text{ kN}$$

and

$$M = \tfrac{1}{8} \times 5 \times 4 = 2.5 \text{ kN m}$$

We may compare this with the single concentrated load (Example 5.4), the two symmetrically arranged concentrated loads, and the series of irregularly arranged concentrated loads (Example 5.7), all of which amount to 5 kN, over a simply supported span of 4 m.

For symmetrical arrangement, the shear force is always the same, namely 2.5 kN. The unsymmetrical arrangement of Example 5.7 makes only a slight difference to the maximum shear force, which is increased by 0.025 kN (25 N).

The bending moment, however, varies considerably. For a single concentrated load it is 5 kN m. For two loads at third points, $M = \tfrac{1}{6} WL = 3.33$ kN m. For the unsymmetrical arrangement in Example 5.7 it is 2.84 kN m. For the uniformly distributed load it is 2.5 kN m.

Example 5.9.   *A beam, 10 m long, carries a uniformly distributed load of 3 kN/m. It is simply supported on two columns, which can be placed anywhere under the beam. Determine the location of the columns that equalizes the maximum positive and negative bending moments.*

Let us call the cantilever overhangs L. Then, using the notation of Fig. 5.20, p. 130:

$$L_1 = L \qquad L_2 = 10 - 2L$$

$$W_1 = 3L \qquad W_2 = 3(10 - 2L)$$

Equalizing the maximum positive and negative bending moments (Fig. 5.20d) requires that

$$\tfrac{1}{8} . W_2 L_2 - \tfrac{1}{2} W_1 L_1 = \tfrac{1}{2} W_1 L_1$$

$$\tfrac{1}{8} \times 3(10 - 2L) \times (10 - 2L) = \left(\tfrac{1}{2} + \tfrac{1}{2}\right) \times 3L^2$$

$$\tfrac{1}{2}(5 - L) \times (5 - L) = L^2$$

$$25 - 10L + L^2 = 2L^2$$

$$L^2 + 10L - 25 = 0$$

$$L = -5 + \sqrt{25 + 25} = 5(\sqrt{2} - 1) = 2.071 \text{ m}$$

We thus obtain $L_1 = 2.071$ m; $L_2 = 5.858$ m; $W_1 = 6.213$ kN; and $W_2 = 17.574$ kN.

The negative moment is $-\frac{1}{2} \times 6.213 \times 2.071 = -6.434$ kN m; and the positive moment is $+\frac{1}{8} \times 17.574 \times 5.858 - 6.434 = 6.434$ kN m.

If we wish to use a beam of uniform section, and the bending moment is the design criterion (see Section 6.4), as it usually is, this is the best location of the columns, because it gives us the lowest possible bending moment. If, on the other hand, the shear force is the design criterion, which may happen in foundation beams (Section 6.6), we require

$$W_1 = \tfrac{1}{2} W_2$$

which evidently means that the columns should be $\frac{1}{4} \times 10$ m, or 2.5 m from the end. This makes $L_1 = 2.5$ m and $L_2 = 5$ m.

Example 5.10. *Sketch the shear force and bending moment diagrams for the beam shown in Fig. 5.22m.a.*

Taking moments about the right-hand support

$$R_1 = \frac{3 \times 10 \times \tfrac{1}{2} \times 10}{7} = 21.43 \text{ kN}$$

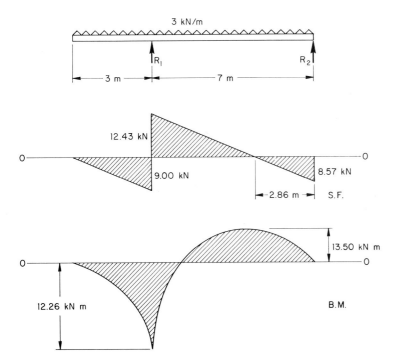

Fig. 5.22m.
Simply supported beam with one cantilever overhang (Example 5.10).

which gives $R_2 = 3 \times 10 - R_1 = 8.57$ kN.

The shear force at the inside end of the cantilever

$$V = 3 \times 3 = 9.0 \text{ kN}$$

and the shear force in the span L_2, just beyond the cantilever, is

$$V = -9 + R_1 = 12.43 \text{ kN}$$

The shear force at the other support is 8.57 kN, and the point of zero shear occurs at $7 \times 8.57/21 = 2.86$ m from the right-hand support.

The maximum cantilever bending moment

$$M = -\tfrac{1}{2} W_1 L_1 = -\tfrac{1}{2} \times 3 \times 3^2 = -13.5 \text{ kN m}$$

The moment $\tfrac{1}{8} W_2 L_2 = \tfrac{1}{8} \times 3 \times 7^2 = 18.375$ kN m, which occurs 3.5 m from the right-hand support. Consequently the moment at the center of the span L_2

$$M = \tfrac{1}{8} W_2 L_2 - \tfrac{1}{4} W_1 L_1 = 18.375 - 6.75 = 11.625 \text{ kN m}$$

The maximum bending moment occurs 2.86 m from the right-hand support:

$$M = R_2 \times 2.86 - \tfrac{1}{2} \times 3 \times 2.86^2 = 34.52 - 12.26 = 12.26 \text{ kN m}$$

This is $5\tfrac{1}{2}\%$ more than the moment at the center of the span L_2.

Example 5.11. *A Gerber beam system in precast concrete carries a load of 6 kN/m over continuous spans of 10 m each, with two pin joints in alternate spans. Determine the location of the pin joints which produce the lowest maximum bending moments.*

Let us call the inset span L, so that the length of the cantilever is $\tfrac{1}{2}(10 - L) = 5 - \tfrac{1}{2}L$. This gives $L_1 = 10$, $L_2 = 5 - \tfrac{1}{2}L$, and $L_3 = L$.

The maximum positive bending moment for equal column spacing and uniformly distributed loading is

$$M_+ = \tfrac{1}{8} w L^2$$

The maximum negative moment is

$$M_- = -\tfrac{1}{2} w (L_2 + L_3) L_2 = -\tfrac{1}{2} w \left(5 - \tfrac{1}{2}L + L\right) \times \left(5 - \tfrac{1}{2}L\right)$$

$$= -\tfrac{1}{8} w (10^2 - L^2) = -\tfrac{1}{8} w \times 10^2 + \tfrac{1}{8} w L^2$$

Equating positive and negative moments, which are of opposite sign

$$\tfrac{1}{8}w \times 10^2 = \tfrac{1}{8}wL^2 + \tfrac{1}{8}wL^2$$

which gives the inset span $L = 10\sqrt{2} = 7.071$ m

The cantilever span $L_2 = \tfrac{1}{2}(10 - 7.071) = 1.465$ m

Consequently the maximum positive moment is

$$M_+ = +\tfrac{1}{8} \times 6 \times 7.071^2 = 37.50 \text{ kN m}$$

and the maximum negative moment is

$$M_- = -\tfrac{1}{2} \times 6(1.465 + 7.071) \times 1.465 = -37.50 \text{ kN m}.$$

The inset span is a little over $\tfrac{2}{3}$ and the cantilever spans are a little less than $\tfrac{1}{6}$ of the column spacing. If the pin joints divided the span exactly in the ratio $1 : 4 : 1$, which would be easier to set out, the maximum moments would not be significantly altered.

Example 5.12. *A rectangular three-pinned portal frame, 3 m high, carries a load of 40 kN over a span of 9 m. Sketch the diagrams for bending moment, shear force, and thrust.*

The horizontal reactions are

$$\tfrac{1}{8} \times \frac{40 \times 9}{3} = 15 \text{ kN}$$

and the maximum bending moment (at the knee of the portal) is

$$15 \times 3 = 45 \text{ kN m}$$

The diagrams for bending moment, shear force, and thrust are shown in Fig. 5.26m.

Example 5.13. *A gabled three-pinned portal frame, with dimensions as shown in Fig. 5.28m carries a load of 40 kN over a span of 9 m. Sketch the diagrams for bending moment, shear force, and thrust.*

The vertical reactions are

$$R_V = \tfrac{1}{2} \times 40 = 20 \text{ kN}$$

and the horizontal reactions are

$$R_H = \tfrac{1}{8} \times \frac{40 \times 9}{7.5} = 6 \text{ kN}$$

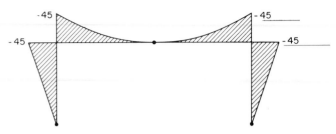

(a) Bending moment diagram

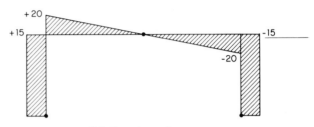

(b) Shear force diagram

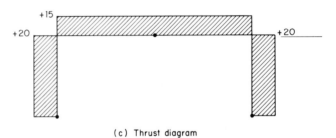

(c) Thrust diagram

Fig. 5.26m.
Three-pin portal (Example 5.12).

The bending moment increases uniformly to the knee of the portal where it is

$$M = -6 \times 3 = -18 \text{ kN m}$$

The bending moment in the gable members, at a distance x from the gable, is

$$M = R_{\text{H}} \times y - \left(\frac{W}{L} \right) x \times \tfrac{1}{2} x$$

Since for a 45° gable $x = y$

$$M = 6x - \tfrac{1}{2} \left(\frac{40}{9} \right) x^2 = (54 - 20x) \frac{x}{9}$$

For an approximate bending moment diagram (Fig. 5.28. m.b) it is sufficient to plot two intermediate values:

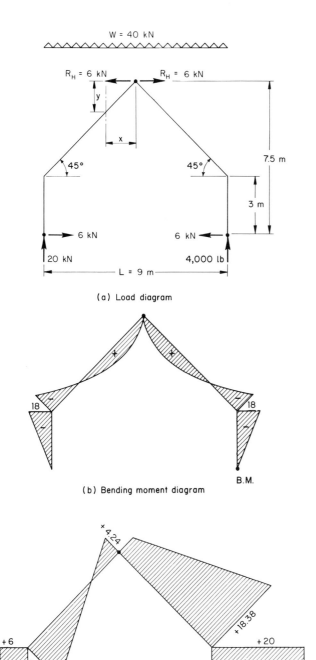

(a) Load diagram

(b) Bending moment diagram

(c) Shear force (left-hand) and thrust (right-hand) diagram

Fig. 5.28m.
Three-pin gabled frame carrying a uniformly distributed vertical load (Example 5.13).
In part (a), the vertical reaction at the top joint is zero by symmetry.

For $x = 0$, $M = 0$
For $x = 1.5$ m, $M = +4.0$ kN m
For $x = 3$ m, $M = -2.0$ kN m
For $x = 4.5$ m, $M = -18.0$ kN m

The shear force in the column is constant and equal to the horizontal reaction, $R_H = 6$ kN. In the gable the shear force is

$$V = R_H \cos 45° - \left(\frac{W}{L} \right) x \sin 45°$$

which varies uniformly from $6/\sqrt{2} = +4.243$ kN at the top pin to $(6 - 20)/\sqrt{2} = -9.90$ kN at the knee (Fig. 5.28 m.c, left-hand side).

The thrust in the column is constant and equal to the vertical reaction $R_V = 20$ kN. At the gable the thrust is

$$P = R_H \sin 45° + \left(\frac{W}{L} \right) x \cos 45°$$

which varies uniformly from $6/\sqrt{2} = 4.243$ kN at the top pin to $(6 + 20)/\sqrt{2} = 18.38$ kN at the knee (Fig. 5.28 m.c, right-hand side).

It should be emphasized that the bending moment diagram is the one which mainly determines the dimensions of the structure, and in a preliminary design it would be sufficient to determine the maximum bending moment, particularly if the frame is to be formed from a single standard section. The maximum bending moment in a three-pin portal almost invariably occurs at the knee, even in this rather extreme example, where the ratio of the span to the height of the knee is 3:1.

The bending moment at the knee is simply

$$M = R_H \times H_1$$

and the basic design is therefore simple, even though a complete determination of all the forces and moments takes time.

Example 5.14. *A semicircular three-pin arch, spanning 9 m, carries a load of 40 kN, uniformly distributed in plan. Sketch the bending moment diagram.*

The solution is shown in Fig. 5.30m. The bending moment along the entire length of the arch is negative; it consists of the difference between the semicircular and the parabolic curves of height $\frac{1}{8} WL$ = 45 kN m. The maximum can be scaled off the curve; it is the longest *vertical* line.

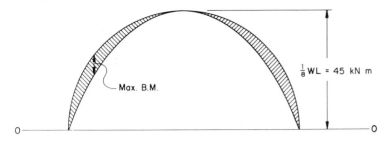

Fig. 5.30m.
Bending moment diagram for semicircular three-pin arch (Example 5.14).

We can also calculate it, although this may be beyond the mathematical competence of some readers. The bending moment

$$M = -R_H \times y + R_V \times x - \left(\frac{W}{L} \right) x \times \tfrac{1}{2} x$$

As we have seen, $R_H = \tfrac{1}{4} W$ and $R_V = \tfrac{1}{2} W$. Furthermore we may calculate the y-coordinate in terms of the radius R, which is equal to the height of the arch, or half the span, so that

$$R = \tfrac{1}{2} L = 4.5 \text{ m}$$

and

$$R^2 = (R - x)^2 + y^2$$

which gives

$$y = (2Rx - x^2)^{\frac{1}{2}}$$

The bending moment therefore becomes

$$M = -\tfrac{1}{4} W (2Rx - x^2)^{\frac{1}{2}} + \tfrac{1}{2} Wx - \left(\frac{\tfrac{1}{4} Wx^2}{R} \right)$$

The maximum bending moment occurs at the value of x for which the differential coefficient equals zero.

$$\frac{dM}{dx} = 0 = -\tfrac{1}{4} W \times \tfrac{1}{2} (2Rx - x^2)^{-\frac{1}{2}} \times (2R - 2x) + \tfrac{1}{2} W - \tfrac{1}{2} \frac{Wx}{R}$$

This gives

$$-\tfrac{1}{2} (2Rx - x^2)^{-\frac{1}{2}} \times (R - x) + 1 - \frac{x}{R} = 0$$

$$-\tfrac{1}{2} (2Rx - x^2)^{-\frac{1}{2}} + \frac{R - x}{R(R - x)} = 0$$

$$2Rx - x^2 = \tfrac{1}{4} R^2$$

$$x = R - \left(R^2 - \tfrac{1}{4}R^2\right)^{\frac{1}{2}} = R\,(1 - 0.866) = 0.134R = 0.603 \text{ m}$$

From geometry, the corresponding value of y is

$$y = \left(R^2 - (R - x)^2\right)^{\frac{1}{2}} = \left(4.5^2 - 3.897^2\right)^{\frac{1}{2}} = 2.250 \text{ m}$$

Substituting these values of x and y, we obtain the maximum bending moment:

$$M = -\tfrac{1}{4} \times 40 \times 2.250 + \tfrac{1}{2} \times 40 \times 0.603 - \left(-\tfrac{1}{2} \times \frac{40}{9} \times 0.603^2\right)$$

$$= 33.75 \text{ kN m}$$

Example 5.15. *Determine the forces and moment in a parabolic three-pin arch, carrying a load of 40 kN uniformly distributed in plan. The arch has a span of 9 m and a rise of 4.5 m.*

Since the arch is parabolic, the bending moment and the shear force are zero along the length of the arch. The maximum thrust occurs at the springings.

The vertical reactions are (Fig. 5.32m)

$$R_V = \tfrac{1}{2} \times 40 = 20 \text{ kN}$$

and the horizontal reactions are

$$R_H = \frac{R_V \times \tfrac{1}{2}L - \tfrac{1}{2}W \times \tfrac{1}{4}L}{H} = \frac{\tfrac{1}{8}WL}{H} = \tfrac{1}{4}W = 10 \text{ kN}$$

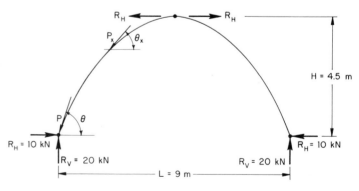

Fig. 5.32m.
Load and thrust diagram for parabolic three-pin arch. The vertical reaction at the top pin joint is zero by symmetry (Example 5.15).

The tangent of the angle θ of a parabola (Fig. 5.33).

$$\tan \theta = \frac{2H}{\frac{1}{2}L} = \frac{4H}{L}$$

This result may be found in appropriate textbooks, see note in text, Example 5.15.

Consequently, $\cos \theta = 1/(1 + \tan^2 \theta)^{\frac{1}{2}} = 1/(1 + 16H^2/L^2)^{\frac{1}{2}}$ $= 1/\sqrt{5}$.

From Fig. 5.32m it is evident that the thrust P at the crown is horizontal and equal to R_{H} for horizontal equilibrium. Some distance down the arch the thrust has increased; but the horizontal component of the maximum thrust is still equal to the horizontal reaction

$$P \cos \theta = R_{\mathrm{H}}$$

Consequently the maximum thrust

$$P = \frac{R_{\mathrm{H}}}{\cos \theta} = 10\sqrt{5} = 22.36 \text{ kN}$$

Problems

5.1. Determine the maximum shear force and the maximum bending moment in the cantilever shown in Fig. 3.36m.

5.2. Draw the bending moment and shear force diagrams for the cantilever shown in Fig. 5.34m, which carries a uniformly distributed load of 4 kN m run over the 2 m nearest to the support, and reactions from two secondary beams and a safety rail, as shown. What are the maximum values?

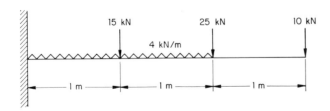

Fig. 5.34m.
Problem 5.2.

5.3. Draw the bending moment and shear force diagram for the beam carrying the concentrated loads shown in Fig. 5.35m, and determine the maximum values.

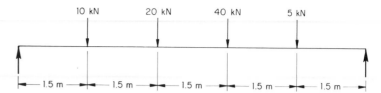

Fig. 5.35m.
Problem 5.3.

5.4. A horizontal beam AB supports three equal loads of P kN each at quarter points. Determine the equation for the bending moment along the beam.

5.5. A simply supported beam carries a uniformly distributed load of 30 kN per metre run over part of its length, and one of 15 kN/m run over the remainder, as shown in Fig. 5.36m. Sketch the bending moment and shear force diagrams, and determine the maximum shear force and bending moment.

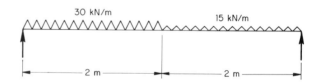

Fig. 5.36m.
Problem 5.5.

5.6. A beam carries a single central load (due to a computer), and uniformly distributed loads on two cantilevers, as shown in Fig. 5.37m. Sketch the bending moment and shear force diagrams, and determine the maximum values.

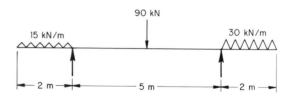

Fig. 5.37m.
Problem 5.6.

5.7. A carport roof is to be supported by edge beams 6 m long, each beam being supported on two columns located at a certain distance from the ends. Calculate the positions for these columns to give the smallest maximum bending moment under uniformly distributed loading, neglecting any effect due to non-uniform or partial loading.

5.8. Fig. 5.38m. shows a ladder resting against a smooth wall, with a 100-kg man at the center. Calculate the reactions, and the maximum bending moment in the ladder.

5.9. A beam, 3.6 m in length and weighing 150 N/m run, is suspended horizontally by three cables as shown in Fig. 5.39m. There is a weight of 500 kN hanging from a position 1.2 m from one end of the beam. Find the force in the cable AE.

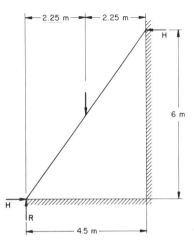

Fig. 5.38m.
Problem 5.8.

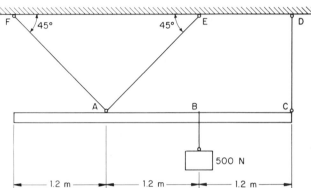

Fig. 5.39m.
Problem 5.9.

5.10. A Gerber beam is continuous over 12 spans of 6 m each, with a uniformly distributed load of 15 kN/m run. Two pins are inserted in alternate spans at a distance of 0.9 m from the supports to make it statically determinate, as shown in Fig. 5.40m. Calculate the bending moment and shear force diagrams for two adjacent spans.

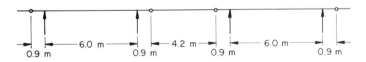

Fig. 5.40m.
Problem 5.10.

5.11. Fig. 5.41m shows a three-hinged frame with a uniformly distributed load of 100 kN on the top member. Calculate the reactions and the maximum bending moment, and draw the bending moment diagram.

397

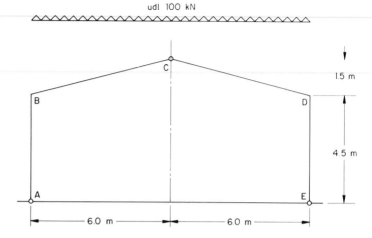

Fig. 5.41m.
Problem 5.11.

5.12. A three-pin frame for a series of carports, shown in Fig. 5.42m has a rigid joint at the top-rear column, and the front column is hinged at both ends. Calculate the reactions, and draw the bending moment and shear force diagrams for a wind load of 5 kN uniformly distributed over the rear column. What are the maximum values of the bending moment and the shear force?

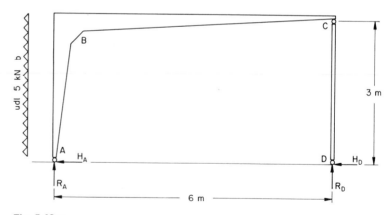

Fig. 5.42m.
Problem 5.12.

5.13. A symmetrical monitor frame shown in Fig. 5.43m is pin-jointed at the supports and the center; it carries a single concentrated load of 150 kN in the position indicated. Neglecting the weight of the frame, sketch the bending moment diagram, and determine the magnitude and position of the maximum bending moment. Indicate the location of the points of zero bending moment.

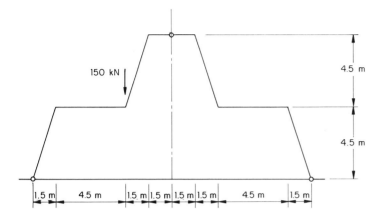

Fig. 5.43m.
Problem 5.13.

5.14. A semicircular arch, spanning 24 m, supports a superstructure which transmits two reactions of 250 kN, as shown in Fig. 5.44m. Draw the bending moment diagram for the arch, neglecting its own weight.

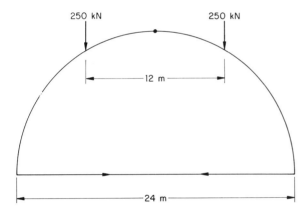

Fig. 5.44m.
Problem 5.14.

A steel cable is to be used to provide the necessary horizontal reaction. If the cable is stressed to 700 MPa, determine the area of the cable required.

Chapter 6

Example 6.1. *Determine the cross-sectional area required for the tie of a steel truss, which is subjected to a force of 225 kN, if the maximum permissible stress is 150 MPa.*

A stress of 150 MPa = 150×10^3 kPa. The cross-sectional area required is $225/150 \times 10^3 = 1.5 \times 10^{-3}$ m^2 = 1,500 mm^2, or 15.00 cm^2.

This is satisfied by:

(i) an Australian equal angle 89 mm × 89 mm × 10 mm (this is a nominal thickness, the actual thickness is 9.5 mm), which weighs 12.50 kg/m, and has a cross-sectional area of 1,590 mm^2;

(ii) a British equal angle 89 mm × 89 mm × 9 mm (actual thickness 9.4 mm), which weighs 12.50 kg/m, and has a cross-sectional area of 15.92 cm^2.

Example 6.2. *Determine the section modulus required for a steel beam subjected to a bending moment of 14 kNm, if the maximum permissible steel stress is 165 MPa.*

The stress of 165 MPa = 165×10^3 kPa. The section modulus required is $14/165 \times 10^3 = 84.848 \times 10^{-6}$ m^3 = 84,848 mm^3, or 84.848 cm^3

This is satisfied by:

(i) an Australian taper flange beam 152 mm × 76 mm, which has an average flange thickness of 9.6 mm, weighs 18 kg/m, and has a section modulus of 115×10^3 mm^3;

(ii) a British joist 152 mm × 89 mm, which has an average flange thickness of 8.3 mm, weighs 17.09 kg/m, and has a section modulus of 115.6 cm^3.

Example 6.3. *A compound steel beam consists of an Australian taper flange beam 152 mm × 76 mm, with a 150 mm × 12mm cover plate on each side. Determine its resistance moment if the maximum permissible stress is 165 MPa.*

The second moment of area (moment of inertia) of the beam (from section tables) without cover plates is $I = 8.75 \times 10^6$ mm^4.

The clear distance between the cover plates is 152 mm, and the center-to-center distance is 152 + 12 = 164 mm. The additional second moment of area (from Table 6.1, line 5) is

$$\frac{2 \times 150 \times 12 \times 164^2}{4} = 24.21 \times 10^6 \text{ mm}^4$$

The total $I = (24.21 + 8.75) \times 10^6 = 32.96 \times 10^6$ mm^4

The overall depth of the symmetrical section is 176 mm, and the section modulus

$$S = \frac{I}{y} = \frac{32.96 \times 10^6}{\frac{1}{2} \times 176} = 374.5 \times 10^3 \text{ mm}^3$$

$$= 374.5 \times 10^{-6} \text{ m}^3$$

Consequently the maximum permissible bending moment

$$M = f\,S = 165 \times 374.5 \times 10^{-6} = 61.80 \times 10^{-3} \text{ MN m}$$

$$= 61.80 \text{ kN m}$$

Example 6.4 *Determine the slope and deflection at the end of a cantilever, 3 m long, carrying a concentrated load of 5 kN at the end (Example 5.1). The cantilever takes the form of an Australian taper flange beam 152 mm × 76 mm (Example 6.2).*

$$W = 5 \text{ kN}, \qquad L = 3 \text{ m}$$

$$M = -W\,L = -15 \text{ kN m}$$

For steel $E = 200$ GPa $= 200 \times 10^6$ kPa.

The second moment of area of the section (from steel section tables) $I = 8.75 \times 10^6 \text{ mm}^4 = 8.75 \times 10^{-6} \text{ m}^4$.

From Table 6.2, the end slope

$$\theta = \tfrac{1}{2}\,\frac{W\,L^2}{EI}$$

$$= +\tfrac{1}{2}\,\frac{5 \times 3^2}{200 \times 10^6 \times 8.75 \times 10^{-6}}$$

$$= 0.0129 \text{ radian}$$

$$= 0.0129 \times \frac{180}{\pi} = 0.739 \text{ degrees}$$

$$= 44 \text{ min of arc}$$

Furthermore, from Table 6.2, the end deflection

$$y = -\tfrac{1}{3}\,\frac{W\,L^3}{E\,I} = -\tfrac{1}{3}\,\frac{5 \times 3^3}{200 \times 10^6 \times 8.75 \times 10^{-6}}$$

$$= 0.0257 \text{ m} = 25.7 \text{ mm} = \frac{L}{116}$$

A cantilever deflection of $1/116$ of the span is too high if the steel section supports a floor or ceiling with a brittle finish (e.g., a plastered ceiling).

Example 6.5. *Determine the end slope and the mid-span deflection of a timber beam carrying a uniformly distributed load of 5 kN over a simply supported span of 3 m (Example 5.8).*

The timber is Douglas fir with a maximum permissible stress of 14 MPa = 14×10^3 kPa, and a modulus of elasticity of 12.5 GPa = 12.5×10^6 kPa.

The bending moment

$$M = \tfrac{1}{8} W L = \tfrac{1}{8} \times 5 \times 3 = 1.875 \text{ kN m}$$

The section modulus required

$$S = \frac{M}{f} = \frac{1.875}{14 \times 10^3} = 133.93 \times 10^{-6} \text{ m}^3 = 133.93 \times 10^3 \text{ mm}^3$$

We can supply the required section modulus by using either a narrow and very deep timber beam or a shallow section, such as a square timber beam. A narrow section needs cross-bracing for stability, and a shallow section uses more material, because S is proportional to b and to d^2. Let us compromise, and select a 140 mm × 45 mm (nominal and actual size), which has (from section tables) a section modulus of 147.0×10^3 mm³. Its second moment of area (from section tables) is 10.29×10^6 mm⁴ = 10.29×10^{-6} m⁴.

From Table 6.2 the end slope

$$\theta = \frac{W L^2}{24 E I} = \frac{5 \times 3^2}{24 \times 12.5 \times 10^6 \times 10.29 \times 10^{-6}}$$

$$= 0.0146 \text{ radian} = 50 \text{ min of arc}$$

and the central deflection

$$y = -\frac{5}{384} \frac{W L^3}{EI} = -\frac{5 \times 5 \times 3^3}{384 \times 12.5 \times 10^6 \times 10.29 \times 10^{-6}}$$

$$= -0.0137 \text{ m} = -13.7 \text{ mm} = \frac{L}{220}$$

This is structurally acceptable but too high if the timber beam supports a plastered ceiling. It would then be necessary to choose a deeper or wider timber beam.

Example 6.6. *Determine the maximum shear stress in the I-beam of Example 6.4.*

The shear force $V = 5$ kN. The section is 152 mm deep, and its web (from steel section tables) is 5.8 mm. Consequently the maximum shear stress is approximately

$$v = \frac{V}{b_w d} = \frac{5}{152 \times 5.8} = 5.672 \times 10^{-3} \text{ kN/mm}^2 = 5.672 \text{ MPa}.$$

Example 6.7. *A rectangular timber section is to be built up from eight laminations, glued together, as shown in Fig. 6.13m. Determine the greatest magnitude of the shear stress in the glued joints, if the shear force at the section is 50 kN.*

The maximum shear stress, which occurs at half depth, is

$$v = \frac{3V}{2b\,d} = \frac{3 \times 50}{2 \times 190 \times 360}$$

$$= 1.096 \times 10^{-3} \text{ kN/mm}^2 = 1.096 \text{ MPa}$$

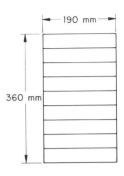

Fig. 6.13m.
Dimension of rectangular timber section
built up from laminations (Example 6.7).

Example 6.8. *A steel bar is tested in tension, and fails when the tensile stress is 350 MPa. Observations show that the bar fails in shear as shown in Fig. 6.21a. Determine the magnitude of the ultimate shear stress.*

From Eq. (6.19) the maximum shear stress

$$v_{max} = \tfrac{1}{2} \times 350 = 175 \text{ MPa}$$

The bar fails at a shear stress of 175 MPa, because the tensile strength of the steel is more than 350 MPa.

Example 6.9. *Determine the magnitude of the diagonal tensile stress in a concrete beam due to a shear stress of 0.9 MPa.*

Let us consider a simply supported beam as shown in Fig. 5.17. The shear force is a maximum just off the support, where the bending moment is zero. Thus the principal stresses are those caused by the shear force alone.
From Eq. (6.25), the principal stresses are both

$$f_{x,\,y} = \pm \sqrt{0.9^2} = \pm 0.9 \text{ MPa}$$

One of the stresses is compressive, and 0.9 MPa is a very small compressive stress for concrete. However, a tensile stress of 0.9 MPa is too high, and shear reinforcement is required (Figs. 6.22c and 6.24).

The direction of the principal tensile stress, from Eq. (6.22) is given by

$$\tan 2\alpha = \frac{0.9}{\frac{1}{2} \times 0} = \text{infinity}$$

so that $2\alpha = 90°$, and $\alpha = 45°$.

The shear reinforcement therefore consists of bars inclined in the direction of the diagonal tension, or of vertical stirrups and horizontal bars, which combine to produce a diagonal tensile force (Fig. 6.24b).

Note that the diagonal tensile stress is numerically equal to the shear force. In example 6.7 the diagonal shear stress was half the tensile stress.

Example 6.10. *The stresses in a shell roof due to the direct membrane forces in two mutually perpendicular directions are 1.5 MPa tension and 2.2 MPa compression, and the membrane shear stress in the same direction is 2.0 MPa. Determine the principal stresses.*

From Eq. (6.24) the principal stresses are

$$f_{x, y} = \tfrac{1}{2}(1.5 - 2.2) \pm \sqrt{\tfrac{1}{4}(1.5 + 2.2)^2 + 2.0^2}$$

$$= -0.35 \pm \sqrt{3.42 + 4.0} = -0.35 \pm 2.72$$

$$f_y = 3.07 \text{ MPa compression}, \qquad f_x = 2.37 \text{ MPa tension}.$$

The angle of inclination of the principal tensile stress is given by Eq. (6.22).

$$\tan 2\alpha = \frac{2.0}{\frac{1}{2}(1.5 + 2.2)} = 1.0811 = 23°37'$$

The angle of inclination to the principal compressive stress is 66° 23'.

Problems

6.2. Draw the bending moment and shear force diagrams for the beam shown in Fig. 6.32m. If the beam is of timber, with a maximum permissible stress of 6.9 MPa, select a suitable cross section.

Examples and Problems in Metric Units

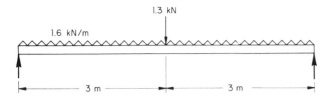

Fig. 6.32m.
Problem 6.2.

6.3. Draw the shear force and bending moment diagrams for the beam shown in Fig. 6.33m, neglecting its own weight.

If the width is 300 mm, and the maximum permissible stress is 8.6 MPa, calculate the minimum depth of the beam.

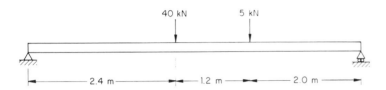

Fig. 6.33m.
Problem 6.3.

6.4. A prestressed concrete box section has an overall depth of 1.2 m and an overall width of 0.9 m. The concrete walls are 150 mm thick on both the horizontal and the vertical parts of the box. Calculate the second moment of area (moment of inertia) and the section modulus.

6.5. A beam AB spans 4 m. At 1 m from the left-hand support an anticlockwise moment of 16 kN m is applied, at 1 m from the right-hand support a 48 kN vertical downward force is applied to the bar. Draw the shear force diagram and the bending moment diagram. If the beam is of timber, with a maximum permissible stress of 22 MPa in bending, select a suitable cross section for the beam to satisfy bending requirements.

6.6. If the beam of Problem 6.5 had to be strengthened to take an increase in load, and all that was available was a length of timber 4 m long and wider than the width determined previously, where should the strengthening piece be placed to obtain maximum benefit from it? Indicate the position by means of a sketch.

6.7. Calculate the deflection of the box section of Problem 6.4 due to its own weight over a simply supported span of 30 m, and the camber required to eliminate it. The weight of concrete may be taken as 2,300 kg/m³, and its modulus of elasticity as 28 GPa.

6.8. The shear force at a section of a laminated timber arch is 125 kN. The arch at that section has a depth of 600 mm, and a width of 250 mm. Calculate the maximum shear stress parallel to

405

the laminations, and also the principal tensile and compressive stresses, if there is no bending moment at that section.

6.9. At a point in the web of a steel beam there is a horizontal tensile stress of 145 MPa due to bending, and a shear stress of 95 MPa. Calculate the magnitude of the principal stresses.

6.10. At a certain position in a floor slab, which acts as a diaphragm, the following stresses occur: in the north-south direction, 1.4 MPa compression; in the east-west direction, 0.7 MPa compression; and in both directions, 1.4 MPa shear.

Calculate the magnitude of the principal stresses, and draw a diagram illustrating the direction of the principal stresses in relation to the direct and shear stresses.

6.11. An element of a reinforced concrete shell roof, 90 mm thick, is subject to a compressive force of 30 kN/m parallel to the span, a tensile force of 50 kN/m perpendicular to the span, and shear forces of 65 kN/m parallel and perpendicular to the span.

Determine the maximum compressive stress in the concrete, the area of reinforcement required, and its inclination to the shell. It may be assumed that the whole of the tension is taken by the steel, and the maximum permissible steel stress is 140 MPa.

Chapter 7

Example 7.1. *Determine the permissible axial load for an Australian 310 UC universal column weighing 118 kg/m; its effective length is 3 m, and the material is structural steel with a minimum yield point of 240 MPa.*

From section tables, the column has a nominal depth and width of 310 mm, an actual depth of 314 mm and an actual width of 307 mm. Its least radius of gyration is $r = 77.6$ mm, and the slenderness ratio is $l/r = 3,000/77.6 = 38.7$. From the tables contained in the Australian Steel Structures Code SA 1250-1975 the permissible stress for a slenderness ratio of 38.7, conforming to the column formula, is 137 MPa.

The cross-sectional area of the section $A = 15,000$ mm^2 = 15 × 10^{-3} m^2, and thus the permissible column load

$$P = 137 \times 15 \times 10^{-3} = 2.06 \text{ MN.}$$

Example 7.2. *Determine the maximum permissible axial load for the column of Example 7.1, if the 310 UC section is strengthened with two 400 × 25 mm plates of the same steel, one on each flange.*

The second moments of area (moments of inertia) of the universal column (from section tables) are 276×10^6 mm^4 about the axis X–X (Fig. 7.6) and 90.0×10^6 mm^4 about the axis Y–Y. To this we must add the second moments of area (moments of inertia) of the plates, which are (from Table 6.1), $A\, d^2/4$ about the axis X–X, where A is the area of the two plates ($2 \times 400 \times 25$ mm^2) and d is the sum of the actual depth of the UC section (314 mm) and the two half-thicknesses of the plate (25 mm); and $2b\, d^3/12$ about the axis Y–Y, where b is the width (25 mm) and d the depth (400 mm) of the plates in that direction.

Thus about the axis X–X

$$I = 276 \times 10^6 + \tfrac{1}{4} \times 2 \times 400 \times 25 \times 339^2 = 850.6 \times 10^6 \text{ mm}^4$$

and about the axis Y–Y

$$I = 90.0 \times 10^6 + \tfrac{1}{12} \times 2 \times 25 \times 400^3 = 356.7 \times 10^6 \text{ mm}^4$$

From section tables the cross-sectional area of the UC section is 15,000 mm^2, so that for the combined section

$$A = 15{,}000 + 2 \times 25 \times 400 = 35{,}000 \text{ mm}^2 = 35 \times 10^{-3} \text{ m}^2$$

The *smaller* radius of gyration of the compound section

$$r = \sqrt{\frac{356.7 \times 10^6}{35{,}000}} = 101.0 \text{ mm}$$

and the slenderness ratio $l/r = 3{,}000/101.0 = 29.7$. The corresponding permissible stress is 140 MPa.

The permissible column load

$$P = 140 \times 35 \times 10^{-3} = 4.90 \text{ MN}$$

Example 7.3. *Determine the maximum permissible load for a 152 mm $\times$ 152 mm square tube column with a wall thickness of 9.5 mm; its effective length is 3 m, and the material is 240 MPa steel.*

Tubes provide neat sections for lightly loaded columns, particularly in conjunction with timber beams or light-gauge steel floors.

From section tables, the radius of gyration is $r = 58.2$ mm, and the slenderness ratio $l/r = 3{,}000/58.2 = 51.5$. The permissible stress is 129 MPa.

The cross-sectional area is 5,400 mm^2, and the permissible column load is $129 \times 5{,}400 \times 10^{-3} = 696.6$ kN.

Example 7.4. *Determine the maximum permissible load for a double-angle strut in a roof truss, consisting of two 102 mm × 76 mm × 10 mm angles, separated by a 10 mm space. The effective length is 1.8 m, and the material is 240 MPa steel.*

Double angles are commonly used in the rafters of roof trusses, because the flanges of a T-section does not offer enough space for connecting the other members (see Section 7.4). A gusset plate is placed between the angles, to which all the members meeting at the joint are connected (see Figs. 4.7 and 7.7). Double angles are therefore a standard form of truss construction, and their properties (acting as a pair) are available in tabular form. The compound section has two radii of gyration: one about the Y–Y axis, and one about the X–X axis. The least value for this compound section (from tables) is $r_x = 31.9$ mm.

The slenderness ratio $l/r = 1,800/31.9 = 56.4$, and the corresponding permissible stress is 126 MPa. The cross-sectional area is 3,180 mm^2, and the permissible column load is 400.7 kN.

Example 7.5. *Determine the maximum permissible load for a 200 mm × 250 mm Douglas fir column, 4 m long.*

If the nominal metric dimensions of the timber section are also its actual dimensions, the ratio $l/d = 4,000/200 = 20$ (where d is the smaller side of the rectangle). The Australian Timber Engineering Code AS 1750–1975 includes a material constant, which for dry Douglas fir is 0.98. The ratio therefore becomes $0.98 \times 20 = 19.6$. For this ratio the Code gives a stability reduction factor of 0.51. The basic stress for Douglas fir in compression is 10.5 MPa, and the maximum permissible stress in accordance with the Australian Code is $0.51 \times 10.5 = 5.355$ MPa.

Consequently the maximum permissible column load is $5.355 \times 200 \times 250 = 267,750$ N $= 267.750$ kN. Note that 1 MPa = 1 N/mm^2.

Example 7.6. *Two steel plates are to be joined by two 6 mm fillet welds, 150 mm long, as shown in Fig. 7.9m.a. Determine the strength of the weld, if the maximum permissible shear stress in the weld metal is 135 MPa.*

In determining the effective size of the weld, the penetration of the weld metal into the structural material and the "reinforcement"

Examples and Problems in Metric Units

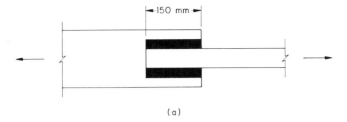

(a)

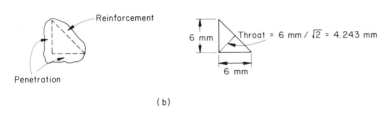

(b)

Fig. 7.9m.
Design of fillet weld (Example 7.6). The dimensions of the weld are its length and its throat, neglecting the reinforcement and the penetration of the weld metal.

of the weld are ignored. The cross-sectional dimensions of the weld are thus its length and its throat (Fig. 7.9m.a and b).

Consequently the strength of the weld is $135 \times 2 \times 150 \times 4.243$ = 171,840 N = 171.84 kN.

This calculation is very simple; but the design of welds is a matter of experience, rather than mechanics. Badly designed welds can distort a thin metal structure. Readers should visit a steel fabricating plant, and talk to people experienced in the practice of welding.

Example 7.7. *Determine the strength of a joint between two 6-mm plates (a) lap-jointed with two 20-mm diameter high-strength bolts, and (b) butt-jointed with two 20 mm diameter high-strength bolts.*

The joint is shown in Fig. 7.10.

According to the Australian High Strength Bolting Code AS 1511, the maximum permissible load for a 20-mm diameter high strength bolt in single shear is 55 kN.

In the lap joints the bolts are in single shear, and thus the strength of the joint is $2 \times 55 = 110$ kN.

In the butt joint with two cover plates, each bolt would have to shear through *twice* before the joint could fail, and the bolts are therefore in *double shear*. Consequently the strength of the joint is $4 \times 55 = 220$ kN.

409

The 20 mm diameter bolts require holes 2 mm larger in the plate, so that $22 \times 6 = 132$ mm^2 must be deducted from the plate to give its effective cross-sectional area as a tensile member.

Example 7.8. *Determine the strength of the same joint if 20 mm diameter hot-driven rivets are used.*

A high-strength bolt holds the plates together by friction, so that bearing of the bolt is not considered; a rivet fills the hole, and failure may occur if the bearing pressure between the plate and the rivet is too high.

Taking the maximum permissible shear stress of rivets as 100 MPa the strength of a 20 mm rivet in single shear is $100 \times \frac{1}{4} \pi \times 20^2$ $= 31,416$ N $= 31.4$ kN, and its strength in double shear is 62.8 kN.

Taking the maximum permissible bearing stress as 324 MPa, its bearing strength on a 6 mm plate is $324 \times 20 \times 6 = 38,880$ N $= 38.9$ kN.

Evidently bearing is not critical for the joint in Fig. 7.10a, which is governed by single shear. Its strength is $2 \times 31.4 = 62.8$ kN.

Bearing is critical for the joint in Fig. 7.10b, since the bearing strength is less than the strength in double shear. Consequently the strength of the joint is $2 \times 38.9 = 77.8$ kN.

Example 7.9. *Determine the strength of the same joint if 20-mm diameter common bolts are used.*

Common bolts do not provide the same friction grip as high-strength bolts, so that bearing must be considered. Taking the maximum permissible shear stress as 80 MPa, and the maximum permissible bearing stress as 324 MPa, the strength of a 20 mm diameter bolt in single shear is 25.1 kN, in double shear it is 50.3 kN, and in bearing on a 6 mm plate it is 38.9 kN.

Consequently the lap joint has a strength of $2 \times 25.1 = 50.2$ kN, and the butt joint has a strength of $2 \times 38.9 = 77.8$ kN.

Example 7.10. *Calculate the live load permissible for a column, 300 mm square, reinforced with four 32-mm bars. The specified ultimate compressive strength of the concrete is 30 MPa, the specified yield stress of the reinforcement is 410 MPa, and the actual dead load is 600 kN.*

Reinforced concrete design in America, Britain, and Australia is now based on ultimate strength. There are, however, slight differences in the load factors used in the three codes. Furthermore, the strength of the concrete is measured in the American Code ACI 318-71 and the Australian Code AS 1480-1974 by testing 6-in (or 150-mm) diameter cylinders 12 in (or 300 mm) long, whereas for the British Code CP 110:1972 150 mm cubes are tested. The results of tests on brittle materials, such as concrete, depend greatly on the cross-sectional area and the length of the test piece. For the same concrete, 150 mm cubes give a higher test result than 150 mm × 300 mm cylinders. The ACI Code has been used in Sections 7.9–7.11, and it is also used in Examples 7.10–7.12 in this appendix. The specified stresses and the dimensions of the concrete sections and the reinforcing bars are in accordance with British and Australian practice.

In this example we have: $P_D = 600$ kN $= 600 \times 10^{-3}$ MN; $f'_c = 30$ MPa; $f_y = 410$ MPa; and A_s (from Table C3, Appendix C) $= 3,200$ mm$^2 = 3.2 \times 10^{-3}$ m^2.

The cross-sectional area of the concrete is the gross cross-sectional area of the column minus the cross-sectional area of the steel:

$$A_c = 300^2 - 3,200 = 86,800 \text{ mm}^2 = 86.8 \times 10^{-3} \text{ m}^2$$

From Eq. (7.13)

$$1.4 \times 600 \times 10^{-3} + 1.7P_L$$

$$= 0.70(0.85 \times 30 \times 86.8 \times 10^{-3} + 410 \times 3.2 \times 10^{-3})$$

This gives the permissible (or service) live load for the column

$$P_L = 957.5 \times 10^{-3} \text{ MN} = 957.5 \text{ kN}.$$

Even when the entire column is too short to buckle, the individual reinforcing bars have very high slenderness ratios. They must thus be restrained by ties at regular intervals (Fig. 7.16).

The size and spacing of these ties is specified in Clause 7.12.3 of the ACI Code, Clause 3.11.4.3 of the British Code, and Clause 6.11.3 of the Australian Code. 10 mm bars at 300 mm centers are appropriate in each case.

Example 7.11. *Determine the amount of reinforcement required for a concrete slab, 150 mm thick, which is subjected to a bending moment due to the dead load of 20 kN m/m width, and to a bending moment due to the live load of 11 kN m/m width. The specified (ultimate) compressive strength of the concrete is 30 MPa, and the specified yield strength of the reinforcement is 410 MPa.*

To determine the effective depth, which is measured to the *center* of the reinforcement, we must assume the bar diameter. However, the error caused by a slightly erroneous assumption is not serious. Let us assume that the bars are 16 mm diameter. The reinforcement requires concrete cover to protect it from rusting, and to ensure proper bond between the concrete and the steel. The minimum cover for a slab is 20 mm for the Australian Code (15 mm for the British Code), or the diameter of the reinforcing bar, whichever is the larger. The effective depth is thus

$$d = 150 - 20 - \tfrac{1}{2} \times 16 = 122 \text{ mm} = 122 \times 10^{-3} \text{m}$$

As we noted in Section 7.9, the load factor for the dead load is 1.4 and the load factor for the live load is 1.7. Therefore the ultimate resistance moment

$$M_u = 1.4 \times 20 + 1.7 \times 11 = 46.7 \text{ kN m/m}$$

$$= 46.7 \times 10^{-3} \text{ MN m/m}$$

Let us assume an ultimate lever arm ratio $j_u = 0.90$. From Eq. (7.16)

$$46.7 \times 10^{-3} = 0.90 \times 410 \times A_s \times 0.90 \times 122 \times 10^{-3}$$

which gives $A_s = 1{,}153 \times 10^{-6} \text{ m}^2/\text{m} = 1{,}153 \text{ mm}^2/\text{m}$.

From Table C4 (Appendix C) we chose 16 mm at 150 mm centers (1,333 mm²/m). 12 mm bars would be at 75 mm centers, and 20 mm bars at 200 mm centers; the first is too close for placing the concrete satisfactorily, and the second might leave a little too much unreinforced concrete between the bars.

We will check our assumption for j_u. Since the reinforcement is stated as an area per m width, $b = 1$ m $= 1{,}000$ mm. From Eq. (7.17) $j_u = 1 - 0.59 \times 410 \times 1{,}333/1{,}000 \times 122 \times 30 = 0.912$.

Example 7.12.   *Determine the reinforcement required for the T-beams of a reinforced concrete floor structure, if the ultimate bending moment at mid-span is 140 kN m. The beams have an overall depth of 300 mm, and an effective width of 1.2 m; the slab is 100 mm thick. The specified yield stress of the reinforcement is 410 MPa.*

The minimum cover required in beams is $1\tfrac{1}{2}$ in (38 mm) in the ACI Code, 25 mm or the bar diameter in the Australian Code, 15 mm or the bar diameter (whichever is the larger) in the British Code. The reinforcing bar diameter is likely to be about 25 mm. Taking the cover as 25 mm, the effective depth is thus

$$d = 300 - 25 - \tfrac{1}{2} \times 25 = 262 \text{ mm}$$

and the length of the lever arm is

$$d - \tfrac{1}{2}t = 262 - \tfrac{1}{2} \times 100 = 212 \text{ mm} = 212 \times 10^{-3} \text{ m}$$

From Eq. (7.19)

$$140 \times 10^{-3} = 0.90 \times 410 \times A_s \times 212 \times 10^{-3}$$

which gives $A_s = 1,790 \times 10^{-6} \text{ m}^2 = 1,790 \text{ mm}^2$. This is satisfied by four 24-mm bars (Australian size) or four 25-mm bars (British size) in each beam (Table C3, Appendix C). The cross section of the beam is shown in Fig. 7.22m.

In actual fact this beam, if analyzed precisely, is a rectangular beam whose neutral axis falls within the flange. The area of reinforcement required is therefore slightly less.

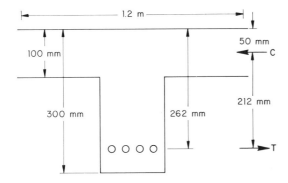

Fig. 7.22m.
Design of mid-span reinforcement for beam (Example 7.12).

Example 7.13. *A 300 mm square column carries a load of 500 kN (a) at its center, (b) 25 mm left of center, (c) 50 mm left of center, and (d) 150 mm left of center, along one center-line of the column. Determine the stresses at the left-hand and right-hand side of the section in each case.*

The cross-sectional area $A = 300^2 = 90 \times 10^3 \text{ mm}^2$, and the section modulus $S = S_t = S_c = 300 \times 300^2/6 = 4.5 \times 10^6 \text{ mm}^3$.

The column load $P = 500 \times 10^3$ N, and the direct compressive stress is

$$\frac{500 \times 10^3}{90 \times 10^3} = 5.556 \text{ N/mm}^2 = 5.556 \text{ MPa}$$

The bending moment varies from 0 to $500 \times 10^3 \times 150 = 75 \times 10^6$ N mm, and the flexural stress varies from 0 to

$$\frac{75 \times 10^6}{4.5 \times 10^6} = 16.667 \text{ N/mm}^2 = 16.667 \text{ MPa}.$$

Thus the left-hand and right-hand stresses from Eqs. (7.20 and 7.21) are (in MPa):

Eccentricity	Left-hand Stress	Right-hand Stress
0	5.56 C	5.56 C
25 mm	8.33 C	2.78 C
50 mm	11.11 C	0
150 mm	22.22 C	11.11 T

C = compression; T = tension.

The column is in compression when the load is in the middle third (100 mm), i.e., when the eccentricity is ± 50 mm. Tension develops when $P/A = M/S$, i.e., when $\epsilon = S/A$.

Example 7.14. *A concrete I-section, with the dimensions shown in Fig. 7.27m, is prestressed with 3,500 mm² of steel, pre-stressed to 700 MPa. Determine the stresses during prestressing and under full load.*

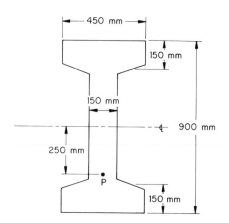

Fig. 7.27m.
Dimension of section, and eccentricity of prestressing force (Example 7.14).

The section has a cross-sectional area $A = 0.225$ m² and a section modulus of $S = 0.0488$ m³.
The prestressing force, from Eq. (7.26) is

$$P = 700 \times 3500 = 2,450,000 \text{ N} = 2.45 \text{ MN}$$

Let us assume that the bending moment due to the weight of the beam has been calculated as $M_G = 0.1$ MNm, and the bending moment due to the superimposed load $M_S, = 1.0$ MN m

At the end of the prestressing operating the concrete stresses are

Examples and Problems in Metric Units

at the top (Eq. 7.22):

$$\frac{2.45}{0.225} - \frac{2.45 \times 0.25}{0.0488} + \frac{0.1}{0.0488}$$

$$= 10.889 - 12.551 + 2.049 = 0.387 \text{ MPa (compression)}$$

at the bottom (Eq. 7.23):

$$10.889 + 12.551 - 2.049 = 21.391 \text{ MPa (compression)}$$

Under full load, at the top (Eq. 7.24):

$$0.387 + \frac{1.0}{0.0488} = 0.387 + 20.492 = 20.879 \text{ MPa (compression)}$$

at the bottom (Eq. 7.25):

$$21.391 - 20.492 = 0.899 \text{ MPa (compression)}$$

Problems

7.1. A sound reflector measuring 6 m by 9 m is to be suspended above a concert platform by four symmetrically placed steel rods. Determine the cross-sectional area of each rod if the combined live and dead load is 3.5 kN/m², and the maximum permissible steel stress is 180 MPa.

7.4. Design a joint between a 76 mm × 76 mm × 10 mm (or 9 mm) angle carrying a tensile force of 150 kN and a 12 mm thick gusset plate (a) using welding, (b) using rivets, (c) using black bolts, and (d) using high-strength bolts. Compare the space occupied by the different joints. An Australian 76 × 76 × 10 mm angle and a British 76 × 76 × 9 mm angle both have an actual thickness of 9.4 mm.

7.5. The open-web joist shown in Fig. 7.28m has a span of 6 m, and a depth of 250 mm between the centers of chords. It is subjected to a uniform loading equivalent of 3 kN/m run. Calculate the forces in the most highly stressed portion of the bottom chord, and the most highly stressed tension and compression diagonals.

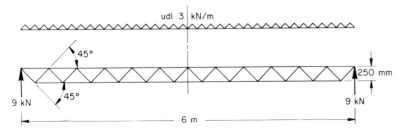

Fig. 7.28m.
Problem 7.5.

7.8. Ignoring in this example the eccentricity of the column load, determine the amount of longitudinal reinforcement required for the 600 mm square columns on the ground floor of a 40-story flat-plate structure if each column supports a floor area of 4.5 m × 4.5 m. The dead load is 6 kN/m², and the live load is 2.4 kN/m². Assume that the columns need only be designed to carry 85% of the full live load. The specified (ultimate) compressive strength of the concrete is 35 MPa, and the specified yield strength of the steel is 410 MPa.

7.9. A 100 mm thick concrete slab cantilevers 2 m beyond the facade of the building. The greatest ultimate bending moment is 38 kN m/m width. Determine the amount of the main reinforcement required if the specified compressive strength of the concrete is 30 MPa, the specified yield strength of the steel is 410 MPa, and $j_u = 0.90$. On what face of the concrete is the reinforcement placed?

7.11. The ultimate negative bending moment at some point in a 125 mm thick flat plate is 48 kN m/m width. Determine the area of reinforcement at that section, using the same constants as in Problem 7.9.

7.12. Fig. 7.29 m shows the plan of a brick pier. The brickwork has a density of 1,900 kg/m³, and the pier is 6 m high. It also carries a point load of 30 kN in the position P. Calculate the properties of the section about X–X (which passes through the center of gravity), and calculate the stress distribution at the base of the pier.

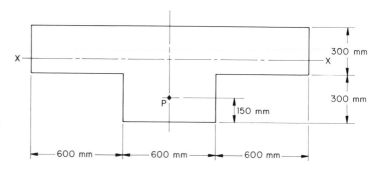

Fig. 7.29m.
Problem 7.12.

7.13 The three-hinged frame shown in Fig. 7.30m is subject to a load of 3.4 kN/m, uniformly distributed in plan. If the actual size of the timber column after dressing is 375 mm × 150 mm, determine the stress distribution in the column immediately below point B.

7.14. A rectangular prestressed concrete beam, 300 mm wide by 150 mm deep, is prestressed with an eccentricity of 50 mm. The initial prestressing force is 320 kN. Ignoring loss of prestress due to

Examples and Problems in Metric Units

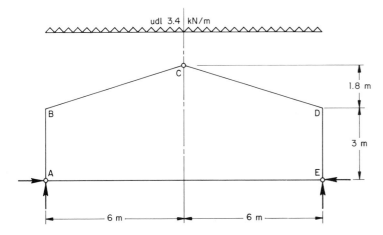

Fig. 7.30m.
Problem 7.13.

creep, etc., calculate the variation of concrete stress at the end of the prestressing operation and under full load. The mid-span bending moments are 8.2 kN m due to the weight of the beam, and 16.4 kN m due to the superimposed load.

Chapter 8

Example 8.1. *Determine the maximum positive and negative bending moments for the beam shown in Fig. 8.6m.*

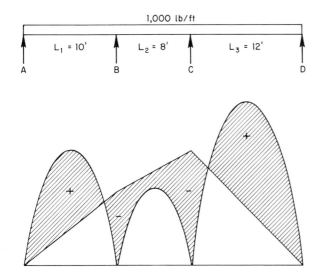

Fig. 8.6m.
Bending moment diagram for continuous beam (Example 8.1).

Let us first consider spans 1 and 2. We have $M_A = 0$, because there is no bending moment at the end of a freely supported beam. From Eq. (8.2)

$$0 + 2 M_B(3.0 + 2.4) + M_C \times 2.4 = -\tfrac{1}{4}(45 \times 3.0^2 + 36 \times 2.4^2)$$

$$10.8 M_B + 294 M_C = -153.09 \qquad \text{(i)}$$

Let us next consider spans 2 and 3. Since $M_D = 0$,

$$M_B \times 2.4 + 2 M_C(2.4 + 3.6) = -\tfrac{1}{4}(36 \times 2.4^2 + 54 \times 3.6^2)$$

$$2.4 M_B + 12.0 M_C = -226.80 \qquad \text{(ii)}$$

Multiplying Eq. (i) by 5

$$54.0 M_B + 12.0 M_C = -765.45 \qquad \text{(iii)}$$

Subtracting Eq. (ii) from Eq. (iii)

$$51.6 M_B = -538.65$$

$$M_B = -10.44 \text{ kN m}$$

From Eq. (i)

$$M_C = \frac{-153.09 + 112.75}{2.4} = -16.81 \text{ kN m}$$

The statically determinate positive moments for the spans 1, 2, and 3 are:

$$M_1 = +\tfrac{1}{8} \times 15 \times 3.0^2 = 16.875 \text{ kN m}$$

$$M_2 = +\tfrac{1}{8} \times 15 \times 2.4^2 = 10.800 \text{ kN m}$$

$$M_3 = +\tfrac{1}{8} \times 15 \times 3.6^2 = 24.300 \text{ kN m}$$

These moments are plotted in Fig. 8.6m. Evidently, M_B and M_C are the maximum negative moments. The maximum positive moment can be obtained by scaling from an accurately drawn diagram. Alternatively, the value of x for which the bending moment is a maximum can be determined from $dM/dx = 0$; the bending moment for this value is then calculated.

This continuous beam is only slightly unsymmetrical; we have merely displaced the support C by 600 mm to the left from a regular 3 m column spacing. However, this is sufficient to eliminate the positive moment in the middle span entirely, and turn it into a negative moment, so that the mid-span reinforcement in a concrete floor is required on top, and not as normally on the bottom. Evidently small irregularities in column spacing can have serious consequences in continuous structures, whereas they have only minor effects in pinned frames.

Examples and Problems in Metric Units

Example 8.2. *From Tables 8.1 and 8.2, determine the maximum positive and negative bending moments for a beam continuous over three equal spans of 3 m, carrying a dead load of 15 kN/m, and a live load of 7.5 kN/m.*

The maximum negative moment occurs at the inner supports:

$$M_- = -0.100 \times 45 \times 3 - 0.117 \times 22.5 \times 3 = -21.40 \text{ kN m}$$

The maximum positive moment occurs in the outer span:

$$M_+ = +0.080 \times 45 \times 3 + 0.101 \times 22.5 \times 3 = +17.62 \text{ kN m}$$

Example 8.3. *Determine the maximum positive and negative bending moments for a primary beam, 9 m long and built in at the ends, which carries two secondary beams at third-points. The reaction of each beam is 50 kN.*

From Table 8.3, line 2, the restraining moment is

$$M_A = - \frac{100 \times 9}{9} = -100 \text{ kN m}$$

The end reactions are 50 kN each, and thus the maximum positive moment (Fig. 8.8m) in a *simply supported* beam carrying the same load over the same span is $50 \times 3 = +150$ kN m. The maximum positive moment in the restrained beam is therefore

$$M_+ = 150 - 100 = +50 \text{ kN m}.$$

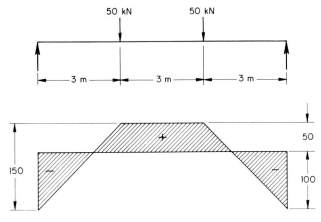

Fig. 8.8m.
Bending moment diagram for built-in beam (Example 8.3).

Example 8.4. *Figure 8.12m shows an interior bay of a steel frame or bent, 15 m high, 6 m wide, and 54 m long. The frame is subject to wind pressure above the ground floor, which is sheltered. The frames are spaced 6 m apart, i.e., the beams which bear on the frame span 6 m in both directions. The horizontal wind pressure is 1 kPa.*

The wind load is distributed as positive pressure acting on the windward face of the building (approximately 60%) and as negative pressure (or suction) on the leeward face (approximately 40%). However, for the purpose of this analysis we may assume that the whole force acts on the windward side, so that

$$W = 1 \times 6 \times 3 = 18 \text{ kN}$$

The shear force due to this wind load on the top-floor columns (Fig. 8.6m.b.) is thus $\frac{1}{2} W = 9$ kN, and it increases by W at each of the two floors below, and finally by $\frac{1}{2} W$. The maximum shear force equals the total horizontal force of 54 kN.

Let us now consider the corresponding column bending moments in the top story. At the hypothetical pin joints (Fig. 8.11), $M = 0$, and under the influence of a constant shear force it increases linearly to a maximum value of

$$M = 9 \times 1.5 = 13.5 \text{ kN m}$$

in the top floor. The rectangular beam-column junction transmits

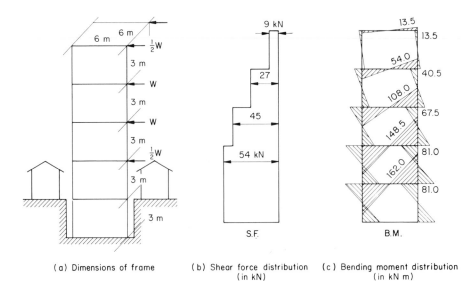

(a) Dimensions of frame   (b) Shear force distribution   (c) Bending moment distribution
                                         (in kN)   (in kN m)

Fig. 8.12m.
Distribution of shear force and bending moment due to wind loads in a steel structure designed by the pinned-frame concept (Example 8.4).

this moment to the roof beam. It is then reduced to 0 at the pin joint, and increases again to a maximum value of 13.5 kN m at the far end of the roof beam, where it is transmitted to the leeward column.

In the columns of the story next to the top, the maximum column moment

$$M = 27 \times 1.5 = 40.5 \text{ kN m}$$

Both this moment and the moment in the column above transmit the curvature to the adjoining floor beams of the top story, which thus has a maximum bending moment

$$M = 13.5 + 40.5 = 54 \text{ kN m}$$

The complete bending moment diagram, shown in Fig. 8.12m.c., is obtained by the successive taking of moments in the columns, descending from top to bottom story, using the increasing shear force from Fig. 8.12m.b. The end moments of *both* adjacent columns are transmitted into each beam.

Example 8.5. *Determine the maximum positive and negative bending moments in the beam and columns of the frame shown in Fig. 8.14m.*

Let us assume that the second moment of area (moment of inertia) of the beam and the columns is the same; then the stiffness of the beam and the columns is the same, since they have the same length.

We will now firmly clamp the joints A, B, C, and D. The beam is then a fully built-in beam. It carries a uniformly distrbuted load of 48 kN, so that the bending moment at B (from Table 8.3, line 5) is

$$M_B = \frac{-WL}{12} = -\frac{48 \times 4}{12} = -16 \text{ kN m}$$

With all joints clamped, we thus have "hogging" bending moments of 16 kN at both ends of the beam, and zero at both ends of each column, where there is no bending moment (Fig. 8.14m.b).

In order to carry the distribution through without error, we propose to adopt the sign convention that is normal for this type of computation. We will call a moment positive when it is applied in a clockwise direction and negative when it is anticlockwise. The value of this convention lies in the interaction of the beam and the columns; if the end of the beam rotates clockwise, it also causes the adjoining end of the column to rotate clockwise. We thus obtain a moment of -16 kN m on the left-hand end of the beam, and $+16$ kN m at the right-hand end of the beam. Please note that this

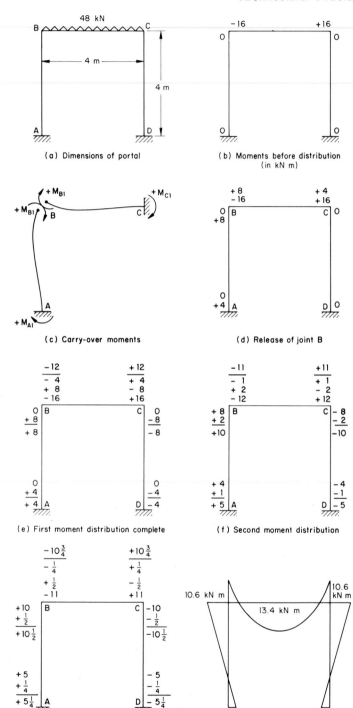

(a) Dimensions of portal

(b) Moments before distribution (in kN m)

(c) Carry-over moments

(d) Release of joint B

(e) First moment distribution complete

(f) Second moment distribution

(g) Third moment distribution

(h) Bending moment diagram

Fig. 8.14m.

The moment distribution method (Example 8.5).

convention differs in character from that used in previous chapters (see Fig. 5.9), where we classified moments by their effect on the deformation of the member.

Let us now release joint B, and thus equalize the beam and column moments at that joint. Since the stiffness of the beam and the column is the same, we distribute the moment equally between both; i.e., the beam and the column each receive a clockwise moment $M_{B1} = +8$ kN m. This is added to $M_B = -16$ kN m in the beam, and to $M_B = 0$ in the column.

In this process we carry over moments to A and C (Fig. 8.14m.c.). The moment M_{B1} acting on the column BA at B produces a moment M_{A1} at A, which is called the *carry-over moment*, because it is produced as a carry-over from the change in moment at B. Since the joint A is still firmly clamped, the slope at A is zero. From Table 6.2, line 12 (using our new sign convention)

$$\theta_A = 0 = \frac{2M_{A1} - M_{B1}}{6EI}$$

Since E and I are constants which cannot be infinity

$$M_{A1} = \tfrac{1}{2} M_{B1} = +4 \text{ kN m}$$

We thus have a *carry-over factor* of $+\tfrac{1}{2}$; i.e., a change in moment at B produces a change of half that amount at A, rotating in the same direction. Similarly $M_{C1} = +\tfrac{1}{2} M_{B1}$.

The initial moments and those produced by the release of joint B are shown in Fig. 8.14m.d. We now clamp joint B, and release joint C. The result is similar, since the structure is symmetrical, and the moments now stand as shown in Fig. 8.14m.e. We cannot release joints A and D, since the structure is built in at those points, and this is therefore the end of the first moment distribution. At joints B and C we started with beam moment of 16 kN m and 0, and error of infinity. We have improved these figures to 12 kN m and 8 kN m, an error of 50%.

We will proceed to the second moment distribution (Fig. 8.1m.f.), using the same procedure. The moments at the beam-column junction are now 11 kN m and 10 kN m, an error of 10%.

After the third distribution (Fig. 8.14m.g.), the moments at the beam-column junctions are 10.75 kN m and 10.5 kN m, an error of 3% which is sufficiently accurate. However, if for some reason high precision is needed, we can make another distribution. For preliminary design, two distributions, with an error of 10%, would be sufficient (see Appendix A6). We now average the remaining difference, and make the bending moments 10.6 kN m and 5.3 kN m respectively (Fig. 8.14m.h.). Readers may have noticed that the distribution procedure provides an immediate check on the accuracy of the result, which must converge at the joint.

We can now draw the bending moment diagram for the portal frame. The statically determinate bending moment varies parabolically from 0 to

$$\frac{WL}{8} = \frac{48 \times 4}{8} = 24 \text{ kN m}$$

Thus the maximum positive bending moment at the mid-span of the beam is $24 - 10.6 = 13.4$ kN m. The bending moment diagram in the columns varies uniformly (Fig. 8.14m.h.).

Example 8.6. *Determine the maximum positive and negative bending moments in the beam and columns of the frame shown in Fig. 8.15m.*

In this frame the beam is subjected to a larger moment, and thus requires a deeper section. It is one of the disadvantages of statically indeterminate design that assumptions must be made about the members *before* they have been designed. This can usually be done from past experience (drawing on reference books and the data of colleagues); but if the initial assumptions are seriously in error we must repeat the calculations with more accurate assumptions.

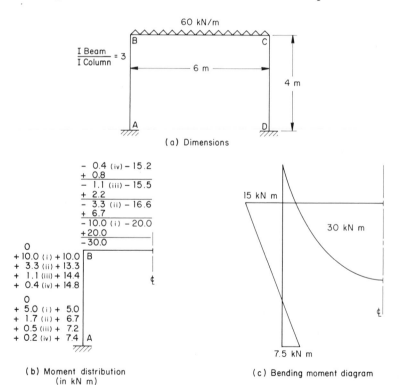

(a) Dimensions

(b) Moment distribution
(in kN m)

(c) Bending moment diagram

Fig. 8.15m.
Moment distribution (Example 8.6).

Examples and Problems in Metric Units

In this example the second moment of area (I) of the beam is three times as large as that of the column. The structural material is the same for both, and one span is 1.5 times the other. Consequently

$$\frac{K_{\text{beam}}}{K_{\text{column}}} = \frac{3}{1.5} = 2$$

$$\frac{K_{\text{beam}}}{\Sigma K} = \frac{2}{3}$$

$$\frac{K_{\text{column}}}{\Sigma K} = \frac{1}{3}$$

The fixing moment for a built-in beam $WL/12 = 60 \times 6/12 = 30$ kN m.

The moment distribution is shown in Fig. 8.15m.b, and four distributions are needed because of the greater stiffness of the beam.

The statically determinate bending moment in the beam is $WL/8 = 60 \times 6/8 = 45$ kN m, and the complete bending moment diagram for half the portal frame is shown in Fig. 8.15m.c.

Problems

8.3. Determine the maximum positive and negative bending moment for a beam continuous over two equal spans of 6 m, carrying a uniformly distributed dead load of 15 kN/m run. Sketch the complete bending moment diagram, and determine approximately the maximum bending moment.

(a) Deformation of continuous beam

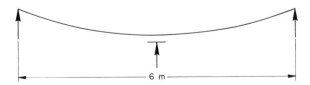

(b) Deformation of continuous beam after large settlement of center support

(c) Deformation of two simply supported beams

Fig. 8.29m.
Effect of large differential settlement on a two-span continuous beam.

8.4. A steel beam is continuous over five equal spans of 5 m each; it carries a uniformly distributed live load of 15 kN/m, and a uniformly distributed dead load of 15 kN/m. Determine the bending moments at all supports, and hence sketch the entire bending moment diagram. What are the maximum negative and positive moments?

8.5. Determine the maximum negative moments for a beam continuous over three spans of 7.5 m, 6 m, and 9 m (6 m being the central span), and carrying a uniformly distributed load of 20 kN/m run. Sketch the complete bending moment diagram, and from it determine the approximate maximum positive moment.

8.7. Derive the maximum positive and negative bending moments for a concrete slab, spanning 3 m built-in at the supports, and carrying a uniformly distributed load of 10 kN/m² (including its own weight).

8.8. A primary reinforced concrete beam, built-in at the supports, carries two secondary beams at third points. Draw the bending moment diagram, neglecting the weight of the primary beam. If the reaction of secondary beams on the primary beam is 25 kN, determine the maximum positive and negative bending moment in the primary beam, which has a span of 7.2 m.

8.9. A rectangular steel portal spans 8 m. The columns are 4 m high, built-in at the supports. The frame is built up from the same 152 mm deep steel section ($I = 18.8 \times 10^6$ mm⁴). Determine the flexural stress at the base of the portal, at the knee-junction, and at mid-span due to a uniformly distributed vertical load of 25 kN.

8.10. A reinforced concrete rectangular portal, built-in at both ends, has a span of 7.2 m and a height of 3.6 m. The beam carries a uniformly distributed load of 200 kN. The second moment of area of the beam (I) is 6×10^{-3}m⁴, and that of the column 1×10^{-3}m⁴. Determine the bending moment at the upper and lower ends of the column.

8.13. At a certain point in a brick shear wall there is a compressive stress of 1.4 MPa and vertical and horizontal shear stresses of 0.7 MPa due to wind load. Calculate the magnitude and direction of the principal stresses.

8.14. An aluminum handrail extrusion has a cross-sectional area of 500 mm²; it is securely fixed on top of a steel handrail, the top member of which has a cross-sectional area of 1,000 mm². Calculate the resulting stresses in the steel and the aluminum, if the temperature falls by 30°C. The moduli of elasticity of steel and aluminum are 200 GPa and 65 GPa, respectively, and the coefficients of linear expansion are 12×10^{-6} and 24×10^{-6} per °C, respectively.

8.15. A 16-story office building is to be built with perimeter walls suspended by hangers at 4 m centers from roof girders which cantilever 9 m from a central service core. Determine the minimum size of the hangers, if the floor construction is to be as light as possible (say, light-gauge steel floor with lightweight concrete top-ping, steel beams with lightweight fireproofing, and a suspended metal ceiling). Conduct your own investigation on the loads, and on the quality of the steel suitable for the hangers.

Having determine the size of the hangers, calculate their elastic deformation under dead and live load, and their temperature move-ment (assuming that they are exposed to sunlight through glass walls in an air-conditioned interior).

Chapter 9

Example 9.1. *Two buildings, 15 m apart, are to be joined by an unstiffened suspension bridge. Determine the cable size re-quired, if each cable carries a load of 15 kN/m run of span.*

Let us assume a ratio of span to sag of 7.5. This gives $L = 15$ m, $s = 2$ m, and $W = 225$ kN. From Eq. (9.1) the cable tension at mid-span

$$T_0 = 225 \times 15/8 \times 2 = 210.94 \text{ kN}$$

From Eq. (9.2) the maximum cable tension

$$T = \sqrt{210.94^2 + 112.50^2} = 239.06 \text{ kN} = 239.06 \times 10^{-3} \text{ MN}$$

If the maximum permissible cable stress is 410 MPa, the steel area is 583 mm^2.

Example 9.2. *Determine the size of cables and the size of the compression ring for a circular suspension roof over an arena 60 m diameter (Figs. 1.17 and 9.2).*

The circumference of the compression ring is $\pi L = 188.5$ m. If we use 62 cables, we obtain a spacing of almost exactly 3 m around the circumference (Fig. 9.2).

The roof needs a fairly large sag to be effective, and this also meets architectural requirements (see Fig. 1.16). Let us make $s = L/5 = 12$ m.

Let us take the weight per unit area of cable roof, allowing for the weight of the cables, the roofing material, and the live load as 5 kN/m^2. Since each cable carries a triangular slice, this varies linearly from a maximum at the compression ring to zero at the tension ring. The total load for a pair of cables (of which there are 31) is

$$W = \frac{5 \times \frac{1}{4}\pi \times 60^2}{31} = 456.0 \text{ kN}$$

Taking moments about the compression ring, we obtain for a single cable (i.e., from the compression ring to the tension ring):

$$T_0 \times s = \tfrac{1}{2} W \times \tfrac{1}{6} L$$

The cable tension at the tension ring

$$T_0 = \frac{456 \times 60}{12 \times 12} = 190.0 \text{ kN}$$

This is also the horizontal reaction of the cable at the compression ring. The vertical reaction is $\tfrac{1}{2} W = 228$ kN, and the combined reaction

$$R = \sqrt{190^2 + 228^2} = 296.8 \text{ kN} = 296.8 \times 10^{-3} \text{ MN}$$

For a maximum permissible steel stress of 410 MPa, the steel area is 723.9 mm^2.

Let us assume that the compression ring is vertically supported, so that the vertical reaction is absorbed by a wall or by columns, The horizontal reaction is 190 kN for each 3 m of the ring, or 63.33 kN/m. This is a condition like that in a thin pipe under suction, and we can solve it in the same way. We thus have to balance a pressure of 63.33 kN/m over a diameter of 60 m (Fig. 9.3m), and this is resisted by the force in two cuts of the ring. Consequently the hoop compression of the ring is

$$\frac{63.33 \times 60}{2} = 1,900 \text{ kN} = 1.9 \text{ MN}$$

Assuming a maximum permissible concrete stress of 11.25 MPa, we require a compression ring of 0.1689 m^2, say 0.3 m wide by 0.6 m deep. In practice we would use a smaller ring, suitably reinforced in compression (see Example 7.10).

The diameter of the tension ring is so small that its dimensions are not primarily determined by structural consideration.

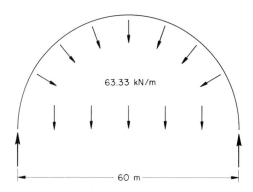

Fig. 9.3m.
The compression ring behaves like a thin tube under suction (Example 9.2).

Examples and Problems in Metric Units

Example 9.3. *Determine the period of the cable in Example 9.1.*

The acceleration due to gravity is 9.81 m per second per second. From Eq. 9.3, the fundamental period $(n = 1)$ is

$$t = \sqrt{\frac{4 \times 225 \times 15}{9.81 \times 1 \times 239.06}} = 2.4 \text{ sec}$$

Example 9.4. *Determine the forces in a hemispherical concrete dome, 75 mm thick, which spans 20 m.*

The load consists of

weight of 75 mm concrete shell	1500 N/m^2
internal and external finishes	450 N/m^2
equivalent uniformly distributed live load	150 N/m^2
equivalent wind load	300 N/m^2
Total load on shell surface	2400 N/m^2 = 2.4 kN/m^2

The membrane shears are zero; the direct membrane forces (Fig. 9.7) are as follows:

Angle from crown of dome	N_ϕ kN/m	N_θ kN/m
0°	12.00 C	12.00 C
30°	7.92 C	12.86 C
45°	2.91 C	14.06 C
60°	4.00 T	16.00 C
90°	24.00 T	24.00 C

C = compression; T = tension.

The maximum compressive stress in the concrete is due to the meridianal force at the springings; it acts on a cross-sectional area of concrete 75 mm thick by 1 m = 1,000 mm wide.

$$f_c = 24 \times 10^3 / 1,000 \times 75 = 0.32 \text{ N/mm}^2 = 0.32 \text{ MPa}$$

The area of reinforcement required for the hoop tension at the springings, for a maximum permissible steel stress of 140 MPa.

$$A_s = \frac{24 \times 10^{-3}}{140} = 171 \times 10^{-6} \text{ m}^2/\text{m} = 171 \text{ mm}^2/\text{m}$$

This is so low (see Table C4) that the actual amount of reinforcement is determined by constructional considerations, both in the hoop and in the meridianal directions.

Evidently 75 mm is a far greater thickness than the structure requires. Whether we can make it thinner depends on waterproofing. If a separate waterproofing membrane is to be used, or the builder and his workmen have the skill to produce a thinner waterproof structure, we can greatly reduce the thickness. However, the main item in the cost of the dome is the cost of the formwork and the cost of placing concrete on a steeply sloping surface. The cost of the concrete materials and of the steel is of less importance.

Example 9.5.   *Design a shallow spherical concrete dome, 75 mm thick, which subtends an angle of 60° at the centre of curvature, and has a span of 20 m.*

Since the dome subtends an angle of 60° at the center, its radius of curvature is

$$\tfrac{1}{2} \times \frac{20}{\sin 30°} = 20 \text{ m}$$

We had a similar dome in Example 9.4, except that its radius of curvature was 10 m. We again take the load as 2.4 kN/m². Both the hoop and the meridianal forces at the crown are 24.0 kN/m (compression), from Eqs. (9.5) and (9.6).

At $\theta = 30°$, i.e., at the springings of the dome, the hoop force is 15.84 kN/m (compression), and the meridianal force is 25.72 kN/m (compression).

The meridianal force is inclined at an angle of 60° to the vertical. The vertical component is absorbed by vertical supports (a wall or columns), but we must absorb the horizontal component $R_H = 25.72 \sin 60° = 22.27$ kN/m with a tension ring (Fig. 9.8), which is designed in precisely the same way as the compression ring for the cable structure in Example 9.2. An alternative, but expensive, solution is to absorb the thrust transmitted by the meridianal forces with buttresses tangential to the shell (Fig. 9.9).

As we explained in Example 9.2, the tension in the ring is

$$\frac{22.27 \times 20}{2} = 222.7 \text{ kN}$$

This is resisted entirely by reinforcement, and since it is essential to avoid large cracks, we limit the stress to 100 MPa. Thus $A_s = 222.7 \times 10^{-3}/100 = 2.227 \times 10^{-3}$ m² $= 2,227$ mm². From Table C3 we select six 25-mm bars (British size) or six 24-mm bars (Australian size).

Since the ring is in tension, it *expands*. However, the hoop forces in the dome are still compressive (15.84 kN/m) so that the edge of the dome *contracts* (Fig. 9.10). The boundary conditions of the dome and the tension ring are incompatible, and thus both are put in bending. We can solve this statically indeterminate problem by a procedure somewhat like that of Example 8.6, where we allowed for the differential stiffness of two joining members. However, this is a complicated and lengthy procedure, and we know from previous designs that the bending stresses are localized in the region of the tie. A proper junction with the tie requires local thickening of the shell, and we introduce top and bottom steel to cope with the local bending stresses. Because bending stresses do not extend too far towards the crown, we can dimension the shell for bending by empirical rules.

Example 9.6. *Determine the forces in a 75 mm thick hypar shell, spanning 7.5 m in both directions, with a rise of 1.5 m.*

The load consists of:

weight of concrete, measured on plan	$1,600 \text{ N/m}^2$
external finish (off the form inside)	150 N/m^2
equivalent uniformly distributed live load	250 N/m^2
equivalent wind load	0
Total load on shell surface	$2,000 \text{ N/m}^2 = 2.00 \text{ kN/m}^2$

The direct forces are zero, and the uniform shear force in the shell

$$V = \frac{2.00 \times 7.5 \times 7.5}{2 \times 1.5} = 37.5 \text{ kN/m}$$

The shear forces produce numerically equal compressive and tensile forces at 45° to the sides of the shell (see Section 6.8). The tensile forces take the form of cable curves, and the compressive forces the shape of arches. Thus we can think of the hypar shell as a series of interconnected arches and suspension cables.

Since the maximum (diagonal) compressive force in the concrete is 37.5 kN/m, the maximum compressive concrete stress is

$$f_c = \frac{37.5 \times 10^3}{1,000 \times 75} = 0.50 \text{ N/mm}^2 = 0.50 \text{ MPa}$$

The area of reinforcement required

$$A_s = 37.5 \times 10^{-3}/100 = 375 \times 10^{-6} \text{ m}^2 = 375 \text{ mm}^2$$

431

It is not advisable to lay the steel in the direction of the principal tension, partly because every bar along the diagonals would have to be cut a different length, and partly because we require reinforcement in both directions, if only to control shrinkage and temperature stresses (see Section 8.9). If we use two sets of bars parallel to the sides, each contributes cos 45° to the tensile force, so that we have $2 A_s \cos 45° = \sqrt{2} A_s$. Consequently the area of steel required parallel to each set of sides is

$$\frac{375}{\sqrt{2}} = 265 \text{ mm}^2$$

This is too low (see Table C4), and the actual amount of reinforcement is determined by constructional considerations.

Problems

9.3. A steel cable spans 110 m between supports, with a sag of 8 m at mid-span. It carries a uniformly distributed load of 1 kN/m. Determine the cable tension at mid-span, the cable tension at the supports, and the reactions transmitted to the supports. If the maximum permissible steel stress is 800 MPa, calculate the cross-sectional area required for the cable.

9.4 Fig. 9.20m. shows a mast for the support of a suspension cable. Calculate the forces in the members AB and AC. Hence, calculate the cross-sectional area required for AB and AC, using permissible stresses of 45 MPa in compression and 140 MPa in tension.

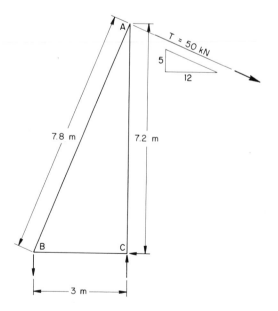

Fig. 9.20m.
Problem 9.4.

Examples and Problems in Metric Units

9.6. A laminated timber three-pin arch of semicircular shape is loaded by two beams as shown in Fig. 9.21m, each beam transmitting a load of 10 kN to the arch. The cross-section of the arch is 150 mm wide by 300 mm deep. Determine the reactions at the support of the arch, and sketch the bending moment diagram. From measurements on this diagram, calculate the approximate maximum tensile and compressive stresses, and indicate which is on the outside of the arch.

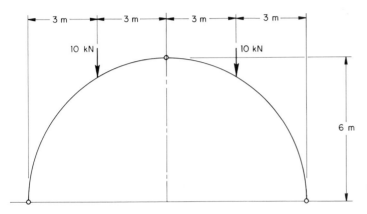

Fig. 9.21m.
Problem 9.6.

Answers to Metric Problems in Appendix G

H

C = compression; T = tension; BM = bending moment; SF = shear force; LH = left hand; RH = right hand; S = section modulus.

Chapter 3

3.1. The force needed is 2.718 kN (horizontal component = 0.860 kN pointing towards the left; vertical component = 2.578 kN pointing downwards).

3.2. 372.15 N; the shortest distance from A is 8.63 mm.

3.3. The LH reaction is 3.5 kN; the RH reaction is 3.0 kN.

3.4. The vertical reaction is 40 kN; the moment reaction is 108.5 kN m.

3.5. The RH reaction is 7.794 kN, acting vertically; the LH reaction is 11.906 kN, inclined at an angle of 49°6′ to the vertical (the vertical component is 7.794 kN, and the horizontal component is 9.000 kN).

3.6. The center of gravity is 183.4 mm to the right and 64.0 mm above the bottom LH corner.

3.7. The answer is 6.890 kN; it is most conveniently obtained by taking moments about E.

3.8. The answer is most conveniently obtained by taking moments about the leeward edge. The vertical force due to the block is 13.244 kN. The block will overturn, since the resultant falls 41 mm outside the base.

3.9. The vertical force due to the building is 26.487 MN. The building will overturn if the resultant falls outside the base. This requires a wind pressure of 795 Pa.

3.10. The resultant must fall within the middle third; this requires a vertical force of 11.25 kPa and a minimum density of 2,548 kg/m³ for the masonry.

Chapter 4

4.4. The stress diagram is shown in Fig. F1 in Appendix F.
AB: 62.5 kN; AE: 387.5 kN; A3: 12.5 kN C; A1: 62.5 kN T;

B1: 88.5 kN C; 12: 53 kN C; C2: 25 kN C; 23: 53 kN T; 34: 265 kN C; D4: 200 kN T; E4: 283 kN C.

4.5. The stress diagram is drawn in Fig. F2 in Appendix F. The forces, scaled from the diagram to the nearest 25 kN, are:
LA 625; DE 575; A1 900 C; A3 925 C; B5 975 C; C7 925 C; D8 925 C; D-10 900 C; D-12 825 C; D-14 800 C; E-14 575 T; F-13 575 T; G-11 775 T; H9 875 T; I6 950 T; J4 875 T; K2 625 T; L1 625 T; 1-2 100 T; 3-4 50 C; 5-6 100 T; 7-8 225 T; 9-10 25 T; 11-12 50 C; 13-14 100 T; 2-3 350 T; 4-5 125 T; 6-7 75 C; 8-9 50 T; 10-11 175 T; 12-13 300 T.

4.6. The *calculated* reactions are: LH 631.25 kN; RH 568.75 kN. The force in the horizontal member is 631.25 kN T; the force in the inclined member is 892.72 kN C.

4.7. 100 kN.

4.8. The reactions are 12.5 kN each.
By resolution at the joints: Joint A: AO 20.833 kN T;
Joint O: OB 20.833 kN C.
By method of sections: Moments about M:
CD 56.667 kN C;
Moments about D:
ML 60.000 kN T.

4.9. Moments about C: BD (and DF by symmetry) 10.0 kN C;
Moments about D: CE 12.5 kN T;
Moments about B: CD (and DE by symmetry) 3.536 kN C.

4.10. The force in AB increases by 10.607 kN.

4.11. The force in XY is 212.13 N C. The reaction at Z is 291.55 N; it makes an angle of 31° to the vertical, and its horizontal component is 150.0 N.

4.12. AB: 9.00 kN C; BC and BD: 4.875 kN T.

4.13. The reactions are, at A: 53.33 kN (down); at B 50.00 kN (up); at C: 53.33 kN (up). The forces are: AB 40.00 kN C; AD 66.67 kN T; BC 40.00 kN C; BD 25.00 kN T; BE 93.94 kN C; CD 66.67 kN C; DE 61.85 kN T.

Chapter 5

5.1. The max. SF occurs at the support $= +40$ kN; the max. BM also occurs at the support $+ -108.5$ kN m.

5.2. The max. SF occurs at the support $= +58$ kN; the max. BM also occurs at the support $= -103$ kN m.

5.3. The max. SF occurs at the RH support $= -38$ kN; the max. BM occurs under the 40 kN load $= +106.5$ kN m.

5.4. The BM varies linearly from 0 at the supports to $+\frac{3}{8}$ PL at the quarter points, and then linearly to $+\frac{1}{2}$ PL at mid-span.

5.5. The SF varies linearly from $+52.5$ kN (max.) at the LH support to -7.5 kN at mid-span, and then linearly to -37.5

kN at the RH support. The BM varies parabolically, with a discontinuity at mid-span.

The maximum occurs where $dM/dx = 0$, which is 1.75 m from the LH support. The max. BM is $+45.94$ kN m.

5.6.   The SF varies linearly from 0 at the LH end to -30 kN at the LH support; then changes to $+39$ kN; remains constant to the concentrated load; changes to -51 kN; remains constant to the RH support; changes to $+60$ kN (max.) and reduces linearly to 0 at the RH end.

The BM varies parabolically from 0 at the LH end to -30 kN m at the LH support; changes linearly to $+67.5$ kN m (max.) under the concentrated load; changes linearly to -60 kN m at the RH support; and reduces parabolically to 0 at the RH end.

5.7.   This problem is basically the same as Example 5.9 (Appendix G). The smallest max BM is obtained when the columns are 1.243 m from the ends of the beam.

5.8.   The man exerts a vertical force of 981 N. The horizontal reactions are 367.9 N, the vertical reaction is 981 N, and the max. BM is 1.1036 kN m.

5.9.   The beam reactions are 655 N at A and 385 N at C: The forces in the cables are 463.2 N at AF and AE, and 385 N at CD, all tensile.

5.10.   The BM and the SF vary in the same way in the cantilevered and in the supported spans. The SF varies linearly from $+45$ kN at the LH supports to -45 kN at the RH supports. The BM varies parabolically from -33.075 kN m at the supports to $+33.075$ at mid-span.

5.11.   The vertical reactions are 50 kN each, and the horizontal reactions are 25 kN each. The BM diagram is shown in Fig. H.1; max. BM $= -112.5$ kN m at the rigid joint.

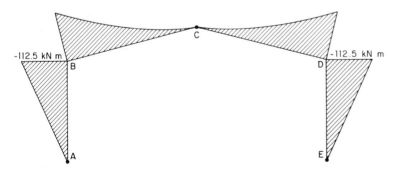

Fig. H.1

5.12. $R_A = -1.25$ kN (uplift): $R_D = +1.25$ kN; $H_A = 5.00$ kN; $H_D = 0$. The BM diagram is shown in Fig. H.2; max. BM $= +7.5$ kN m at the rigid joint.

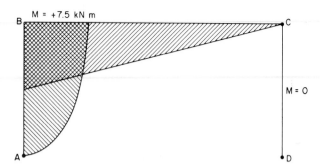

Fig. H.2

5.13. The vertical reactions are 100 kN (LH) and 50 kN (RH); the horizontal reactions are 50 kN each. The BM diagram is shown in Fig. H.3; max. BM = +375 kN m under the concentrated load.

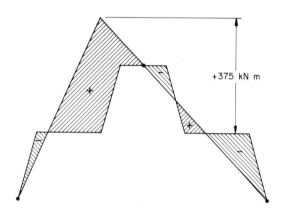

Fig. H.3

5.14. Vertical reactions: 250 kN; horizontal reactions: 125 kN. The BM diagram is shown in Fig. F6. The cross-sectional area of the cable is 178.6 mm².

Chapter 6

6.2. Max. SF = 5.45 kN; Max. BM = 9.15 kN m; S required = 1.326×10^6 mm³. 100 mm × 300 mm (actual size) gives S = 1.5×10^6 mm³.

6.3. Max. SF = 24.64 kN; Max. BM = 59.136 kN m; S = 6. 876×10^6 mm⁴; Depth required = 371 mm, say 400 mm.

6.4. $I = 93.15 \times 10^9$ mm⁴; S = 155.25×10^6 mm³.

6.5. Max. SF = 32 kN; Max. BM = 32 kN m (see Fig. H.4). S = 1.455×10^6 mm³. 150 mm × 250 mm (actual size) gives S = 1.56×10^6 mm³.

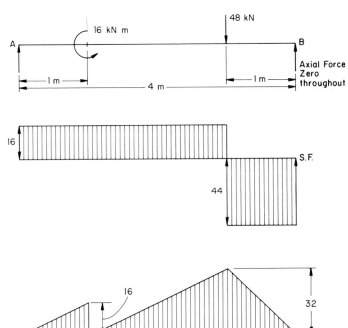

Fig. H.4

6.6. Either above or below the existing beam, not adjacent to it, and connected to it by sufficient bolts to resist the horizontal shear. If appearance is no objection, a greater advantage can be obtained by cutting the length of timber in half, and fixing one 2-m length above and another below the existing beam covering the range of maximum bending moment.

6.7. The deflection is 49.27 mm downwards. The camber required to eliminate is 49.27 mm upwards.

6.8. The horizontal shear stress and both principal stresses are 1.25 MPa.

6.9. The principal stresses are 192 MPa T and 47 MPa C.

6.10. The principal stresses are 2.493 MPa C and 0.393 MPa T. The angle of inclination is $38°0'$.

6.11. The max. forces are 86.32 kN/m T and 66.32 kN/m C. The max. compressive stress in the concrete is 0.737 MPa. The area of reinforcement required is 161.6 mm^2/m.

Chapter 7

7.1. 262.5 mm^2.

7.4. Using the same maximum permissible stresses as in Section 7.4: (a) a 100-mm long 9.4-mm fillet weld (83.6 mm required); (b) five 20-mm rivets (4.78 required); (c) six 30-mm black bolts (5.98 required); (d) three 20-mm high-strength bolts (2.72 required).

7.5. By resolution at the joints, the most highly stressed diagonal tension member is the diagonal closest to the support (12.728 kN); the most highly stressed diagonal compression member is the adjacent diagonal (12.728 kN).

By method of sections, or by taking the bending moment (13.5 kN m) and dividing by the resistance arm (0.250 m), the most highly stressed portion of the bottom chord is at mid-span (54 kN T).

7.8. The (factored) ultimate load is 9.613 MN. The area of reinforcement required is 7,950 mm^2. Twelve 32-mm bars (10 would do, but we require 12 for symmetry).

7.9. Assuming 16 mm bars, the effective depth is 72 mm, and the area of reinforcement required is 1,589 mm^2/m. Use 16 mm bars at 125 mm centers (1,600 mm^2/m) on the top of the slab.

7.11. The flat slab requires bars in both directions, and one set of bars lies inside the other. Considering the interior set (which gives the smaller effective depth), and assuming 20 mm bars, the effective depth is 75 mm. The area of reinforcement required is 1,927.1 mm^2/m. Use 20-mm bars at 150-mm centers (2,067 mm^2/m).

7.12. From Fig. 6.8: A = 0.72 m^2; $I = 17.55 \times 10^{-3}$ m^4; S_t(wall side) = 78×10^{-3} m^3; S_b(pier side) = 46.8×10^{-3} m^3. Vertical load = 110.52 kN; BM = 6.75 kN m; stress on pier side = 297.73 kPa C; stress on wall side = 66.96 kPa C.

7.13. Vertical reactions = 20.4 kN each; horizontal reactions = 12.75 kN each; Max. BM = 38.25 kN m; stress in column varies from 10.942 MPa C to 10.216 MPa T.

7.14. After prestressing the concrete stress varies from 0.178 MPa C (top) to 14.044 MPa C (bottom). Under full load it varies from 14.756 MPa C (top) to 0.534 MPa T (bottom).

Chapter 8

8.3. Using the Theorem of Three Moments, or Table 8.2, the negative BM at the central support is 67.5 kN m. This is the max. negative BM. The moments M_1 and M_2 are $+67.5$ kN m, and the BM diagram is sketched in Fig. H.5. Scaling from this diagram, or by calculation from Table 8.2, the max. positive BM is 37.8 kN m.

8.4. Using Table 8.2, the positive BM in the first span = 66.75 kN m (max.); second span = 43.375 kN m; center span = 49.5 kN m. The negative BM at the outer support = 0; at the first interior support = 82.875 kN m; at the support next to the center span = 69.75 kN m.

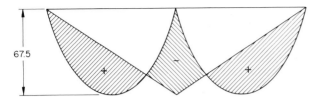

67.5

Fig. H.5

8.5. The problem is similar to Example 8.1. Using the notation of Fig. 8.6m, $M_A = 0$; $M_B = -87.00$ kN m; $M_C = -140.10$ kN m (max. negative BM); $M_D = 0$; $M_1 = +140.625$ kN m; $M_2 = +90.00$ kN m; $M_3 = +202.50$ kN m. The net BM diagram is drawn in Fig. 8.6m, and the positive BM can be scaled.

8.7. The max. negative BM can be calculated from theory, or using Table 8.3; it is 7.5 kN m/m. The max. positive BM is the difference between this and WL/8; it is 3.75 kN m/m.

8.8. The problem is similar to Example 8.3. Max. Negative BM = 40 kN m; max. positive BM = 20 kN m. The BM diagram is sketched in Fig. H.6.

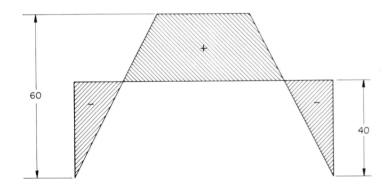

60

40

Fig. H.6

8.9. The stiffness of the beam is half the stiffness of the column because it is twice as long. After three moment distributions, the BM at the base of the portal = +6.615 kN m, at the knee-junction = -13.28 kN m. From the latter the BM at mid-span is obtained = +11.72 kN m. S = 247 × 10³ mm³. The flexural stresses at the base = 26.78 MPa; at the knee-junction = 53.76 MPa; at mid-span = 47.49 MPa.

8.10. The BM at the upper end of the column = -47.0 kN m; at the lower end of the column = +23.6 kN m.

8.13. The angle is 22°30′, and the principal stresses are 1.690 MPa C and 0.290 MPa T.

8.14. The difference in strain due to temperature change is 360×10^{-6}, and this is the sum of the tensile strain in the aluminum and the compressive strain in the steel, since they are fixed to one another. Furthermore, the tensile force in the aluminum is equal to the compressive force in the steel for equilibrium. The aluminum stress is 20.13 MPa T, and the steel stress is 10.06 MPa C.

Chapter 9

9.3. Cable tension at mid-span = 189.06 kN; reaction at support = 196.90 kN (horizontal component = 189.06 kN, vertical component = 55 kN); max. cable tension (at support) = 196. 90 kN; cross-sectional area of cable = 246.13 mm^2.

9.4. Resolving horizontally and vertically at A: force in AB = 120 kN T; force in AC = 130 kN C. Cross-sectional areas required for AB = 857 mm^2; for AC = 2889 mm^2.

9.6. Vertical reactions = 10 kN; horizontal reactions 5 kN. BM diagram is sketched in Fig. F10. Max. positive BM under concentrated load. Max. negative BM (and absolute max. BM) occurs approx. 1.2 m from supports and is approx. 7.4 kN m (scaled from Fig. F10). The angle of inclination of the arch to the horizontal at that point is approx. 54°. Hence the thrust in the arch is 5 cos 54° + 10 sin 54° = 11.0 kN. The max. stresses are 3.53 MPa C on the inside of the arch, and 3.04 MPa T on the outside of the arch.

Index